Universitext

Universitext

Universitext is a series of textbooks that presents material from a wide variety of mathematical disciplines at master's level and beyond. The books, often well class-tested by their author, may have an informal, personal even experimental approach to their subject matter. Some of the most successful and established books in the series have evolved through several editions, always following the evolution of teaching curricula, to very polished texts.

Thus as research topics trickle down into graduate-level teaching, first textbooks written for new, cutting-edge courses may make their way into *Universitext*.

More information about this series at http://www.springer.com/series/223

Gabor Toth

Measures of Symmetry for Convex Sets and Stability

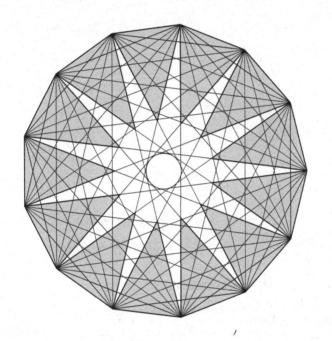

 Springer

Gabor Toth
Department of Mathematical Sciences
Rutgers University
Camden, NJ, USA

ISSN 0172-5939   ISSN 2191-6675 (electronic)
Universitext
ISBN 978-3-319-23732-9   ISBN 978-3-319-23733-6 (eBook)
DOI 10.1007/978-3-319-23733-6

Library of Congress Control Number: 2015950621

Springer Cham Heidelberg New York Dordrecht London

Printed on acid-free paper

Springer International Publishing AG Switzerland is part of Springer Science+Business Media (www.springer.com)

Preface

> *"Two roads diverged in a wood, and I–*
> *I took the one less traveled by,*
> *And that has made all the difference."*
> from THE ROAD NOT TAKEN
> by Robert Frost, 1920.

About the Title

CONVEXITY is one of the most fundamental and multifaceted concepts in mathematics. It can be traced back to antiquity, especially to Archimedes of Syracuse who gave the first precise definitions of convex curve and surface. The thorough and systematic development of *convexity theory* began at the dawn of the twentieth century by the monumental work of Minkowski, and it has been expanding ever since. At present, this field comprises several subjects. Based on the ideas, methods, and the arsenal of tools employed, it can be roughly divided into three major areas: *convex geometry*, *convex analysis*, and *discrete* or *combinatorial convexity*. Each of these areas is intricately interconnected with well-established and central subjects in mathematics such as computational geometry, functional analysis, calculus of variations, integral geometry, and linear programming.

This diversity is also apparent from the long list of eminent mathematicians who contributed to convexity theory, and also from the even longer list of applications beyond mathematics in areas such as econometrics, finance, computer science, crystallography, physics, and engineering.

MEASURES OF SYMMETRY is a collective name of a wide variety of geometric and analytic constructions in convex geometry that quantify various concepts of symmetry for convex sets.

In 1963, in a seminal paper, Branko Grünbaum gave a detailed account on literally hundreds of interrelated problems on measures of symmetry. Moreover, this work is of fundamental importance in that it established a framework of basic definitions and problems pertaining to this area of convex geometry.

According to his most general and abstract definition, an affine measure of symmetry is an affine invariant continuous function on the space of all convex bodies (with a suitable metric topology). In addition, he also required that the minimum value of a measure of symmetry must be attained precisely on symmetric bodies.

With this in mind, it turns out that measures of symmetry are frequently lurking amidst many geometric situations. In addition to the classical *Minkowski measure* and its relatives, simple volume comparisons (partitioning a convex body with hyperplanes) give rise to the *Winternitz measure* and its variants; and, in more complex geometric settings, one considers volume quotients such as the classical *Rogers–Shephard volume ratio*, or the more recent *Hug–Schneider measure* which, for a convex body, takes the volume ratio of the minimum volume circumscribed *Löwner ellipsoid* and maximum volume inscribed *John ellipsoid*.

STABILITY is a desirable and often valid property of a *geometric inequality* concerning a specific set of geometric objects. The question of stability arises in an *inequality* if the geometric objects for which the *equality* sign holds form a "well-characterized" class. Objects belonging to this class are called "extremal." Stability holds if whenever, for an object, the inequality sign is close to equality, then the deviation (or distance) of the object is also close to one of the extremal objects.

For example, the well-known *Bonnesen inequality*, relating the length, the area, the radii of the incircle and the circumcircle of a planar Jordan curve, can be viewed as an early stability estimate for the *isoperimetric problem*: if, for a Jordan curve, the isoperimetric inequality is close to equality then the two radii are also close so that the Jordan curve is contained in an annulus of small width. This width can be considered as the deviation of the Jordan curve from a circle, the extremal object of the isoperimetric problem.

In 1993, in yet another seminal survey, Helmut Groemer gave a general and comprehensive account on the known stability estimates for geometric inequalities concerning various classes of convex sets.

Why This Book?

Since the survey articles of Grünbaum and Groemer, many papers have been written on measures of symmetry and stability, shedding new lights on classical topics, and developing new branches. In contrast to this proliferation of literature, there is no single graduate-level textbook that starts from the beginning, guides the reader through the necessary background, and ends up in recent research. This book wishes to make a humble contribution to this end.

The goal of this book is to serve the novice who seeks an introduction to measures of symmetry assuming only basic mathematics that can be found in undergraduate curricula, and also to serve mathematicians who, bypassing the preliminary material in Chapters 1 and 2, wish to have a quick path to the most recent developments of the subject.

This book is not a comprehensive survey, nor does it claim completeness; rather, we make a careful selection of measures of symmetry covering a variety of geometric settings, and follow with a thorough treatment and detailed proofs.

Unifying Theme

The apparent diversity of the notion of convexity is amply reflected in the occurrence of measures of symmetry throughout the convex geometry literature. In this book this multiformity of measures of symmetry is elevated to be the main guiding light to unify seemingly disparate results to a single source, and thereby bringing coherence to the subject. Led by this, throughout the text, we see no harm, indeed, we feel justified pointing out various connections to other mathematical disciplines.

About the Reader

Throughout the writing of this book, utmost attention was paid for gradual development of the subject; to assume as little as possible, and, at the same time, to be self-contained as much as possible. The reader is assumed to be familiar with basic topology, analysis, group theory, and geometry.

Since we believe in progressive learning through continuous challenge, this book contains 70 problems assembled in groups at the end of each chapter. For a newcomer to graduate studies, about half of these problems are difficult, and oftentimes require novel ideas. It would have been possible to incorporate them into the main text, but this would have been at the expense of the size of the book, and at the expense of losing the main focus. To aid the reader, the more difficult problems are marked with asterisk and have detailed hints and references.

Synopsis

The purpose of Chapter 1 is to give a gentle but rapid introduction to the basic constructions and results in convexity theory, and therefore to lay a solid foundation for the rest of the more advanced material. Highlights of some preparatory topics treated in full detail are: (1) the *Hausdorff distance* leading to *Blaschke's selection theorem*; (2) the *Hahn–Banach theorem* giving a variety of *separation theorems* for convex sets; (3) the *Minkowski–Krein–Milman theorem*; and (4) a brief introduction to Minkowski's *mixed volumes*. This chapter culminates in the three pillars of combinatorial geometry of convex sets; the classical theorems of *Carathéodory*, *Radon*, and *Helly*.

Paving the way to subsequent developments and applications, Helly's theorem and some of its applications and extensions receive special care. As an application, we use the Yaglom–Boltyanskiĭ approach to Helly's theorem to derive an a priori upper bound for the *distortion* of a convex set; this is a prelude to the *Minkowski–Radon inequality* treated in the next chapter. As an extension, an extra section is devoted to *Klee's extension* of Helly's theorem to infinite non-compact families, along with the necessary topological arsenal, the *Painlevé–Kuratowski convergence* for sets.

Finally, still in this chapter, we introduce the four basic metric invariants of a convex body; the *inradius*, the *circumradius*, the *diameter*, and the *minimal width*. We derive the classical *Jung–Steinhagen estimates* for their two non-trivial ratios.

In Chapter 2 we introduce the notions of *critical set* and *Minkowski measure* of a convex body. The critical set is usually defined as the minimal level set associated with the Minkowski measure. However appealing and concise this may seem, here we chose "the road not taken" (as Robert Frost wrote in the epitaph above) through *Hammer's decomposition* of a convex body not only because it reveals structural details of the convex body (including an insight to the Yaglom–Boltyanskiĭ method alluded to above), but also because of its connection to affine diameters.

As a quick by-product of the work in the previous chapter, we derive the Minkowski –Radon inequality for the Minkowski measure. (Two additional classical proofs of this fundamental inequality are in Chapter 3.)

The bulk of the material in this chapter is taken up by Klee's dimension estimate of the critical set (sharpening the Minkowski–Radon upper bound). As before and particularly here, no effort is spared to give full details of Klee's brilliant analysis.

Convex bodies of constant width have a steadily growing literature; they provide a wealth of explicitly computable examples for measures of symmetry. This chapter ends with an extended treatise on them; our main purpose is to provide transparent as well as beautiful illustrations of the general theory.

Note that, as the brief summary above shows, the first two chapters have sufficient material for an introductory graduate course in convex geometry (as it was done at Rutgers—Camden in Fall 2014).

Armed with the necessary arsenal, in Chapter 3, we begin in earnest to study measures of symmetry for convex bodies, and stability of the corresponding geometric inequalities. The Minkowski measure is reintroduced here in dual settings. The corresponding Minkowski–Radon inequality is archetypal in understanding one of the principal problems in dealing with stability estimates: the choice of distance on the space of convex bodies by which we measure the deviation of a convex body from an extremal one.

We start with the Hausdorff distance, and first derive *Groemer's stability estimate* for the lower bound of the Minkowski measure.

In seeking a distance better fit to stability, *affine invariance* (a fundamental property of the Minkowski measure) as a desirable requirement comes to the fore. This leads directly to the *Banach–Mazur "metric."* This proves to be a better choice because it almost immediately gives a simple and strong stability result for the lower bound of the Minkowski measure.

We keep the Banach–Mazur metric to obtain stability for the upper bound in the Minkowski–Radon inequality. From a few competitors in the literature, we choose to derive *Schneider's stability estimate* not only because it is the sharpest but also because it uses the Yaglom–Boltyanskiĭ approach to Helly's theorem treated in Chapter 1.

The affine invariant Banach–Mazur metric is sub-multiplicative, and therefore its logarithm induces a distance on the (extended) *Banach–Mazur compactum*, the space of (not necessarily symmetric) convex bodies modulo affine transformations.

As pointed out by Schneider, compactness of this space can be shown directly, but it is an easy consequence of *John's ellipsoid theorem* (along with Blaschke's selection theorem discussed in Chapter 1). This striking and celebrated result of Fritz John in 1948 is well-worth a detour. John's ellipsoid theorem asserts that a maximal volume ellipsoid inscribed in a convex body is *unique*, and its scaled copy (from the center of the ellipsoid) with scaling ratio equal to the dimension of the ambient space *contains* the convex body. A significant portion of this chapter is devoted to this result. We first discuss John's original approach as an optimization problem with KKT (Karush–Kuhn–Tucker) conditions. Then, we give complete details of a more recent proof by Guo–Kaijser based on constructing volume increasing affine transformations. This method not only recovers John's ellipsoid theorem (along with Lassak's theorem), but also gives an interesting upper estimate for the Banach–Mazur distance between convex bodies in terms of their Minkowski measures.

After this detour, we return to the main line with another highlight of this chapter; a complete proof of *Groemer's stability* of the *Brunn–Minkowski inequality*.

At this juncture, we come to the main point of the book: Grünbaum's general notion of measure of symmetry. We illustrate this with a host of classical and recent examples such as the Winternitz measure (with a fairly complete treatment), the Rogers–Shephard volume ratio (with connection to the Brunn–Minkowski inequality), and Guo's L^p-Minkowski measure.

In closing this chapter, we return to the four classical metric invariants of a convex body. Since in the Jung–Steinhagen estimates equalities are not attained by well-defined classes of extremal convex bodies, stability results for these ratios cannot be expected. The situation radically changes, however, if we relax the Euclidean scalar product on the ambient space by weaker Minkowski norms, and take the supremum of the respective two ratios of metric invariants with respect to all norms. We adopt here Schneider's approach to this topic which not only recovers the analogues of the Jung–Steinhagen universal estimates in Minkowski space (due to Bohnenblust and Leichtweiss) but also gives stability estimates for these ratios.

In the last chapter, Chapter 4, we introduce and study a new (infinite) sequence of *mean Minkowski measures* associated with a convex body. It is a sequence of functions on the interior of the convex body, and, in addition to subtle functional behavior, it has a host of arithmetic properties such as sub-arithmeticity and monotonicity. Roughly speaking, at an interior point, the kth member of the sequence measures how far the convex body is from having an affine k-dimensional simplicial intersection across the interior point. Each member of the sequence is a measure of symmetry with universal lower and upper bounds. The corresponding classes of extremal convex bodies are those of simplices (of the respective dimension) and symmetric bodies. In this chapter we give a full analysis of this sequence including stability estimates.

As a seemingly remote connection to convexity, Appendix A deals with *eigenmaps*, maps of a compact Riemannian manifold (as the domain) to spheres whose component functions belong to a given eigenspace corresponding to an eigenvalue of the Laplace operator (acting on functions of the domain). The *moduli (space)* of such maps is a convex body. If the Riemannian manifold is homogeneous, then the acting Lie group induces a linear representation on the linear span of the moduli, and leaves the moduli itself invariant. Although the dimension of the moduli increases fast with the increase of the eigenvalue, in certain instances, a measure of symmetry of the moduli can be explicitly calculated. This therefore gives a rough measure of symmetry of the otherwise intractable moduli.

Acknowledgments

A large part of this book was written during my sabbatical leave in Spring 2014 while traveling extensively in China and Taiwan. It is a pleasure to record my thanks to several mathematicians and institutions for their support, including Prof. Hui Ma of Tsinghua University, Beijing; Prof. H.-J. Cheng of the Academia Sinica, Taiwan; and Profs. H. Li and Q. Wang of Xiamen University.

I am particularly indebted to Prof. Qi Guo of Suzhou University of Science and Technology, China, for my stay in Suzhou in Winter 2015 (supported by the NSF-China, No. 11271282) for marathon discussions and seminars on convexity.

The original draft of the manuscript underwent several major revisions. In the final phase, many explanatory paragraphs have been added, innumerable typos were corrected, and, to increase readability, several proofs have been expanded. In this effort, an invaluable aid was provided by the reviewers. I wish to convey my thanks to them for their time and dedication.

Finally, I also thank Christopher Lim who has read the manuscript and made many helpful suggestions, and to Catherine Meehan who spent countless hours of painstaking work in developing the illustrations.

Camden, NJ, USA   Gabor Toth

Contents

Chapter 1
First Things First on Convex Sets

1.1 Preliminaries

A. Convex Sets. We begin here with some basic definitions and elementary facts. Throughout this book we will work in an n-dimensional real *vector space* $\mathcal{X}$. The choice of a basis in $\mathcal{X}$ amounts to a linear isomorphism $\mathcal{X} \cong \mathbb{R}^n$, but we will make this identification only in examples and some explicit computations.

We do not distinguish between the linear structure of a vector space $\mathcal{X}$ and the underlying *affine space*, where the latter amounts to disregarding a distinguished point that serves as the origin.

Several key concepts in convexity depend on the affine structure of $\mathcal{X}$ only. While we will occasionally indicate this by using the term "affine," we will leave it to the reader to make this finer distinction to avoid a detailed and somewhat redundant introduction to affine geometry.

Using the additive structure of $\mathcal{X}$, for $V \in \mathcal{X}$, the *translation* $T_V : \mathcal{X} \to \mathcal{X}$ with *translation vector* V is defined by $T_V(X) = X + V$, $X \in \mathcal{X}$. The vector space $\mathcal{X}$ can be naturally identified with the set $\mathcal{T}(\mathcal{X}) = \{T_V \,|\, V \in \mathcal{X}\}$ of all translations of $\mathcal{X}$ (via $V \mapsto T_V$, $V \in \mathcal{X}$), and this identification makes $\mathcal{T}(\mathcal{X})$ a vector space (of dimension n). We call $\mathcal{T}(\mathcal{X})$ the (additive) *group of translations* of $\mathcal{X}$. Due to their different roles, we will usually keep $\mathcal{T}(\mathcal{X})$ and $\mathcal{X}$ separate.

Since the primary role of translations is to displace points (such as the origin in a vector space), in affine geometry $\mathcal{T}(\mathcal{X})$ is postulated to be a vector space that acts on $\mathcal{X}$ (regarded only as a set) by transformations satisfying certain axioms. This defines the affine structure on $\mathcal{X}$.

An *affine combination* of a finite set $\{A_1, \ldots, A_m\} \subset \mathcal{X}$, $m \geq 1$, is a sum $\sum_{i=1}^{m} \lambda_i A_i \in \mathcal{X}$ with coefficients $\{\lambda_1, \ldots, \lambda_m\} \subset \mathbb{R}$ satisfying $\sum_{i=1}^{m} \lambda_i = 1$.

As noted above, the term *affine* refers to the fact that affine combinations depend only on the underlying affine structure of $\mathcal{X}$. This means that they do not depend on the specific location of the origin, or, in other words, they are *invariant* under translations in the sense that

© Springer International Publishing Switzerland 2015
G. Toth, *Measures of Symmetry for Convex Sets and Stability*,
Universitext, DOI 10.1007/978-3-319-23733-6_1

$$T_V \left(\sum_{i=1}^{m} \lambda_i A_i \right) = \sum_{i=1}^{m} \lambda_i T_V(A_i), \quad V \in \mathcal{X}.$$

A subset $\mathcal{X}' \subset \mathcal{X}$ is an *affine subspace* if, along with any finite set of points $\{A_1, \ldots, A_m\} \subset \mathcal{X}'$, $m \geq 1$, any affine combinations $\sum_{i=1}^{m} \lambda_i A_i$, $\sum_{i=1}^{m} \lambda_i = 1$, $\{\lambda_1, \ldots, \lambda_m\} \subset \mathbb{R}$, also belong to $\mathcal{X}'$.

An affine subspace of $\mathcal{X}$ is a *linear* subspace if and only if it contains the origin. Consequently, an affine subspace in $\mathcal{X}$ is a translated copy of a linear subspace of $\mathcal{X}$.

The concepts of *line, plane, hyperplane*, etc. (corresponding to affine subspaces of dimensions $1, 2, n - 1$, etc.) in $\mathcal{X}$ are readily understood.

An *affine map* $\phi : \mathcal{X}' \rightarrow \mathcal{X}$ between vector spaces is a map that preserves affine combinations, that is, we have

$$\phi \left(\sum_{i=1}^{m} \lambda_i A_i \right) = \sum_{i=1}^{m} \lambda_i \phi(A_i),$$

where $\sum_{i=1}^{m} \lambda_i = 1$, $\{A_1, \ldots, A_m\} \subset \mathcal{X}'$, $\{\lambda_1, \ldots, \lambda_m\} \subset \mathbb{R}$, $m \geq 1$.

The composition of affine maps is affine, and the inverse of a bijective affine map is also affine.

An affine map $\phi : \mathcal{X}' \rightarrow \mathcal{X}$ is linear if and only if it sends the origin of $\mathcal{X}'$ to the origin of $\mathcal{X}$.

The *affine group* $\text{Aff}(\mathcal{X})$ of $\mathcal{X}$ is the group of *invertible affine* self-maps of $\mathcal{X}$. The affine group contains two distinctive subgroups. First, the *general linear group* $GL(\mathcal{X})$, the group of all *linear* transformations of $\mathcal{X}$, is a subgroup of $\text{Aff}(\mathcal{X})$ consisting of all affine transformations that keep the origin of $\mathcal{X}$ fixed. Second, the group of translations $\mathcal{T}(\mathcal{X})$ is a normal subgroup in the affine group: $\mathcal{T}(\mathcal{X}) \lhd \text{Aff}(\mathcal{X})$.

For an affine transformation $\phi \in \text{Aff}(\mathcal{X})$, there exist unique $A \in GL(\mathcal{X})$ and $T_V \in \mathcal{T}(\mathcal{X})$, $V \in \mathcal{X}$, such that $\phi(X) = T_V(A \cdot X) = A \cdot X + V$, $X \in \mathcal{X}$. Thus, we have $\text{Aff}(\mathcal{X}) = \mathcal{T}(\mathcal{X}) \times GL(\mathcal{X})$ *as sets*. The product of two elements $(T_V, A), (T_{V'}, A') \in \mathcal{T}(\mathcal{X}) \times GL(\mathcal{X})$ is given by $(T_V, A) \cdot (T_{V'}, A') = (T_{A \cdot V' + V}, A \cdot A')$. This is the archetype of a *semidirect product*: $\text{Aff}(\mathcal{X}) \cong \mathcal{T}(\mathcal{X}) \rtimes GL(\mathcal{X})$.

A subset $\mathcal{C} \subset \mathcal{X}$ is said to be *convex* if, for any $C_0, C_1 \in \mathcal{C}$ and $\lambda \in [0, 1]$, the *convex combination* $(1 - \lambda)C_0 + \lambda C_1$ (affine combination with non-negative coefficients) belongs to $\mathcal{C}$. Using geometric language, for any two distinct points $C_0, C_1 \in \mathcal{C}$, the *line segment*

$$[C_0, C_1] = \{(1 - \lambda)C_0 + \lambda C_1 \mid \lambda \in [0, 1]\} \subset \mathcal{X}$$

(with endpoints C_0 and C_1) is contained in $\mathcal{C}$.

It follows by a simple induction that $\mathcal{C} \subset \mathcal{X}$ is convex if and only if, for any *finite* subset $\{C_1, \ldots, C_m\} \subset \mathcal{C}$ and $\{\lambda_1, \ldots, \lambda_m\} \subset [0, 1]$ with $\sum_{i=1}^{m} \lambda_i = 1$, $m \geq 1$, the *convex combination* $\sum_{i=1}^{m} \lambda_i C_i$ belongs to $\mathcal{C}$. ($m = 1$ is a tautology and $m = 2$ recovers the definition of convexity.) Clearly, convexity depends on the affine structure of $\mathcal{X}$ only.

For an *arbitrary* subset $\mathcal{A} \subset \mathcal{X}$, we define the *convex hull* $[\mathcal{A}]$ of $\mathcal{A}$ by

$$[\mathcal{A}] = \left\{ \sum_{i=1}^{m} \lambda_i A_i \in \mathcal{X} \,\middle|\, \sum_{i=1}^{m} \lambda_i = 1, \{A_1, \ldots, A_m\} \subset \mathcal{A}, \{\lambda_1, \ldots, \lambda_m\} \subset [0, 1], m \geq 1 \right\}.$$

(1.1.1)

The name is justified since $\mathcal{A} \subset [\mathcal{A}]$ ($m = 1$), $[\mathcal{A}]$ is convex, and, for any convex set $\mathcal{C} \subset \mathcal{X}$, $\mathcal{A} \subset \mathcal{C}$ implies $[\mathcal{A}] \subset \mathcal{C}$. In other words, $[\mathcal{A}]$ is the *smallest* convex set which contains $\mathcal{A}$, or equivalently, it is the intersection of all convex sets that contain $\mathcal{A}$.

The frontispiece depicts a 13-sided regular polygon. It is the convex hull of the inscribed (unshaded) star-polygon $\{\frac{13}{6}\}$.

Remark. Many authors use the notation $\mathrm{conv}(\mathcal{A})$ for the convex hull of $\mathcal{A} \subset \mathcal{X}$. Our notation conforms with that of the line segment $[C_0, C_1]$, $C_0, C_1 \in \mathcal{X}$, $C_0 \neq C_1$, which is the convex hull of its endpoints C_0 and C_1. To improve clarity we will occasionally write: "the convex hull $[\mathcal{A}]$ of $\mathcal{A}$."

Some of the elementary (and easily verified) properties of convexity are as follows:

(1) The (non-empty) convex sets in a (one-dimensional) line are the intervals (including singletons).
(2) The images and inverse images of convex sets by *affine maps* are convex. In particular, affine subspaces are convex.
(3) Any intersection of convex sets is convex.

Removing the bounds on the parameters in the definition of the convex hull in (1.1.1), for $\mathcal{A} \subset \mathcal{X}$, we define the *affine span* (or *affine hull*) $\langle \mathcal{A} \rangle$ of $\mathcal{A}$ by

$$\langle \mathcal{A} \rangle = \left\{ \sum_{i=1}^{m} \lambda_i A_i \in \mathcal{X} \,\middle|\, \sum_{i=1}^{m} \lambda_i = 1, \{A_1, \ldots, A_m\} \subset \mathcal{A}, \{\lambda_1, \ldots, \lambda_m\} \subset \mathbb{R}, m \geq 1 \right\}.$$

(1.1.2)

Clearly $\mathcal{A} \subset [\mathcal{A}] \subset \langle \mathcal{A} \rangle$, $\langle \mathcal{A} \rangle \subset \mathcal{X}$ is an affine subspace, and for any affine subspace $\mathcal{X}' \subset \mathcal{X}$, $\mathcal{A} \subset \mathcal{X}'$ implies $[\mathcal{A}] \subset \langle \mathcal{A} \rangle \subset \mathcal{X}'$. In other words, $\langle \mathcal{A} \rangle$ is the smallest affine subspace that contains $\mathcal{A}$ (or $[\mathcal{A}]$), or equivalently, it is the intersection of all affine subspaces that contain $\mathcal{A} \subset [\mathcal{A}]$.

The concepts of *relative interior* and *relative boundary* of a set will be understood with respect the affine span of the set.

We need a *distance function* or *metric* (inducing a metric topology) on $\mathcal{X}$. To do this we endow $\mathcal{X}$ with a *norm* $|\cdot| : \mathcal{X} \to \mathbb{R}$, and assume that $\mathcal{X}$ is a *Minkowski space*, a finite dimensional *Banach space*. The norm has the following properties:

(1) $|X| \geq 0$, $X \in \mathcal{X}$, and equality holds if and only if $X = 0$;
(2) $|\lambda \cdot X| = |\lambda| \cdot |X|$, $X \in \mathcal{X}$, $\lambda \in \mathbb{R}$;
(3) $|X + X'| \leq |X| + |X'|$, $X, X' \in \mathcal{X}$.

The *distance function* $d : \mathcal{X} \times \mathcal{X} \to \mathbb{R}$ derived from a norm $| \cdot |$ on $\mathcal{X}$ is given by $d(X, X') = |X - X'|$, $X, X' \in \mathcal{X}$. As direct consequences of the definitions, the distance function satisfies the following properties:

(1) (Positivity) $d(X, X') \geq 0$, $X, X' \in \mathcal{X}$, and equality holds if and only if $X = X'$;
(2) (Symmetry) $d(X, X') = d(X', X)$, $X, X' \in \mathcal{X}$;
(3) (Triangle inequality) $d(X, X'') \leq d(X, X') + d(X', X'')$, $X, X', X'' \in \mathcal{X}$;
(4) (Translation invariance) $d(T_V(X), T_V(X')) = d(X + V, X' + V) = d(X, X')$, $V \in \mathcal{X}, X, X' \in \mathcal{X}$.

Endowed with this norm-induced distance, $\mathcal{X}$ is called an n-dimensional *Minkowski* space. Note that, since $\mathcal{X}$ is finite dimensional, completeness of $\mathcal{X}$ in the induced metric topology is automatic.

A norm on $\mathcal{X}$, in turn, can be derived from a *scalar product* $\langle \cdot, \cdot \rangle : \mathcal{X} \times \mathcal{X} \to \mathbb{R}$, a symmetric positive definite bilinear form on $\mathcal{X}$. The associated norm is defined by $|X| = \langle X, X \rangle^{1/2}$, $X \in \mathcal{X}$. Endowed with this structure, $\mathcal{X}$ is called an n-dimensional *Euclidean* space.

Although some of the constructions require only a norm in Minkowski space setting, unless stated otherwise (as in Section 3.9), in this book we will work in *Euclidean spaces*. Whenever available, we will make specific remarks about possible generalizations to Minkowski spaces.

The largest possible affine span of a finite set $\{C_0, \ldots, C_m\} \subset \mathcal{X}$ of $(m + 1)$ points is m-dimensional. This is the case if and only if the set $\{C_0, \ldots, C_m\}$ is *affinely independent* (that is, $\{C_i - C_0\}_{1 \leq i \leq m} \subset \mathcal{X}$ is *linearly* independent). In this case the convex hull $\Delta = [C_0, \ldots, C_m]$ is said to be an m-*simplex* with *vertices* $\{C_0, \ldots, C_m\}$. Thus, a 2-simplex is a triangle, a 3-simplex is a tetrahedron, a 4-simplex is a pentatope, etc.

As a simple example, taking the vertices to be the elements of the standard basis in $\mathbb{R}^{n+1}$ as vertices, we obtain the *standard* (*regular*) n-*simplex*:

$$\Delta_n = \left\{ (\lambda_0, \ldots, \lambda_n) \in \mathbb{R}^{n+1} \,\middle|\, \sum_{i=0}^{n} \lambda_i = 1, \{\lambda_0, \ldots, \lambda_n\} \subset [0, 1] \right\}.$$

Any two n-simplices $\Delta, \Delta' \subset \mathcal{X}$, $\dim \mathcal{X} = n$, are affine equivalent, that is, there exists $\phi \in \text{Aff}(\mathcal{X})$ such that $\phi(\Delta) = \Delta'$. Moreover, ϕ is uniquely determined by its action on the respective sets of vertices of Δ and Δ'.

A compact convex set $\mathcal{C} \subset \mathcal{X}$ is called a *convex body* (in $\mathcal{X}$) if it has non-empty interior *in* $\mathcal{X}$. Clearly, a compact convex set $\mathcal{C}$ is a convex body in its affine span $\langle \mathcal{C} \rangle$.

Motivated by this, we define the *dimension* of a convex set $\mathcal{C} \subset \mathcal{X}$ as the dimension of its affine span: $\dim \mathcal{C} = \dim \langle \mathcal{C} \rangle$. It follows that a compact convex set $\mathcal{C} \subset \mathcal{X}$ is a convex body (in $\mathcal{X}$) if and only if $\dim \mathcal{C} = n (= \dim \mathcal{X})$.

Remark. Several authors define a convex body as a compact convex subset in $\mathcal{X}$ without requiring a non-empty interior within $\mathcal{X}$.

The distance function d on $\mathcal{X}$ defines the *open metric ball*

$$\mathcal{B}_r(C) = \{X \in \mathcal{X} \mid d(X, C) < r\}$$

with *center* at $C \in \mathcal{X}$ and *radius* $r > 0$. (The triangle inequality implies that the open metric balls are open in the metric topology defined by d.) Replacing the strict inequality in the definition with $\leq$ we obtain the definition of the *closed metric ball* $\bar{\mathcal{B}}_r(C)$, where the overline is justified as this is the topological closure of $\mathcal{B}_r(C)$. Since $\mathcal{X}$ is finite dimensional, the closed metric balls are *compact*. Clearly, metric balls (open or closed) are convex (even in Minkowski space).

The boundary

$$\mathcal{S}_r(C) = \partial \mathcal{B}_r(C) = \partial \bar{\mathcal{B}}_r(C) = \{X \in \mathcal{X} \mid d(X, C) = r\}$$

is called the (metric) *sphere* with center C and radius $r > 0$. The sphere $\mathcal{S}_r(C)$ is closed in $\bar{\mathcal{B}}_r(C)$, and hence it is compact. The *unit* radius is suppressed from the notation, so that $\mathcal{B}(C)$ denotes the open *unit* ball with center at C, etc. Similarly, the origin as the center is not indicated, so that $\bar{\mathcal{B}}_r$ stands for the closed ball with center at the origin and radius r, and $\mathcal{S}$ is the *unit sphere*, etc.

An *ellipsoid* $\mathcal{E} \subset \mathcal{X}$ is the image of the closed unit ball $\bar{\mathcal{B}}$ (or any other closed ball) under an affine transformation $\phi \in \mathrm{Aff}(\mathcal{X})$, that is, we have $\mathcal{E} = \phi(\bar{\mathcal{B}})$. In particular, any two ellipsoids are affine equivalent.

For $\mathcal{A}, \mathcal{A}' \subset \mathcal{X}$ we define $d(\mathcal{A}, \mathcal{A}') = \inf\{d(A, A') \mid A \in \mathcal{A}, A' \in \mathcal{A}'\}$. In particular, we set $d(X, \mathcal{A}) = d(\{X\}, \mathcal{A})$, $X \in \mathcal{X}$. By the triangle inequality, we have

$$|d(X, \mathcal{A}) - d(X', \mathcal{A})| \leq d(X, X'), \quad X, X' \in \mathcal{X},$$

in particular, $d(\cdot, \mathcal{A}) : \mathcal{X} \to \mathbb{R}$ is a continuous function.

In addition, once again by the triangle inequality, we have $d(\cdot, \mathcal{A}) = d(\cdot, \bar{\mathcal{A}})$. Finally, due to the compactness of closed metric balls, given $X \in \mathcal{X}$, there exists $A \in \bar{\mathcal{A}}$ such that $d(X, A) = d(X, \bar{\mathcal{A}}) = d(X, \mathcal{A})$. In particular, $d(X, \mathcal{A}) = 0$ if and only if $X \in \bar{\mathcal{A}}$.

Given $r > 0$, the *open r-neighborhood* of a subset $\mathcal{A} \subset \mathcal{X}$ is defined by

$$\mathcal{A}_r = \bigcup_{A \in \mathcal{A}} \mathcal{B}_r(A) = \{X \in \mathcal{X} \mid d(X, \mathcal{A}) < r\} = \mathcal{A} + \mathcal{B}_r.$$

The equivalence of the first two definitions, the second equality, follows from the definitions. In the last definition we used the *Minkowski sum* of two subsets $\mathcal{A}, \mathcal{A}' \subset \mathcal{X}$ defined by

$$\mathcal{A} + \mathcal{A}' = \{X + X' \mid X \in \mathcal{A}, X' \in \mathcal{A}'\}.$$

With this, the third definition follows from translation invariance of the distance function d.

We have $(\mathcal{A}_r)_{r'} = \mathcal{A}_{r+r'}$, $r, r' > 0$. By the first definition (as a union of open balls), $\mathcal{A}_r$, $r > 0$, is *open* for any $\mathcal{A} \subset \mathcal{X}$. Finally, we also have $\bar{\mathcal{A}} = \bigcap_{r>0} \mathcal{A}_r$.

For $r \geq 0$, the *closed r-neighborhood* is defined for a *closed* subset $\mathcal{A} \subset \mathcal{X}$ (a natural assumption in view of $d(\cdot, \mathcal{A}) = d(\cdot, \bar{\mathcal{A}})$) by

$$\bar{\mathcal{A}}_r = \bigcup_{A \in \mathcal{A}} \bar{B}_r(A) = \{X \in \mathcal{X} \mid d(X, \mathcal{A}) \leq r\} = \mathcal{A} + \bar{B}_r.$$

As before, we have $(\overline{\bar{\mathcal{A}}_r})_{r'} = \bar{\mathcal{A}}_{r+r'}$, $r, r' > 0$. The second definition implies that $\bar{\mathcal{A}}_r$, $r \geq 0$, is closed.

A function $f : \mathcal{D} \to \mathbb{R}$ with domain $\mathcal{D} \subset \mathcal{X}$, is called *convex* if, for any line segment $[X_0, X_1] \subset \mathcal{D}$ and $\lambda \in [0, 1]$, we have

$$f((1 - \lambda)X_0 + \lambda X_1) \leq (1 - \lambda)f(X_0) + \lambda f(X_1).$$

Concavity is defined by reversing the inequality sign.

If $f : \mathcal{D} \to \mathbb{R}$ is a convex function defined on a *convex* set $\mathcal{D} \subset \mathcal{X}$ then the *level-sets* $\{X \in \mathcal{D} \mid f(X) < r\}$ and $\{X \in \mathcal{D} \mid f(X) \leq r\}$, $r \in \mathbb{R}$, are convex.

As a simple but important example, we now claim that, for a convex set $\mathcal{C} \subset \mathcal{X}$, the distance function $d(., \mathcal{C}) : \mathcal{X} \to \mathbb{R}$ is convex.

Let $X_0, X_1 \in \mathcal{X}$. Given $\epsilon > 0$, choose $C_0, C_1 \in \mathcal{C}$ such that $d(X_0, C_0) \leq d(X_0, \mathcal{C}) + \epsilon$ and $d(X_1, C_1) \leq d(X_1, \mathcal{C}) + \epsilon$. Then $(1 - \lambda)C_0 + \lambda C_1 \in \mathcal{C}$, $\lambda \in [0, 1]$, and we have

$$
\begin{aligned}
d((1 - \lambda)X_0 + \lambda X_1, \mathcal{C}) &\leq d((1 - \lambda)X_0 + \lambda X_1, (1 - \lambda)C_0 + \lambda C_1) \\
&= |(1 - \lambda)(X_0 - C_0) + \lambda(X_1 - C_1)| \\
&\leq (1 - \lambda)|X_0 - C_0| + \lambda|X_1 - C_1| \\
&\leq (1 - \lambda)d(X_0, \mathcal{C}) + \lambda d(X_1, \mathcal{C}) + \epsilon.
\end{aligned}
$$

Letting $\epsilon \to 0$, the claim follows. (Note that, as the proof shows, this holds in Minkowski space.)

In particular, we see that the open r-neighborhood of a convex set is convex, and the closed r-neighborhood of a closed (compact) convex set is also a closed (compact) convex set.

Example 1.1.1. In $\mathcal{X}$ the (affine) *hyperplane* containing $X_0 \in \mathcal{X}$ and having normal vector $0 \neq N \in \mathcal{X}$ is given by $\mathcal{H} = \{X \in \mathcal{X} \mid \langle X - X_0, N \rangle = 0\}$. The *closed half-space* $\mathcal{G} = \{X \in \mathcal{X} \mid \langle X - X_0, N \rangle \leq 0\}$ then has boundary $\partial \mathcal{G} = \mathcal{H}$ and *outward* normal vector N. Replacing $\leq$ with strict inequality in the definition of $\mathcal{G}$, we obtain the *open half-space* int $\mathcal{G}$. Switching the sign of the normal vector N gives the complementary closed half-space $\mathcal{G}' = \mathcal{X} \setminus \text{int}\,\mathcal{G}$ with $\partial \mathcal{G}' = \mathcal{H}$. We then have $\mathcal{G} \cap \mathcal{G}' = \mathcal{H}$ and $\mathcal{X} = \text{int}\,\mathcal{G} \cup \text{int}\,\mathcal{G}' \cup \mathcal{H}$, a disjoint union. Clearly, $\mathcal{G}$, $\mathcal{G}'$, int $\mathcal{G}$, int $\mathcal{G}'$, and $\mathcal{H}$ are convex sets.

In Minkowski space setting the construction of these objects requires the replacement of the scalar product $\langle \cdot, N \rangle$ by a non-zero linear functional $\phi : \mathcal{X} \to \mathbb{R}$, an element of the dual $\mathcal{X}^*$.

Further properties of convexity are as follows:

(1) The intersection of closed half-spaces is a closed convex set.
(2) The closure and the interior of a convex set are convex.

The converse of (1) is a classical fact (to be proved in the next section): A closed convex set is the intersection of closed half-spaces. In particular, taking only finitely many closed half-spaces one arrives at the concept of *polyhedron*.

B. Distances Between Convex Sets. In this subsection we prove the celebrated *Blaschke selection theorem* about bounded sequences of compact convex sets.

Let $\mathfrak{C} = \mathfrak{C}_\mathcal{X}$ denote the set of all *compact* subsets of $\mathcal{X}$. (Whenever convenient, the subscript will be suppressed.) We define the (*Pompéiu–*)*Hausdorff(–Blaschke*) *distance function* $d_H : \mathfrak{C} \times \mathfrak{C} \to \mathbb{R}$ by

$$d_H(C, C') = \inf\{r \geq 0 \,|\, C \subset \overline{C'}_r, C' \subset \bar{C}_r\}$$

$$= \max\left(\sup_{X \in C} d(X, C'), \sup_{X' \in C'} d(X', C)\right)$$

$$= \inf\{r \geq 0 \,|\, C \subset C' + \bar{B}_r, C' \subset C + \bar{B}_r\}, \quad C, C' \in \mathfrak{C}.$$

The equivalence of these definitions is clear by noting that

$$C \subset \overline{C'}_r \quad \Leftrightarrow \quad \sup_{X \in C} d(X, C') \leq r, \quad r \geq 0.$$

As the name suggests, this notion of distance was first studied by [Hausdorff], and somewhat earlier, a non-symmetric version, by [Pompéiu]. For convex sets it was first used by Blaschke. By compactness, the infima and suprema are attained.

The Hausdorff distance d_H is obviously symmetric:

$$d_H(C, C') = d_H(C', C), \quad C, C' \in \mathfrak{C},$$

and satisfies the triangle inequality:

$$d_H(C, C'') \leq d_H(C, C') + d_H(C', C''), \quad C, C', C'' \in \mathfrak{C}.$$

Clearly, $d_H(C, C') \geq 0, C, C' \in \mathfrak{C}$; and, by compactness of C and C', $d_H(C, C') = 0$ if and only if $C = C'$.

From now on, unless stated otherwise, we use the Hausdorff distance d_H as the natural metric (inducing the natural metric topology) on $\mathfrak{C}$.

Theorem 1.1.2. $\mathfrak{C}$ *is a complete metric space.*

Proof. We first claim that if $\{C_k\}_{k \geq 1} \subset \mathfrak{C}$ is a *decreasing* sequence of compact sets then its (Hausdorff) limit exists and is equal to the (non-empty) intersection $C = \bigcap_{k \geq 1} C_k$.

Indeed, otherwise, for some $\epsilon > 0$ and then for each $k \geq 1$, the kth member C_k is not contained in the (open) ϵ-neighborhood C_ϵ of C. Then, the sequence of (non-empty and compact) differences $\{A_k\}_{k \geq 1} \subset \mathfrak{C}$, $A_k = C_k \setminus C_\epsilon$, $k \geq 1$, is decreasing. Therefore it has a non-empty intersection $A = \bigcap_{k \geq 1} A_k$. By construction, A is disjoint from C, but also $A = \bigcap_{k \geq 1} A_k \subset \bigcap_{k \geq 1} C_k = C$, a contradiction. The claim follows.

Turning to the proof of the theorem, let $\{C_k\}_{k \geq 1} \subset \mathfrak{C}$ be a *Cauchy sequence*. For $k \geq 1$, by the Cauchy property, the set $A_k = \overline{\bigcup_{l \geq k} C_l}$ is (closed and) bounded, hence compact. The sequence $\{A_k\}_{k \geq 1} \subset \mathfrak{C}$ is decreasing, so that, by the first claim, it (Hausdorff) converges to the intersection $A = \bigcap_{k \geq 1} A_k \in \mathfrak{C}$. We claim that $\lim_{l \to \infty} C_l = A$.

Indeed, let $\epsilon > 0$, and choose $k_0 \geq 1$ such that $A_k \subset A_\epsilon$, $k \geq k_0$. By construction, we also have $C_l \subset A_\epsilon$, $l \geq k_0$. Using the Cauchy property again, there exists $K_0 \geq k_0$ such that $C_l \subset (C_m)_\epsilon$, $l, m \geq K_0$. Now, taking unions we obtain $\bigcup_{l \geq k} C_l \subset (C_m)_\epsilon$, $k, m \geq K_0$. Taking closure (with a possibly larger k_0) we arrive at $A_k \subset (C_m)_\epsilon$. Thus, $A \subset (C_m)_\epsilon$, or equivalently, $d_H(C_m, A) \leq \epsilon$, $m \geq K_0$. The theorem follows.

Remark. Theorem 1.1.2 is classical and contained in several books; see, for example, [Schneider 2, 1.8] and [Berger, 9.11.2]. The proof above shows that this is true in *any metric space* with the property that the closed and bounded subsets are compact (such as the finite dimensional vector space $\mathcal{X}$).

We now introduce the following convenient terminology: a sequence *subconverges* if it has a convergent subsequence.

Theorem 1.1.3. *In $\mathfrak{C}$ every bounded sequence subconverges. Consequently, every closed and bounded subset of $\mathfrak{C}$ is compact; in particular, $\mathfrak{C}$ is locally compact.*

Proof. The proof is via a diagonalization method in selecting an infinite sequence of nested subsequences from a given *bounded* sequence $\{C_k^0\}_{k \geq 1} \subset \mathfrak{C}$ contained in a cube $Q \subset \mathcal{X}$ of edge length a. (To be specific, a cube can be chosen with faces parallel to the coordinate hyperplanes with respect to a fixed isomorphism $\mathcal{X} \cong \mathbb{R}^n$.)

First, systematically halving Q by hyperplanes parallel to the faces, for each $m \geq 1$, we can split Q into $(2^n)^m = 2^{mn}$ subcubes. Each subcube has edge length $a/2^m$ (and diagonal length $a\sqrt{n}/2^m$). For any $C \in \mathfrak{C}$, let $A_m(C)$ denote the union of the cubes in the mth subdivision which intersect C.

To begin the selection process, consider the sequence $\{A_1(C_k^0)\}_{k \geq 1}$. Now, the crux is that there are only finitely many subcubes in a given (here the first) level, so that this sequence must have a *constant subsequence* $\{A_1 = A_1(C_k^1)\}_{k \geq 1}$ with $\{C_k^1\}_{k \geq 1} \subset \{C_k^0\}_{k \geq 1}$.

Continuing inductively, using the mth subdivision, from $\{C_k^{m-1}\}_{k \geq 1}$ we can select a subsequence $\{C_k^m\}_{k \geq 1}$ with constant $\{A_m = A_m(C_k^m)\}_{k \geq 1}$.

Summarizing, we have

$$\{C_k^0\}_{k \geq 1} \supset \{C_k^1\}_{k \geq 1} \supset \ldots \supset \{C_k^{m-1}\}_{k \geq 1} \supset \{C_k^m\}_{k \geq 1} \supset \ldots \qquad (1.1.3)$$

with

$$\mathcal{A}_m = \mathcal{A}_m(\mathcal{C}_k^m), \quad k \geq 1.$$

By construction, given $m \geq 1$, any two members $\mathcal{C}_k^m$ and $\mathcal{C}_{k'}^m$, $k, k' \geq 1$, share a common $\mathcal{A}_m$ so that we have

$$d_H(\mathcal{C}_k^m, \mathcal{C}_{k'}^m) \leq \frac{a\sqrt{n}}{2^m}.$$

Using (1.1.3), we obtain

$$d_H(\mathcal{C}_k^m, \mathcal{C}_{k'}^{m'}) \leq \frac{a\sqrt{n}}{2^m}, \quad m' \geq m, \ k, k' \geq 1.$$

We now select the "diagonal" $\mathcal{C}_m = \mathcal{C}_m^m$ and have

$$d_H(\mathcal{C}_m, \mathcal{C}_{m'}) \leq \frac{a\sqrt{n}}{2^m}, \quad m' \geq m.$$

This means that $\{\mathcal{C}_m\}_{m \geq 1}$ is a Cauchy sequence, thereby convergent (Theorem 1.1.2). Since $\{\mathcal{C}_m\}_{m \geq 1}$ is a (convergent) subsequence of the original sequence $\{\mathcal{C}_k^0\}_{k \geq 1}$, the theorem follows.

We let $\mathfrak{K} = \mathfrak{K}_{\mathcal{X}} \subset \mathfrak{C}_{\mathcal{X}}$ denote the subspace of all compact *convex* subsets of $\mathcal{X}$.

Theorem 1.1.4. $\mathfrak{K}$ *is closed in* $\mathfrak{C}$.

Proof. We show that $\mathfrak{C} \setminus \mathfrak{K}$ is open. Let $\mathcal{C} \in \mathfrak{C}$ be a *non-convex* (compact) subset in $\mathcal{X}$. Choose $C_\lambda = (1 - \lambda)C_0 + \lambda C_1 \notin \mathcal{C}$ with $C_0, C_1 \in \mathcal{C}$ and $0 < \lambda < 1$. Let $\epsilon > 0$ be such that the open metric ball $\mathcal{B}_\epsilon(C_\lambda)$ is disjoint from $\mathcal{C}$.

We claim that the *Hausdorff* metric ball with center at $\mathcal{C}$ and radius $\epsilon/2$ is disjoint from $\mathfrak{K}$. Indeed, let $\mathcal{C}' \in \mathfrak{C}$ be such that $d_H(\mathcal{C}, \mathcal{C}') < \epsilon/2$. We need to show that $\mathcal{C}'$ is *not* convex. By the definition of the Hausdorff distance, we have $C_0' \in \mathcal{B}_{\epsilon/2}(C_0)$ and $C_1' \in \mathcal{B}_{\epsilon/2}(C_1)$, for some $C_0', C_1' \in \mathcal{C}'$. Setting $C_\lambda' = (1 - \lambda)C_0' + \lambda C_1'$, we then have $C_\lambda' \in \mathcal{B}_{\epsilon/2}(C_\lambda)$. But $\mathcal{B}_\epsilon(C_\lambda)$ does not intersect $\mathcal{C}$ while $d_H(\mathcal{C}, \mathcal{C}') < \epsilon/2$. Thus C_λ' cannot be in $\mathcal{C}'$. We obtain that $\mathcal{C}'$ is not convex. The theorem follows.

As an immediate corollary of Theorems 1.1.3 and 1.1.4, we obtain Blaschke's selection theorem:

Theorem 1.1.5 ([Blaschke 1]). *A bounded sequence in* $\mathfrak{K}$ *subconverges within* $\mathfrak{K}$.

Given $\mathcal{A} \subset \mathcal{X}$, we let

$$\mathfrak{K}_\mathcal{A} = \{\mathcal{C} \in \mathfrak{K} \mid \mathcal{C} \subset \mathcal{A}\}.$$

Note that this notation conforms with our earlier $\mathfrak{K}_\mathcal{X} = \mathfrak{K}$.

Corollary 1.1.6. *Let* $C_0 \in \mathfrak{K}$. *Then* $\mathfrak{K}_{C_0} \subset \mathfrak{K}$ *is compact; in particular, any sequence in* $\mathfrak{K}_{C_0}$ *subconverges within* $\mathfrak{K}_{C_0}$.

Proof. $\mathfrak{K}_{C_0}$ is bounded since

$$\sup_{C \in \mathfrak{K}_{C_0}} d_H(C, C_0) = \sup\{d(C, C') \mid C, C' \in C_0\}.$$

(The right-hand side is the *diameter* of C_0; see Section 1.5.) The rest follows from Theorems 1.1.3 and 1.1.4.

The space of all *convex bodies* in $\mathcal{X}$ will be denoted by $\mathfrak{B} = \mathfrak{B}_{\mathcal{X}}$. We thus have the inclusions

$$\mathfrak{B} \subset \mathfrak{K} \subset \mathfrak{C}.$$

We claim that $\mathfrak{B}$ is *dense* and *open* in $\mathfrak{K}$. Density is clear since, for $C \in \mathfrak{K}$ and $r > 0$, we have $\bar{C}_r \in \mathfrak{B}$ and $d_H(C, \bar{C}_r) = r$.

To show openness, let $C \in \mathfrak{B}$ with $B_r(O) \subset C$, $O \in \text{int}\,C$ and $r > 0$. Let $\epsilon > 0$ and $C' \in \mathfrak{K}$ such that $d_H(C, C') < \epsilon$. Assume, on the contrary, that the interior of C' is empty. Then C' is contained in a hyperplane $\mathcal{H}$. Since C is ϵ-close to C', we obtain that C is contained in a slab of width 2ϵ bounded by two hyperplanes parallel to $\mathcal{H}$. For ϵ small enough this cannot happen. (For a more general statement, see [Schneider 2, Lemma 1.8.14].)

In particular, $\mathfrak{B} \subset \mathfrak{K}$ is not a closed subspace. (This can also be seen directly: A sequence of nested closed metric balls converging to a point is a sequence in $\mathfrak{B}$ but its limit is in $\mathfrak{K} \setminus \mathfrak{B}$.)

Remark 1. Once again, Theorems 1.1.3–1.1.6 are classical; for recent expositions, see [Schneider 2, 1.8] and [Gruber, 6.1]. The proofs can easily be fitted to Minkowski spaces as in [Valentine, Part III] which itself is adapted from a more general approach of [Whyburn] for compact metric spaces.

Remark 2. Given $C \in \mathfrak{K}_{\mathcal{X}}$ and $\epsilon > 0$, C can be covered by finitely many open metric balls of radius ϵ and *centers in* C. The convex hull of these centers is a polyhedron $\mathcal{P}$ which, by the definition of the Hausdorff distance, satisfies $d_H(C, \mathcal{P}) \leq \epsilon$. We obtain that compact convex sets can be approximated by polyhedra.

In addition, choosing a specific isomorphism $\mathcal{X} \cong \mathbb{R}^n$ and selecting rational points (in $\mathbb{Q}^n$) for the centers of the balls, we see that $\mathfrak{K}_{\mathcal{X}}$ is *separable* (with respect to the Hausdorff metric).

Another concept of distance is the *symmetric difference distance* $d_\Delta : \mathfrak{B} \times \mathfrak{B} \to \mathbb{R}$ defined by using the *n-dimensional volume* $\text{vol} = \text{vol}_n$ (the n-dimensional Lebesgue measure). It is given by

$$d_\Delta(C, C') = \text{vol}\,(C \Delta C') = \text{vol}\,(C \cup C') - \text{vol}\,(C \cap C') = \text{vol}\,(C \setminus C') + \text{vol}\,(C' \setminus C), \quad C, C' \in \mathfrak{B},$$

where we used the symmetric difference

$$\mathcal{C}\Delta\mathcal{C}' = (\mathcal{C} \cup \mathcal{C}') \setminus (\mathcal{C} \cap \mathcal{C}') = (\mathcal{C} \setminus \mathcal{C}') \cup (\mathcal{C}' \setminus \mathcal{C}).$$

We obviously have symmetry:

$$d_\Delta(\mathcal{C},\mathcal{C}') = d_\Delta(\mathcal{C}',\mathcal{C}), \quad \mathcal{C},\mathcal{C}' \in \mathfrak{B}.$$

The triangle inequality is a direct consequence of the set-theoretical identity

$$(\mathcal{C}\Delta\mathcal{C}')\Delta(\mathcal{C}'\Delta\mathcal{C}'') = \mathcal{C}\Delta\mathcal{C}'', \quad \mathcal{C},\mathcal{C}',\mathcal{C}'' \in \mathfrak{B}.$$

Finally, $d_\Delta(\mathcal{C},\mathcal{C}') \geq 0$, $\mathcal{C},\mathcal{C}' \in \mathfrak{B}$, clearly holds. For the non-trivial direction in proving positivity, that is

$$d_\Delta(\mathcal{C},\mathcal{C}') = 0 \quad \Rightarrow \quad \mathcal{C} = \mathcal{C}', \quad \mathcal{C},\mathcal{C}' \in \mathfrak{B},$$

one needs to use an elementary property of convex bodies as in Proposition 1.1.7 below; see Problem 6.

Remark 1. The symmetric difference metric was introduced and studied by [Dinghas]. It can be generalized to any *measure space* $(\mathcal{X},\mathfrak{A},\mu)$, where $\mathfrak{A}$ is a σ-algebra of subsets of $\mathcal{X}$, and μ is a σ-finite measure on $\mathfrak{A}$. Adopting the definition as above: $d_\Delta(\mathcal{A},\mathcal{A}') = \mu(\mathcal{A}\Delta\mathcal{A}')$, $\mathcal{A},\mathcal{A}' \in \mathfrak{A}$, we immediately see that d_Δ is a *pseudometric* (non-negative, symmetric, and satisfies the triangle inequality). It becomes a metric on the quotient of $\mathfrak{A}$ by the equivalence relation: $\mathcal{A} \sim \mathcal{A}'$ if and only if $\mu(\mathcal{A}\Delta\mathcal{A}') = 0$. Finally, note that d_Δ on this quotient defines a separable metric space if and only if $L^2(\mathcal{X},\mathfrak{A},\mu)$ is separable.

Remark 2. As shown by [Shephard–Webster], the Hausdorff distance and the symmetric difference distance induce the same topology on $\mathfrak{B}$. Explicit inequalities asserting this have been obtained by [Groemer 1] as follows. We have

$$d_\Delta(\mathcal{C},\mathcal{C}') \leq c \cdot d_H(\mathcal{C},\mathcal{C}'), \quad \mathcal{C},\mathcal{C}' \in \mathfrak{B},$$

where

$$c = \frac{2\kappa_n}{2^{1/n} - 1}\left(\frac{\max(D_\mathcal{C},D_{\mathcal{C}'})}{2}\right)^{n-1} \quad \text{and} \quad \kappa_n = \text{vol}(\mathcal{B}) = \frac{\pi^{n/2}}{\Gamma\left(\frac{n}{2}+1\right)}.$$

($D_\mathcal{C}$ and $D_{\mathcal{C}'}$ are the diameters of $\mathcal{C}$ and $\mathcal{C}'$; see Section 1.5 below.)

For the reverse estimate, we need to assume that $\text{int}(\mathcal{C} \cap \mathcal{C}') \neq \emptyset$. We then have

$$d_H(\mathcal{C},\mathcal{C}') \leq c' \cdot d_\Delta(\mathcal{C},\mathcal{C}')^{1/n},$$

where

$$c' = \left(\frac{n}{\kappa_{n-1}}\right)^{1/n} \left(\frac{\max(D_C, D_{C'})}{r_{C \cap C'}}\right)^{(n-1)/n},$$

and $r_{C \cap C'}$ is the *inradius* of the convex body $C \cap C' \in \mathfrak{B}$ (see again Section 1.5).

C. Topology of Convex Sets. In this section we derive several geometric and topological properties of convex sets; in particular, we prove that the boundary of a compact convex set is homeomorphic with a sphere.

For $\lambda \in \mathbb{R}$, $\lambda \neq 0$, and $C \in \mathcal{X}$, we denote by $S_{\lambda,C} \in \text{Aff}(\mathcal{X})$ the *(central) similarity* or *homothety* in $\mathcal{X}$ with center C and ratio λ:

$$S_{\lambda,C}(X) = C + \lambda(X - C) = (1 - \lambda)C + \lambda X, \quad X \in \mathcal{X}. \tag{1.1.4}$$

Within the affine group $\text{Aff}(\mathcal{X})$, the central similarities $S_{\lambda,C}$, $\lambda \in \mathbb{R}$, $\lambda \neq 0$, $C \in \mathcal{X}$, and the translation group $\mathcal{T}(\mathcal{X})$ form the *group of dilatations* $\text{Dil}(\mathcal{X})$. We have the chain of normal subgroups

$$\mathcal{T}(\mathcal{X}) \lhd \text{Dil}(\mathcal{X}) \lhd \text{Aff}(\mathcal{X}).$$

In the forthcoming chapters we will see many applications of central similarities. As for now, we use this to prove the following simple result:

Proposition 1.1.7. *Any ray emanating from a (relative) interior point of a convex set $C \subset \mathcal{X}$ meets the (relative) boundary of C in at most one point.*

Proof. We may assume that $C \in \mathfrak{B}$. Let $O \in \text{int}\, C$ so that, for some $r > 0$, we have $\mathcal{B}_r(O) \subset C$. Being convex, the intersection of a ray emanating from O with C is a (connected) line segment with one endpoint at O.

First, assume that this line segment is finite with other endpoint $C \in \partial C$. We have $[O, C) \subset C$. (We adopt the usual interval notation for line segments: replacing a square bracket with a parenthesis means that the respective endpoint is not counted in.) Let $O' \in [O, C)$. By convexity, for $0 \leq \lambda < 1$, we have

$$S_{1-\lambda,O'}(\mathcal{B}_r(O)) = \lambda O' + (1 - \lambda)\mathcal{B}_r(O) = \mathcal{B}_{(1-\lambda)r}(O_\lambda) \subset C,$$

where $O_\lambda = (1 - \lambda)O + \lambda O'$. Hence $O_\lambda, 0 \leq \lambda < 1$, is contained in the interior of C. Since $O' \in [O, C)$ was arbitrary, we obtain that the entire line segment $[O, C)$ is contained in the interior of C. (See Figure 1.1.1.)

Applying this argument for the points in $\mathcal{B}_r(O)$, we see that the entire cone $[\mathcal{B}_r(O), C] \setminus \{C\}$ (without the vertex C) is contained in the interior of C. To complete this case, we claim that the line extension of $[O, C]$ beyond C cannot meet the boundary of C again. Otherwise, let $C' \in \partial C$ be a point on this extension beyond C.

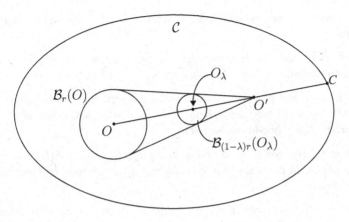

Fig. 1.1.1

Applying the construction above, we obtain that $[\mathcal{B}_r(O), C'] \setminus \{C'\}$ is contained in the interior of C. Since $C \in [O, C')$, this means that $C \in \text{int}\, C$, a contradiction.

Second, assume that the ray emanating from $O \in \text{int}\, C$ is contained in C. Sliding a point O' along this ray away from O and applying the construction above, we see that the entire ray is contained in the interior of C. The proposition follows.

Remark. As in the first case of the proof, let $\mathcal{B}_r(O) \subset \text{int}\, C$ and $C \in \partial C$. The union of all rays emanating from points in $\mathcal{B}_r(O)$ and passing through C is a *double cone* with vertex at C. Deleting the vertex, it falls into two open connected components: the open bounded cone $[\mathcal{B}_r(O), C] \setminus \{C\}$ and an open infinite cone $\mathcal{K}$ (the union of the part of the rays beyond C). By the proof, $\mathcal{K}$ is contained in the *exterior* of C. (See Figure 1.1.2.)

Corollary 1.1.8. *For a convex set C, we have* $\text{int}\, C = \text{int}\, \bar{C}$. *If* $\text{int}\, C \neq \emptyset$ *then* $\overline{\text{int}\, C} = \bar{C}$.

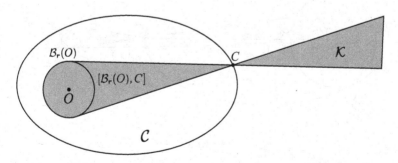

Fig. 1.1.2

The following immediate corollary of Proposition 1.1.7 will be used repeatedly
in the future:

Corollary 1.1.9. *Any ray that emanates from an interior point of a convex body
intersects the boundary of the convex body at a single point.*

Given $C \in \mathfrak{B}$ and $O \in \operatorname{int} C$, this corollary asserts that the radial projection
$\pi_O : S(O) \to \partial C$ of the *unit* sphere $S(O) = \partial \mathcal{B}(O)$ with center at O is well
defined. π_O is clearly one-to-one and onto.

Proposition 1.1.10. *For $C \in \mathfrak{B}$, the radial projection $\pi_O : S(O) \to \partial C$ is a
homeomorphism. In particular, the boundary of a convex body is homeomorphic
with the sphere.*

Proof. Since $S(O)$ is compact, we need only to show that π_O is (sequentially)
continuous.

Let $\{X_k\}_{k \geq 1} \subset S(O)$ converge to $X \in S(O)$ and $C = \pi_O(X)$. (We may assume
that $X_k \neq X, k \geq 1$.) Choose $r > 0$ such that $\mathcal{B}_r(O) \subset \operatorname{int} C$ and consider the double
cone $[\mathcal{B}_r(O), C] \cup \mathcal{K}$ as in the remark after Proposition 1.1.7. For k sufficiently large,
only a finite segment of the ray emanating from O and passing through X_k will
be outside of this double cone, in fact, as $k \to \infty$, this segment converges to the
vertex C. (See Figure 1.1.3.)

By the proof of Proposition 1.1.7, the double cone meets the boundary of C only
at the vertex C. Thus, for large k, $\pi_O(X_k) \in \partial C$ must be contained in this line
segment, and as $k \to \infty$, must therefore converge to $C = \pi_O(X)$. The proposition
follows.

Remark. Note that the construction also implies that $\pi_O : S(O) \to \partial C$ is Lipschitz
continuous. Therefore, by the classical Rademacher theorem, π_O is differentiable
almost everywhere.

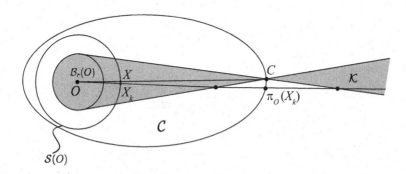

Fig. 1.1.3

We now briefly return to the Hausdorff metric and give a characterization of
convergence (of compact sets) in terms of convergent sequences of points. This
will give an insight to the Painlevé–Kuratowski convergence to be discussed in
Section 1.6.

Theorem 1.1.11. *The Hausdorff convergence* $\lim_{k \to \infty} C_k = C$ *in* $\mathfrak{K}$ *is equivalent to the following two conditions:*

(1) Each point $C \in C$ *is the limit of a sequence* $\{C_k\}_{k \geq 1}$, $C_k \in C_k$, $k \geq 1$;
(2) The limit of a convergent sequence $\{C_{k_l}\}_{l \geq 1}$, $C_{k_l} \in C_{k_l}$, $l \geq 1$, *is contained in* C.

Proof. ($\Rightarrow$) Assume that in $\mathfrak{K}$ we have $d_H(C_k, C) \to 0$ as $k \to \infty$. Let $C \in C$. For each $k \geq 1$, select $C_k \in C_k$ such that $d(C, C_k) = d(C, C_k)$. (By convexity of C_k, the point C_k is actually unique; see Problem 7.) Since $d(C, C_k) \leq d_H(C, C_k)$, $k \geq 1$, we obtain $\lim_{k \to \infty} C_k = C$. (1) follows.

Next, assume that the sequence $\{C_{k_l}\}_{l \geq 1}$ is as in (2), but its limit C is not in C. Since C is compact, for some $\epsilon > 0$, the open ϵ-neighborhood C_ϵ and the open metric ball $\mathcal{B}_\epsilon(C)$ are disjoint. For large $l \geq 1$, however, we have $d(C_{k_l}, C) < \epsilon$ and $C_{k_l} \in C_{k_l} \subset C_\epsilon$. This is a contradiction. (2) follows.

($\Leftarrow$) Let $\epsilon > 0$. Assuming (1)–(2), to prove $\lim_{k \to \infty} d_H(C_k, C) = 0$ we need to show

$$C \subset (\overline{C_k})_\epsilon \quad \text{and} \quad C_k \subset \bar{C}_\epsilon, \tag{1.1.5}$$

each for sufficiently large $k \geq 1$.

The failure of the first inclusion in (1.1.5) (for large $k \geq 1$) means the existence of a sequence $\{A_{k_l}\}_{l \geq 1} \subset C$, converging to some $C \in C$, say, such that $d(A_{k_l}, C_{k_l}) \geq \epsilon$ for all $l \geq 1$. For this particular $C \in C$, let $\{C_k\}_{k \geq 1} \subset \mathcal{X}$ be as in (1). Then $d(A_{k_l}, C_{k_l}) \to 0$ as $l \to \infty$, a contradiction. Thus, the first inclusion in (1.1.5) holds.

The failure of the second inclusion in (1.1.5) means the existence of a sequence $\{C_{k_l}^0\}_{l \geq 1} \subset \mathcal{X}$ with $C_{k_l}^0 \in C_{k_l}$ but $C_{k_l}^0 \notin \bar{C}_\epsilon$, $l \geq 1$. Choose any $C \in C$, and use (1) to obtain a sequence $\{C_k^1\}_{k \geq 1}$ converging to C such that $C_k^1 \in C_k$, $k \geq 1$. Then, for $l \geq l_0$ (with l_0 large), we have $C_{k_l}^1 \in \bar{C}_\epsilon$ (and $C_{k_l}^1 \in C_{k_l}$). Finally, use Corollary 1.1.9 to select $C_{k_l} \in [C_{k_l}^0, C_{k_l}^1] \cap \partial \bar{C}_\epsilon \subset C_{k_l}$, $l \geq l_0$. (Note that $\bar{C}_\epsilon$ is a convex *body* in $\mathcal{X}$ even though C may not be.) By (2), the sequence $\{C_{k_l}\}_{l \geq l_0}$ subconverges to a point in C, a contradiction. The theorem follows. $\blacksquare$

1.2 Separation Theorems for Convex Sets

The concepts of *separation* and *support* are fundamental notions in convex geometry. In this section we study under what topological conditions can two convex sets be separated. We prove the *Hahn–Banach theorem*, the cornerstone of all separation theorems. In addition, we derive two fundamental results of Minkowski about the *extremal points* of a convex set and *mixed volumes*.

Given two convex sets C and C', a hyperplane $\mathcal{H}$ is said to *separate* them if C and C' lie in different *closed* half-spaces determined by $\mathcal{H}$. Using the notations of Example 1.1.1 and with a suitable choice of the normal vector N for $\mathcal{H}$, we have $C \subset \mathcal{G}$ and $C' \subset \mathcal{G}'$. Clearly, if $\mathcal{H}$ separates C and C' then $C \cap C' \subset \mathcal{H}$.

We say that $\mathcal{H}$ *strictly separates* $\mathcal{C}$ and $\mathcal{C}'$ if $\mathcal{C} \subset \operatorname{int} \mathcal{G}$ and $\mathcal{C}' \subset \operatorname{int} \mathcal{G}'$. In this case $\mathcal{C}$ and $\mathcal{C}'$ are disjoint.

As an instructive example, the (open) convex set

$$\mathcal{C} = \{(x, y) \in \mathbb{R}^2 \mid xy > 1, x, y > 0\}$$

and either of the coordinate axes of $\mathbb{R}^2$ can be separated (by a line), but cannot be strictly separated.

The following result is usually termed as Hahn–Banach theorem. It was proved independently by H. Hahn and S. Banach in the late 1920s.

Theorem 1.2.1. *Given an open convex set $\mathcal{C} \subset \mathcal{X}$, an affine subspace $\mathcal{E} \subset \mathcal{X}$ disjoint from $\mathcal{C}$ can be extended to a hyperplane $\mathcal{H}$ still disjoint from $\mathcal{C}$.*

Remark. Note that $\mathcal{E}$ may consist of a singleton on the boundary of $\mathcal{C}$; in particular, openness of $\mathcal{C}$ cannot be dispensed with.

Proof. We begin with the planar case $n = \dim \mathcal{X} = 2$ as it has a simple geometric interpretation. We may assume that $\mathcal{E}$ is zero dimensional, $\mathcal{E} = \{C\}$, $C \notin \mathcal{C}$, since otherwise there is nothing to prove. Let $\mathcal{S}(C) = \partial \mathcal{B}(C)$ be the unit circle with center C and $\pi_C : \mathcal{X} \setminus \{C\} \to \mathcal{S}(C)$ the radial projection. The image $\mathcal{A} = \pi_C(\mathcal{C})$ of $\mathcal{C}$ is connected and (relatively) open in $\mathcal{S}(C)$. (See Figure 1.2.1.)

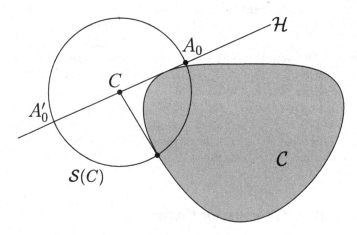

Fig. 1.2.1

Hence $\mathcal{A} \subset \mathcal{S}(C)$ is an open arc. $\mathcal{A}$ cannot contain antipodal points (with respect to C). Indeed, if $A = \pi_C(B)$ and $A' = \pi_C(B')$, $B, B' \in \mathcal{C}$, were antipodal in $\mathcal{A}$ then we would have $C \in [B, B'] \subset \mathcal{C}$, a contradiction.

In particular, the arc $\mathcal{A}$ has two endpoints; let A_0 be one of them. Now, the antipodal A_0' of A_0 cannot be in (the open) $\mathcal{A}$ since otherwise a point in $\mathcal{A}$ close to A_0 would have an antipodal in $\mathcal{A}$. We obtain that $A_0, A_0' \notin \mathcal{C}$ so that the line $\mathcal{H}$ passing through them is the required extension of $\mathcal{E}$.

For the general case, let $\mathcal{H} \supset \mathcal{E}$ be a *maximal* affine subspace of $\mathcal{X}$ disjoint from $\mathcal{C}$. We may assume that codim $\mathcal{H} \geq 2$ since otherwise we are done. Translate $\mathcal{H}$ to a linear subspace $\mathcal{H}_0 \subset \mathcal{X}$, and consider the projection $\pi : \mathcal{X} \to \mathcal{X}/\mathcal{H}_0$. In the quotient $\mathcal{X}/\mathcal{H}_0$ (of dimension ≥ 2), $\pi(\mathcal{H})$ reduces to a single point $\bar{C}$ and $\pi(\mathcal{C})$ is an open convex set disjoint from $\bar{C}$. Let $\bar{\mathcal{X}} \subset \mathcal{X}/\mathcal{H}_0$ be a two-dimensional affine subspace containing $\bar{C}$ and intersecting $\pi(\mathcal{C})$ in an open convex set $\bar{\mathcal{C}}$.

The first part of the proof now applies to $\bar{C} \notin \bar{\mathcal{C}} \subset \bar{\mathcal{X}}$ yielding a line $\mathcal{L} \subset \bar{\mathcal{X}}$ through $\bar{C}$ disjoint from $\bar{\mathcal{C}}$. The affine subspace $\pi^{-1}(\mathcal{L}) \supset \mathcal{H}$ in $\mathcal{X}$ has one less codimension than $\mathcal{H}$, and it is still disjoint from $\mathcal{C}$. This contradicts to the maximality of $\mathcal{H}$. The theorem follows.

Corollary 1.2.2. *Let C and C' be disjoint convex sets in $\mathcal{X}$. If C is open then it can be separated from C'.*

Proof. The algebraic difference $C_0 = C - C'$ is convex and also open since $C_0 = \bigcup_{C' \in C'}(C - C')$. Since C and C' are disjoint, C_0 does not contain the origin 0 and the Hahn–Banach theorem applies. Let $\mathcal{H}_0$ be a *linear* hyperplane disjoint from C_0. Let $N \in \mathcal{X}$ be a normal vector of $\mathcal{H}_0$ pointing to the half-space that contains C_0. Then, for all $C \in \mathcal{C}$ and $C' \in \mathcal{C}'$, we have $\langle C, N \rangle > \langle C', N \rangle$. Let $X_0 \in \mathcal{X}$ be such that $\inf_C \langle \cdot, N \rangle \geq \langle X_0, N \rangle \geq \sup_{C'} \langle \cdot, N \rangle$. Then the hyperplane $\mathcal{H} = \{X \in \mathcal{X} \mid \langle X, N \rangle = \langle X_0, N \rangle\}$ (containing X_0 and parallel to $\mathcal{H}_0$) separates C and C'.

An immediate consequence of this corollary is that two disjoint *open* convex sets can be *strictly* separated by a hyperplane.

In particular, a disjoint pair of a *closed* convex set and a *compact* convex set can be *strictly* separated (since, by compactness of the second, they have *disjoint* open r-neighborhoods for some $r > 0$).

Finally, note that two disjoint *closed* convex sets can be separated. (By the closure of the example at the beginning of this section, in general they cannot be strictly separated.)

Indeed, if C and C' are closed convex disjoint sets then, for a fixed $C \in \mathcal{C}$ and any $k \geq 1$, the compact convex set $C \cap \bar{B}_k(C)$ can be (strictly) separated from C' by a hyperplane $\mathcal{H}_k$. Then the sequence $(\mathcal{H}_k)_{k \geq 1}$ of hyperplanes subconverges to a hyperplane which automatically separates C and C'.

We have now come to the fundamental concept of *support* in convex geometry. Let $\mathcal{A} \subset \mathcal{X}$ be an arbitrary subset and $A \in \mathcal{A}$. A hyperplane $\mathcal{H}$ in $\mathcal{X}$ is *supporting* $\mathcal{A}$ at A if $\mathcal{H}$ separates $\{A\}$ and $\mathcal{A}$. Sometimes the role of the point of support A is irrelevant and therefore may be suppressed.

If $\mathcal{A}$ is compact and $\mathcal{H} \subset \mathcal{X}$ is a hyperplane then there exists *at least one* hyperplane $\mathcal{H}'$ parallel to $\mathcal{H}$ and supporting $\mathcal{A}$.

Indeed, let $\mathcal{H}_0$ be the *linear* hyperplane parallel to $\mathcal{H}$, and consider the projection $\pi : \mathcal{X} \to \mathcal{X}/\mathcal{H}_0$ to the line $\mathcal{X}/\mathcal{H}_0$. If C is an outermost boundary point of (the compact) $\pi(\mathcal{A})$ (that is, $\pi(\mathcal{A})$ is on one side of C in $\mathcal{X}/\mathcal{H}_0$) then the hyperplane $\mathcal{H}' = \pi^{-1}(C)$ supports $\mathcal{A}$.

This proof also shows that if, in addition, $\mathcal{A}$ has non-empty interior, then there are at least two hyperplanes supporting $\mathcal{A}$ and parallel to $\mathcal{H}$.

Finally, if $C \in \mathcal{B}$ is a *convex body* then there are *exactly two* hyperplanes supporting C and parallel to a given hyperplane $\mathcal{H}$.

Corollary 1.2.3. *Let $\mathcal{C} \subset \mathcal{X}$ be a closed convex set and $C \in \partial\mathcal{C}$. Then there exists a hyperplane supporting $\mathcal{C}$ at C. Moreover, $\mathcal{C}$ is the intersection of closed half-spaces whose boundaries support $\mathcal{C}$.*

Proof. We may assume that $\mathcal{C}$ has non-empty interior (since otherwise a hyperplane extension of the affine span $\langle \mathcal{C} \rangle$ supports all points of $\mathcal{C}$). Given $C \in \partial\mathcal{C}$, the first statement now follows from Corollary 1.2.2 applied to $\{C\}$ and $\operatorname{int}\mathcal{C}$.

For the second statement, once again we can assume that $\mathcal{C}$ has non-empty interior. Let $\mathcal{C}'$ be the intersection of all closed half-spaces containing $\mathcal{C}$ whose boundary hyperplanes support $\mathcal{C}$. Then $\mathcal{C}'$ is a closed convex set and contains $\mathcal{C}$. Now, there cannot be a point $X \in \mathcal{C}' \setminus \mathcal{C}$ since then $\{X\}$ could be strictly separated from $\mathcal{C}$ by a hyperplane, contradicting the definition of $\mathcal{C}'$. Thus $\mathcal{C}' = \mathcal{C}$ and the second statement also follows.

Remark. The *boundary* of *any* convex body has supporting hyperplanes at all its points yet it is not convex. Thus the converse of the first statement of Corollary 1.2.3 is clearly false.

On the other hand, if $\mathcal{A} \subset \mathcal{X}$ is a closed subset with *non-empty* interior such that $\mathcal{A}$ has supporting hyperplane at each of its boundary points then $\mathcal{A}$ is convex. This is an easy exercise in the use of a plane intersection of $\mathcal{A}$ containing an interior point of $\mathcal{A}$ and a line segment $[C_0, C_1]$, $C_0, C_1 \in \mathcal{A}$, that *fails* the convexity property.

Corollary 1.2.4. *If an open convex set $C \neq \mathcal{X}$ contains an entire hyperplane $\mathcal{H}$ then ∂C consists of one or two hyperplanes parallel to $\mathcal{H}$.*

Proof. A supporting hyperplane $\mathcal{H}'$ at any point of the boundary must be parallel to $\mathcal{H}$ and, by convexity, the entire region between $\mathcal{H}$ and $\mathcal{H}'$ must belong to $\mathcal{C}$.

We now briefly digress here and discuss an alternative approach to the concept of support.

In convexity theory it is customary to call an affine map $f : \mathcal{X} \to \mathbb{R}$ an *affine functional* (on $\mathcal{X}$). We let $\operatorname{aff}(\mathcal{X})$ denote the vector space of affine functionals. Since an affine functional is a linear functional plus a constant, we have $\dim \operatorname{aff}(\mathcal{X}) = \dim \mathcal{X} + 1 = n + 1$.

An affine functional $f : \mathcal{X} \to \mathbb{R}$ is non-constant if and only if it is onto ($\mathbb{R}$), and, in this case, for any $t \in \mathbb{R}$, the level-set $f^{-1}(t) = \{X \in \mathcal{X} \mid f(X) = t\}$ is a hyperplane in $\mathcal{X}$. Moreover, level-sets corresponding to different range values, being disjoint, must be parallel hyperplanes.

Conversely, given two parallel hyperplanes $\mathcal{H}$ and $\mathcal{H}'$ in $\mathcal{X}$, there exists a unique affine functional $f : \mathcal{X} \to \mathbb{R}$ such that $\mathcal{H} = f^{-1}(0) = \{X \in \mathcal{X} \mid f(X) = 0\}$ and $\mathcal{H}' = f^{-1}(1) = \{X \in \mathcal{X} \mid f(X) = 1\}$. (Indeed, in the spirit of our previous discussions, let $\mathcal{H}_0 \subset \mathcal{X}$ be the *linear* hyperplane parallel to $\mathcal{H}$, and consider the projection

$\pi : \mathcal{X} \to \mathcal{X}/\mathcal{H}_0$ to the line $\mathcal{X}/\mathcal{H}_0$. Since $\mathcal{H}'$ is parallel to $\mathcal{H}$, the projections $\pi(\mathcal{H})$ and $\pi(\mathcal{H}')$ are distinct points on the line $\mathcal{X}/\mathcal{H}_0$. Define the affine transformation $g : \mathcal{X}/\mathcal{H}_0 \to \mathcal{X}/\mathcal{H}_0$ by $g(\pi(\mathcal{H})) = 0$ and $g(\pi(\mathcal{H}')) = 1$. Then, the composition $f = g \circ \pi$ is an affine functional possessing the stated level-sets.)

Given a convex body $\mathcal{C} \in \mathcal{B}$, we say that an affine functional $f \in \text{aff}(\mathcal{X})$ is *normalized* for $\mathcal{C}$ if $f(\mathcal{C}) = [0, 1]$. The set of all normalized functionals for $\mathcal{C}$ is denoted by $\text{aff}_\mathcal{C} \subset \text{aff}(\mathcal{X})$. Clearly, if $f \in \text{aff}_\mathcal{C}$ then the level-sets $\mathcal{H} = f^{-1}(0) = \{X \in \mathcal{X} \,|\, f(X) = 0\}$ and $\mathcal{H}' = f^{-1}(1) = \{X \in \mathcal{X} \,|\, f(X) = 1\}$ are parallel hyperplanes both supporting $\mathcal{C}$.

Conversely, if $\mathcal{H}$ is a hyperplane supporting $\mathcal{C}$ then there is a *unique* $f \in \text{aff}_\mathcal{C}$ such that $\mathcal{H} = f^{-1}(0)$. (Indeed, by the discussion before Corollary 1.2.3, there exists a unique hyperplane $\mathcal{H}'$ parallel to $\mathcal{H}$ and supporting $\mathcal{C}$. By what we have just shown, $\mathcal{H}$ and $\mathcal{H}'$ determine an affine functional $f \in \text{aff}(\mathcal{X})$ uniquely. Having the right level-sets, f is normalized for $\mathcal{C}$.) It follows that the set $\text{aff}_\mathcal{C}$ can be identified with the unit sphere $\mathcal{S}$ (by associating to a hyperplane supporting $\mathcal{C}$ the unit vector pointing outward $\mathcal{C}$); in particular, $\text{aff}_\mathcal{C}$ carries a natural compact topology.

Finally, once again our discussion above implies that, for *any* non-constant affine functional $f \in \text{aff}(\mathcal{X})$, there is an affine transformation $g : \mathbb{R} \to \mathbb{R}$ such that the composition $g \circ f$ is normalized for $\mathcal{C}$.

Many of the concepts discussed so far can be conveniently expressed in terms of the *support function* $h_\mathcal{C} : \mathcal{X} \to \mathbb{R}$, $\mathcal{C} \in \mathfrak{K}$. We define

$$h_\mathcal{C}(X) = \sup_{C \in \mathcal{C}} \langle C, X \rangle, \quad X \in \mathcal{X}. \tag{1.2.1}$$

Remark. If $\mathcal{X}$ is a Minkowski space, the support function is naturally defined on the dual $\mathcal{X}^*$ of $\mathcal{X}$ as $h_\mathcal{C} : \mathcal{X}^* \to \mathbb{R}$ given by

$$h_\mathcal{C}(\phi) = \sup_{C \in \mathcal{C}} \phi(C), \quad \phi \in \mathcal{X}^*.$$

Clearly, $h_\mathcal{C}$ is a degree-one positively homogeneous function: $h_\mathcal{C}(tX) = t\, h_\mathcal{C}(X)$, $t \geq 0$, $X \in \mathcal{X}$. Hence, its restriction to the unit sphere $\mathcal{S} \subset \mathcal{X}$ uniquely determines $h_\mathcal{C}$.

Given $N \in \mathcal{S}$, the hyperplane $\mathcal{H}(N) = \{X \in \mathcal{X} \,|\, \langle X, N \rangle = h_\mathcal{C}(N)\}$ supports $\mathcal{C}$ at the points where the supremum in (1.2.1) is attained. In particular, $h_\mathcal{C}(N)$ is the signed distance of this supporting hyperplane $\mathcal{H}(N)$ from the origin. The corresponding closed half-space $\mathcal{G}(N) = \{X \in \mathcal{X} \,|\, \langle X, N \rangle \leq h_\mathcal{C}(N)\}$ contains $\mathcal{C}$, and, by Corollary 1.2.3, we have

$$\mathcal{C} = \bigcap_{N \in \mathcal{S}} \mathcal{G}(N).$$

Being the supremum of linear functions, $h_\mathcal{C}$ is *convex*. In addition, it is obviously *sub-additive*

$$h_\mathcal{C}(X + X') \leq h_\mathcal{C}(X) + h_\mathcal{C}(X'), \quad X, X' \in \mathcal{X}.$$

Therefore, h_C is a *Lipschitz function*

$$|h_C(X) - h_C(X')| \leq |h_C(X - X')| = \sup_{C \in \mathcal{C}} \langle X - X', C \rangle \leq \sup_{C \in \mathcal{C}} |C| \cdot |X - X'|, \; X, X' \in \mathcal{X},$$

with Lipschitz constant $\sup_{C \in \mathcal{C}} |C|$. In particular, h_C is *continuous*.

The algebraic properties of the support function are as follows:

(1) $h_{\lambda C} = \lambda h_C, \lambda \geq 0, C \in \mathfrak{K}$;
(2) $h_{C+C'} = h_C + h_{C'}, C, C' \in \mathfrak{K}$, where $C + C'$ is the Minkowski sum of C and C' (as in Section 1.1/A).

In addition, as a direct consequence of the definition, complementing (1), we have $h_{-C}(X) = h_C(-X), X \in \mathcal{X}, C \in \mathfrak{K}$.

The proofs of (1)–(2) are straightforward. For $X \in \mathcal{X}$ and $\lambda > 0$, we have

$$h_{\lambda C}(X) = \sup_{C \in \mathcal{C}} \langle \lambda C, X \rangle = \lambda \sup_{C \in \mathcal{C}} \langle C, X \rangle = \lambda h_C(X),$$

and (1) follows. Similarly, using the definition of the Minkowski sum $C + C'$, for $X \in \mathcal{X}$, we calculate

$$\begin{aligned} h_{C+C'}(X) &= \sup_{C \in \mathcal{C}, C' \in \mathcal{C}'} \langle C + C', X \rangle \\ &= \sup_{C \in \mathcal{C}, C' \in \mathcal{C}'} \left(\langle C, X \rangle + \langle C', X \rangle \right) \\ &= \sup_{C \in \mathcal{C}} \langle C, X \rangle + \sup_{C' \in \mathcal{C}'} \langle C', X \rangle \\ &= h_C(X) + h_{C'}(X). \end{aligned}$$

Hence, (2) follows.

Finally, the Hausdorff distance between two compact convex sets is equal to the *maximum norm* of the difference of the corresponding support functions:

$$d_H(\mathcal{C}, \mathcal{C}') = \sup_S |h_C - h_{C'}|, \quad \mathcal{C}, \mathcal{C}' \in \mathfrak{K}. \tag{1.2.2}$$

Indeed, if $d_H(\mathcal{C}, \mathcal{C}') \leq r$ then $C \subset C' + \bar{B}_r = C' + r\bar{B}$, and, by (1)–(2) above, for $N \in S$, we have

$$h_C(N) \leq h_{C'+\bar{B}_r}(N) = h_{C'}(N) + r h_{\bar{B}}(N) = h_{C'}(N) + r.$$

Switching the roles of C and C', this gives

$$|h_C(N) - h_{C'}(N)| \leq r.$$

Taking the supremum for $N \in \mathcal{S}$, we obtain that $d_H(\mathcal{C}, \mathcal{C}')$ is at most the maximum norm in (1.2.2). This argument can clearly be reversed, and the opposite inequality also follows.

Remark. The discussion above shows that associating to a compact convex set its support function (restricted to the unit sphere) gives rise to an embedding of $\mathfrak{K}_{\mathcal{X}}$ into the space $C^0(\mathcal{S})$ of continuous functions on the unit sphere $\mathcal{S} \subset \mathcal{X}$. This embedding is an *isometry* with respect to the Hausdorff metric on $\mathcal{X}$ and the maximum norm on $C^0(\mathcal{S})$. In addition, it is an *algebraic* isomorphism with respect to scalar multiplication and addition in the respective spaces.

A boundary point C of a convex set $\mathcal{C}$ is said to be *extremal* if $\mathcal{C} \setminus \{C\}$ is (still) convex. Clearly, $C \in \partial\mathcal{C}$ is *not* extremal if and only if there exists a line segment $[C_0, C_1] \subset \partial\mathcal{C}$, $C_0 \neq C_1$, such that $C \in (C_0, C_1)$. We denote by $\mathcal{C}^\wedge$ the set of extremal points of $\mathcal{C}$. Clearly, $\mathcal{C}^\wedge$ is contained in the (relative) boundary of $\mathcal{C}$.

Example 1.2.5. Let $\dim \mathcal{X} \geq 2$. Given any (non-empty) subset $\mathcal{A} \subset \mathcal{S}$ (of the unit sphere $\mathcal{S} \subset \mathcal{X}$), the convex hull $\mathcal{C} = [\mathcal{S} \times \{0\} \cup \mathcal{A} \times \{\pm 1\}]$ in $\mathcal{X} \times \mathbb{R}$ is a convex set with non-empty interior. (It is a convex body if and only if $\mathcal{A} \subset \mathcal{S}$ is closed.) We have

$$\mathcal{C}^\wedge = (\mathcal{S} \setminus \mathcal{A}) \times \{0\} \cup \mathcal{A} \times \{\pm 1\}.$$

We see that the extremal set is not necessarily closed in dimension ≥ 3. (Note that $\mathcal{C}^\wedge$ is closed in dimension 2; see Problem 9.)

Theorem 1.2.6 ([Minkowski 2, Krein-Milman]). *A compact convex set is the convex hull of its extremal points: $\mathcal{C} = [\mathcal{C}^\wedge]$, $\mathcal{C} \in \mathfrak{K}$.*

Proof. We first note that if $\mathcal{H}$ is a supporting hyperplane for $\mathcal{C}$ then, as a consequence of the definition of extremal point, we have

$$(\mathcal{C} \cap \mathcal{H})^\wedge = \mathcal{C}^\wedge \cap \mathcal{H}. \tag{1.2.3}$$

Turning to the proof, we clearly have $[\mathcal{C}^\wedge] \subset \mathcal{C}$ and $[\partial\mathcal{C}] = \mathcal{C}$ (Corollary 1.1.9) so that we need only to show that

$$\partial\mathcal{C} \subset [\mathcal{C}^\wedge].$$

We proceed by induction with respect to the dimension of $\mathcal{C}$. The one-dimensional case is obvious. Let $\mathcal{C} \in \mathfrak{K}$, $\dim\mathcal{C} \geq 2$, and assume that the inclusion above is valid in dimensions $< \dim\mathcal{C}$. Let $C \in \partial\mathcal{C}$ and $\mathcal{H}$ a hyperplane supporting $\mathcal{C}$ at C (such that $\dim\mathcal{C} \cap \mathcal{H} < \dim\mathcal{C}$). Applying the induction hypothesis to $\mathcal{C} \cap \mathcal{H}$ (a convex body in its affine span), by (1.2.3), we have $C \in \mathcal{C} \cap \mathcal{H} \subset [\partial(\mathcal{C} \cap \mathcal{H})] \subset [(\mathcal{C} \cap \mathcal{H})^\wedge] = [\mathcal{C}^\wedge \cap \mathcal{H}] \subset [\mathcal{C}^\wedge]$. This completes the general induction step. The theorem follows.

Remark 1. Theorem 1.2.6 is often called the "Krein–Milman theorem." For finite dimensional convex sets (that we have here), it was actually proved by [Minkowski 2]; see also [Price].

Remark 2. A boundary point C of a closed convex set C is called *exposed* if $\{C\} = C \cap \mathcal{H}$ for a hyperplane $\mathcal{H}$ supporting C (at C). Clearly, exposed points are extremal but the converse is false. (As an example, consider a closed square with an open semi-disk surmounted on one of its sides.)
By a result of Strasziewicz, the set of exposed points is dense in the set of extremal points; see [Grünbaum 1].

If C is a compact *polyhedron*, an intersection of finitely many closed hyperplanes, then it follows by an easy induction in the use of (1.2.3) that C has only finitely many extremal points. We obtain the following:

Corollary 1.2.7. *A compact polyhedron is the convex hull of finitely many points.*

Let $C \subset \mathcal{X}$ be a closed convex set. Given a boundary point $C \in \partial C$, the intersection of *all* supporting hyperplanes for C at C is an affine subspace of $\mathcal{X}$ whose dimension is called the *order* of C.

The two extreme cases are as follows: $C \in \partial C$ with the highest possible order $n - 1$ is called a *smooth* point, and with the lowest possible order 0 is called a *vertex*.

By definition, a non-extremal point $C \in \partial C$ must have positive order so that a vertex is always extremal. On the other hand any point on the boundary of a metric ball is smooth and extremal so that the converse is false.

A convex polyhedral body is called a *polytope*. A polytope is therefore a compact convex polyhedron with non-empty interior.

Representing a polytope $\mathcal{P}$ as a non-redundant intersection of closed half-spaces, one easily arrives at the stratification

$$\partial \mathcal{P} = \mathcal{P}^{n-1} \supset \ldots \supset \mathcal{P}^1 \supset \mathcal{P}^0,$$

where $\mathcal{P}^j \setminus \mathcal{P}^{j-1}$, $1 \le j < n$, is the set of boundary points of order j. The latter is a disjoint union of components, the *open j-faces* of $\mathcal{P}$. Clearly, $\mathcal{P}^0 = \mathcal{P}^\wedge$, that is, the set of vertices and the set of extremal points coincide.

Finally, the closure of a *j*-face, a *closed j-face*, is a polyhedron of dimension *j*.

The corollary above can be restated as follows: A polytope is the convex hull of its vertices. (For more details, see [Berger, Vol. II, 12.1.5–12.1.9], and for a full analysis, see [Schneider 2, 2].)

As an application of these ideas, we briefly discuss the volume functional vol $=$ $\mathrm{vol}_n : \mathcal{R} \to \mathbb{R}$ introduced in Section 1.1/B. By definition, it is the (n-dimensional) Lebesgue measure restricted to the space of all compact convex subsets of $\mathcal{X}$ (metrized by the Hausdorff distance d_H). (The subscript n indicating the dimension will often be suppressed from the notation. Note also that, by Proposition 1.1.10, the

Lebesgue measure of a compact convex set is equal to its Peano–Jordan volume.) The proofs of the following properties are straightforward:

(1) The volume functional is invariant under isometries of $\mathcal{X}$;
(2) $\mathrm{vol}\,(\lambda\,\mathcal{C}) = \lambda^n\,\mathrm{vol}\,(\mathcal{C})$, for any $\mathcal{C} \in \mathfrak{K}$ and $\lambda \geq 0$;
(3) For $\mathcal{C} \in \mathfrak{K}$, $\mathrm{vol}\,(\mathcal{C}) = 0$ if and only if $\mathcal{C}$ is contained in a hyperplane of $\mathcal{X}$;
(4) If $\mathcal{C} \subset \mathcal{C}', \mathcal{C}, \mathcal{C}' \in \mathfrak{K}$, then $\mathrm{vol}\,(\mathcal{C}) \leq \mathrm{vol}\,(\mathcal{C}')$, and equality holds if and only if $\mathrm{vol}\,(\mathcal{C}') = 0$ or $\mathcal{C} = \mathcal{C}'$;
(5) The volume functional is continuous (with respect to the Hausdorff metric on $\mathfrak{K}$).

We note a simple but useful explicit formula for the volume of a *polytope* $\mathcal{P} \in \mathfrak{B}$. If $\mathcal{Q}_1, \ldots \mathcal{Q}_m \subset \partial \mathcal{P}$ are the closed $(n-1)$-faces of $\mathcal{P}$, and $N_k \in \mathcal{S}$ is the *outer* unit normal vector of $\mathcal{P}$ of the affine span $\langle \mathcal{Q}_k \rangle$, $k = 1, \ldots, m$, then we have

$$\mathrm{vol}\,(\mathcal{P}) = \frac{1}{n} \sum_{k=1}^{m} h_{\mathcal{P}}(N_k)\,\mathrm{vol}_{n-1}\,(\mathcal{Q}_k), \qquad (1.2.4)$$

where $h_{\mathcal{P}}$ is the support function of $\mathcal{P}$, and vol_{n-1} is the $(n-1)$-dimensional volume functional. (This holds since $h_{\mathcal{P}}(N_k)$ is the signed distance of $\langle \mathcal{Q}_k \rangle$ from the origin, so that the right-hand side in (1.2.4) is the sum of signed volumes of pyramids with respective bases $\mathcal{Q}_k$, $k = 1, \ldots, m$.)

The following result is of fundamental importance:

Theorem 1.2.8 ([Minkowski 2]). *There is a symmetric function $V : \mathfrak{K}^n \to \mathbb{R}$ such that, for $\mathcal{C}_i \in \mathfrak{K}$ and $\lambda_i \geq 0$, $i = 1, \ldots, r$, we have*

$$\mathrm{vol}\left(\sum_{i=1}^{r} \lambda_i \mathcal{C}_i\right) = \sum_{i_1, \ldots, i_n = 1}^{r} \lambda_{i_1} \ldots \lambda_{i_n} V(\mathcal{C}_{i_1}, \ldots, \mathcal{C}_{i_n}). \qquad (1.2.5)$$

Minkowski's theorem asserts that the volume of the linear combination $\sum_{i=1}^{r} \lambda_i \mathcal{C}_i$ is a degree n homogeneous polynomial in the variables λ_i, $i = 1, \ldots, r$. The symmetric function V (whose values are the coefficients of the polynomial) is called the *mixed volume*. The obvious properties of V are as follows:

(1) The mixed volume is invariant under translations in each variable separately, and under any simultaneous isometry (in all the variables);
(2) $\mathrm{vol}\,(\mathcal{C}) = V(\mathcal{C}, \ldots, \mathcal{C})$ for $\mathcal{C} \in \mathfrak{K}$;
(3) The mixed volume is continuous (with respect to the Hausdorff metric on $\mathfrak{K}$).

Remark. A less trivial property of V (which we will not use) is:

(4) If $\mathcal{C}_1 \subset \mathcal{C}_1', \mathcal{C}_1, \mathcal{C}_1' \in \mathfrak{K}$, then we have $V(\mathcal{C}_1, \mathcal{C}_2, \ldots, \mathcal{C}_n) \leq V(\mathcal{C}_1', \mathcal{C}_2, \ldots, \mathcal{C}_n)$, $\mathcal{C}_i \in \mathfrak{K}$, $i = 2, \ldots, n$.
   Note that this implies that V is non-negative. (See [Bonnesen-Fenchel, 28].)

Proof of Theorem 1.2.8. We first prove (1.2.5) for polyhedra.

We proceed by induction with respect to $n \geq 1$. The first step is obvious. For the general induction step $n - 1 \Rightarrow n$, assume that (1.2.5) holds for compact convex polyhedra of dimension $\leq n - 1$. Let $\dim \mathcal{X} = n$, $\mathcal{P}_i \in \mathfrak{K}$ and $\lambda_i \geq 0, i = 1 \ldots, r$. We may assume that the linear combination

$$\mathcal{P} = \sum_{i=1}^{r} \lambda_i \mathcal{P}_i \tag{1.2.6}$$

is n-dimensional (that is, it belongs to $\mathfrak{B}$) since otherwise the induction hypothesis applies.

Let $\mathcal{H} \subset \mathcal{X}$ be a hyperplane supporting $\mathcal{P}$ with outer unit normal $N \in \mathcal{S}$ such that $\mathcal{Q} = \mathcal{P} \cap \mathcal{H}$ is a closed $(n - 1)$-dimensional face of $\mathcal{P}$. For $i = 1, \ldots, r$, let $\mathcal{H}_i$ be the hyperplane supporting $\mathcal{P}_i$ with the *same* outer unit normal N for $\mathcal{P}_i$ as for $\mathcal{H}$. For $i = 1, \ldots, r$, the intersection $\mathcal{Q}_i = \mathcal{P}_i \cap \mathcal{H}_i$ is a closed face of $\mathcal{P}_i$ of dimension $\leq n - 1$. We claim that

$$\mathcal{Q} = \sum_{i=1}^{r} \lambda_i \mathcal{Q}_i. \tag{1.2.7}$$

To prove this, we let $X \in \mathcal{P}$ be decomposed according to (1.2.6), that is, we write $X = \sum_{i=1}^{r} \lambda_i X_i$ with $X_i \in \mathcal{P}_i, i = 1, \ldots, r$. Using the respective support functions, we have

$$\langle X, N \rangle \leq h_{\mathcal{P}}(N) \quad \text{and} \quad \langle X_i, N \rangle \leq h_{\mathcal{P}_i}(N), \quad i = 1, \ldots, r. \tag{1.2.8}$$

Now the crux is that equality holds in the first inequality in (1.2.8) if and only if $X \in \mathcal{Q}$, and similarly, equalities hold in the second set of inequalities if and only if $X_i \in \mathcal{Q}_i, i = 1, \ldots, r$. Furthermore, using the algebraic properties of the support function, the first inequality in (1.2.8) can be expanded as

$$\sum_{i=1}^{r} \lambda_i \langle X_i, N \rangle \leq \sum_{i=1}^{r} \lambda_i h_{\mathcal{P}_i}(N).$$

Finally, since $\lambda_i \geq 0, i = 1, \ldots, r$, these imply that $X \in \mathcal{Q}$ is equivalent to $X_i \in \mathcal{Q}_i$, $i = 1, \ldots, r$. The decomposition in (1.2.7) follows.

For $i = 1, \ldots, r$, the orthogonal projection of $\mathcal{Q}_i \subset \mathcal{H}_i$ to $\mathcal{H}$ does not change its $(n - 1)$-dimensional volume (since $\mathcal{H}_i$ is parallel to $\mathcal{H}$), and (1.2.7) stays intact for the projections. Since all convex sets in $\mathcal{H}$ are of dimension $\leq n - 1$ the induction hypothesis applies. We obtain that $\mathrm{vol}_{n-1}(\mathcal{Q}) = \mathrm{vol}_{n-1}\left(\sum_{i=1}^{r} \lambda_i \mathcal{Q}_i\right)$ is a degree $(n - 1)$ homogeneous polynomial in the variables $\lambda_i, i = 1, \ldots, r$.

Applying this to *all* closed $(n - 1)$-faces $\mathcal{Q}_k$, $k = 1, \ldots, m$, of $\mathcal{P}$ and using (1.2.4), we obtain (1.2.5) since the support function of $\mathcal{P}$ is a degree 1 homogeneous polynomial (in $\lambda_i, i = 1, \ldots, r$) via $h_{\mathcal{P}} = \sum_{i=1}^{r} \lambda_i h_{\mathcal{P}_i}$. Thus the theorem follows for polyhedra.

We now show (1.2.5) in full generality for $C_i \in \mathfrak{K}$ and $\lambda_i \geq 0$, $i = 1, \ldots, r$. We set $C = \sum_{i=1}^{r} \lambda_i C_i$. Recall from the second remark after Corollary 1.1.6 that a compact convex set can be approximated by polyhedra with arbitrary precision with respect to the Hausdorff distance d_H. Thus, for each $i = 1, \ldots, r$, there exists a sequence $\{\mathcal{P}_i^l\}_{l \geq 1}$ of polyhedra such that $\lim_{l \to \infty} \mathcal{P}_i^l = C_i$ in the Hausdorff metric. Taking linear combinations, we obtain $\lim_{l \to \infty} \mathcal{P}^l = C$, where $\mathcal{P}^l = \sum_{i=1}^{r} \lambda_i \mathcal{P}_i^l$. Since the volume functional is continuous, the sequence $\{\text{vol}(\mathcal{P}^l)\}_{l \geq 1}$ converges to $\text{vol}(C)$.

Now, by what we proved above, every element of this sequence is a degree n homogeneous polynomial in λ_i, $i = 1, \ldots, r$. The convergence being pointwise in these variables, it follows that the limit $\text{vol}(C)$ is also a degree n polynomial in λ_i, $i = 1, \ldots, r$. The theorem follows.

We now discuss an important special case when $C \in \mathfrak{B}$ and $\bar{B} \subset \mathcal{X}$ is the unit ball. Then, for the sum $C + \epsilon \bar{B}$, $\epsilon > 0$, (1.2.5) reduces to

$$\text{vol}(C + \epsilon \bar{B}) = \sum_{i=1}^{n} \binom{n}{i} \epsilon^i V(\bar{B}[i], C[n - i])$$

$$= V(C, \ldots, C) + \epsilon \, n \, V(\bar{B}, C, \ldots, C) + O(\epsilon^2),$$

where the numerals in the square brackets indicate repetitions. As noted above, $V(C, \ldots, C) = \text{vol}(C)$. The coefficient of the linear term is

$$n \, V(\bar{B}, C, \ldots, C) = \lim_{\epsilon \to 0^+} \frac{\text{vol}(C + \epsilon \bar{B}) - \text{vol}(C)}{\epsilon}. \tag{1.2.9}$$

We claim that, for $C = \mathcal{P} \in \mathfrak{B}$ a polytope, the right-hand side in (1.2.9) is the *surface area* of $\mathcal{P}$. (We tacitly assume that $n \geq 2$ with the area for $n = 2$ interpreted as perimeter.)

Indeed, if $\mathcal{Q}_1, \ldots, \mathcal{Q}_m \subset \partial \mathcal{P}$ are the closed $(n-1)$-faces of $\mathcal{P}$ then $(C + \epsilon \bar{B}) \setminus \text{int} \, C$ consists of the union of (closed) cylinders with bases $\mathcal{Q}_i$, $i = 1, \ldots, m$, and height $\epsilon > 0$, and a *remainder* which is contained in the union of closed balls of radius ϵ and centers in $\mathcal{P}^{n-2}$. The latter has combined volume $O(\epsilon^2)$ (since $n \geq 2$). We obtain

$$\text{vol}(C + \epsilon \bar{B}) = \text{vol}(C) + \epsilon \sum_{i=1}^{m} \text{vol}_{n-1}(\mathcal{Q}_i) + O(\epsilon^2).$$

The claim follows.

We now *define* the surface area of $C \in \mathfrak{B}$ as given in (1.2.9). We obtain that the surface area of a convex body always exists, and it is a continuous functional on $\mathfrak{B}$ with respect to d_H; in particular, the surface area of $C \in \mathfrak{B}$ is the limit of the surface areas of convex polytopes converging to C in the Hausdorff metric.

Remark. The discussion of mixed volumes above follows closely [Bonnesen-Fenchel, 28]. For a recent treatment, see [Schneider 2, 5.1].

1.3   The Theorems of Carathéodory and Radon

The three pillars of combinatorial convexity are the classical theorems of *Carathéodory*, *Radon*, and *Helly*. In this section we discuss the first two. A detailed treatment of Helly's theorem will be given in the next section.

Recall from Section 1.1/A the definition of the convex hull $[\mathcal{A}]$ of a subset $\mathcal{A} \subset \mathcal{X}$: $B \in [\mathcal{A}]$ if there exists a finite subset $\{A_1, \ldots, A_m\} \subset \mathcal{A}$, $m \geq 1$, such that $B = \sum_{i=1}^m \lambda_i A_i$, for some $\{\lambda_1, \ldots, \lambda_m\} \subset [0, 1]$ with $\sum_{i=1}^m \lambda_i = 1$.

A natural problem is to find a *universal* upper bound for m. The simple answer, $m \leq n + 1$, is given by the following:

Theorem 1.3.1 ([Carathéodory]). *Given $\mathcal{A} \subset \mathcal{X}$, $B \in [\mathcal{A}]$ if and only if there exists $\{A_1, \ldots, A_{n+1}\} \subset \mathcal{A}$ such that $B \in [A_1, \ldots, A_{n+1}]$.*

Proof. Let $B = \sum_{i=1}^m \lambda_i A_i$, $\{A_1, \ldots, A_m\} \subset \mathcal{A}$, $\{\lambda_1, \ldots, \lambda_m\} \subset (0, 1]$ with $\sum_{i=1}^m \lambda_i = 1$, and assume that m is *minimal*, and $m > n + 1$.

We will make a linear change in the coefficients of the representation of B above to reduce the number of non-zero terms.

To do this, it is natural to fix a *non-trivial* solution $\{\mu_1, \ldots, \mu_m\} \subset \mathbb{R}$ of the system $\sum_{i=1}^m \mu_i A_i = 0$, $\sum_{i=1}^m \mu_i = 0$ (which exists since there are m variables and $n + 1(< m)$ equations). We consider the set

$$\mathcal{T} = \{\tau \in \mathbb{R} \mid \lambda_i + \tau\mu_i \geq 0, i = 1, \ldots, m\}.$$

Clearly, $0 \in \mathcal{T} \neq \mathbb{R}$, and $\mathcal{T}$ is closed and convex. Let $\tau_0 \in \partial\mathcal{T}$ and choose $j \in \{1, \ldots, m\}$ such that $\lambda_j + \tau_0\mu_j = 0$. With this, we have

$$B = \sum_{i=1}^m \lambda_i A_i + \tau_0 \sum_{i=1}^m \mu_i A_i = \sum_{i=1}^m (\lambda_i + \tau_0\mu_i)A_i.$$

The last sum is a representation of B in $[A_1, \ldots, A_m]$ since the coefficients are non-negative (by the definition of $\mathcal{T}$) and they add up to 1 (by the choice of $\{\mu_1, \ldots, \mu_m\}$). In addition, the jth term is zero. This contradicts to the minimal choice of m. Thus, $m \leq n + 1$, and Carathéodory's theorem follows.

Remark. Splitting the index-set $\mathcal{I} = \{1, \ldots, m\}$ as $\mathcal{I}^+ = \{i \in \mathcal{I} \mid \mu_i \geq 0\}$ and $\mathcal{I}^- = \{i \in \mathcal{I} \mid \mu_i < 0\}$, it is easy to see that $\mathcal{T}$ is a (finite) closed interval containing 0 in its interior. The positive boundary is at $-\min_{i\in\mathcal{I}^-}(\lambda_i/\mu_i)$ which can be taken as τ_0.

Corollary 1.3.2. *The convex hull of a compact set is compact.*

Proof. Recall from Section 1.1/A the standard n-simplex $\Delta_n \subset \mathbb{R}^{n+1}$. Given a compact subset $\mathcal{A} \subset \mathcal{X}$, by Carathéodory's theorem, the convex hull $[\mathcal{A}]$ is the image of the *continuous* map $\Delta_n \times \mathcal{A}^{n+1} \to \mathcal{X}$ given by $(\lambda_1, \ldots, \lambda_{n+1}; A_1, \ldots, A_{n+1}) \mapsto \sum_{i=1}^{n+1} \lambda_i A_i$. Thus, $[\mathcal{A}]$ is compact and the corollary follows.

Remark. A subset $\mathcal{A} \subset \mathcal{X}$ is *bounded* if it is contained in a metric ball. Since the latter is convex, it immediately follows that the convex hull of a bounded set is bounded. On the other hand, the convex hull of a closed set is not necessarily closed. For example, we have

$$[\{(x,y) \in \mathbb{R}^2 \mid xy \geq 1, x, y > 0\} \cup \{(0,0)\}] = \{(x,y) \in \mathbb{R}^2 \mid x, y > 0\} \cup \{(0,0)\}.$$

(Compare this with the example at the beginning of Section 1.2.) This shows that Corollary 1.3.2 does not immediately follow from the definitions. For a proof that does not use Carathéodory's theorem, see [Eggleston 1, p. 22].

Corollary 1.3.3. *Let $C \in \mathfrak{B}$. Then any point of C is contained in an m-simplex, $m \leq n$, with vertices being extremal points of C.*

Proof. Let $C \in \mathcal{C}$. By Theorem 1.2.6, C is in the convex hull of $C^\wedge$, the set of extremal points of C. Carathéodory's theorem (applied to $C^\wedge$) then guarantees the existence of $\{C_0, \ldots, C_m\} \subset C^\wedge$, $m \leq n$, such that $C \in [C_0, \ldots, C_m]$.

Assuming that m is minimal, we claim that $[C_0, \ldots, C_m]$ is a simplex. Indeed, otherwise $\dim[C_0, \ldots, C_m] < m$, and applying Carathéodory's theorem to C and $\langle C_0, \ldots, C_m \rangle$ again, we would get contradiction to the minimality of m. The corollary follows.

Theorem 1.3.4 ([Radon]). *Let $\mathcal{A} \subset \mathcal{X}$ consist of at least $n + 2$ points. Then $\mathcal{A} = \mathcal{A}^+ \cup \mathcal{A}^-$, $\mathcal{A}^+ \cap \mathcal{A}^- = \emptyset$, such that $[\mathcal{A}^+] \cap [\mathcal{A}^-] \neq \emptyset$.*

Proof. Choose a subset $\{A_i\}_{i \in \mathcal{I}} \subset \mathcal{A}$ of $m \geq n + 2$ points indexed by $\mathcal{I} = \{1, \ldots, m\}$. In the spirit of the proof of Carathéodory's theorem and the subsequent remark, we consider a *non-trivial* solution $\{\mu_i\}_{i \in \mathcal{I}}$ of the system $\sum_{i \in \mathcal{I}} \mu_i A_i = 0$, $\sum_{i \in \mathcal{I}} \mu_i = 0$, and let $\mathcal{I}^+ = \{i \in \mathcal{I} \mid \mu_i \geq 0\}$ and $\mathcal{I}^- = \{i \in \mathcal{I} \mid \mu_i < 0\}$.

Non-triviality implies that $\sum_{i \in \mathcal{I}^+} \mu_i = \sum_{i \in \mathcal{I}^-}(-\mu_i) = \mu > 0$. Letting $\{A_i \mid i \in \mathcal{I}^+\} \subset \mathcal{A}^+$ and $\{A_i \mid i \in \mathcal{I}^-\} \subset \mathcal{A}^-$ (with the rest of the points in $\mathcal{A}$ distributed arbitrarily), a stated point in the intersection of the convex hulls is $\sum_{i \in \mathcal{I}^+}(\mu_i/\mu)A_i = \sum_{i \in \mathcal{I}^-}(-\mu_i/\mu)A_i$.

Remark 1. A generalization of Radon's theorem to general partitions is due to [Tverberg]: Given a set $\mathcal{A} \subset \mathcal{X}$ of at least $(n + 1)(k + 1) + 1$ points, $\mathcal{A}$ can be partitioned into k subsets whose convex hulls have a non-trivial intersection.

Remark 2. Let $f : \mathcal{X} \to \mathcal{X}_0$, $n = \dim \mathcal{X} > \dim \mathcal{X}_0$, be an affine map. Radon's theorem implies that, given a polytope $C \subset \mathcal{X}$, the set of vertices $C^\wedge$ of C can be split into two disjoint subsets $\mathcal{A}^\pm$ such that $f([\mathcal{A}^+]) \cap f([\mathcal{A}^-]) \neq \emptyset$.

Since *any* subset of at least $n + 1$ points in $\mathcal{X}_0$ is the affine image of the vertices of a polytope in $\mathcal{X}$ by an *affine* map $f : \mathcal{X} \to \mathcal{X}_0$, this statement is actually equivalent to Radon's theorem.

It is a surprising fact proved by [Bajmóczy–Bárány] that in this statement the word "affine" can be replaced by "continuous": Given a *continuous* map $f : \mathcal{X} \to \mathcal{X}_0$, $\dim \mathcal{X} > \dim \mathcal{X}_0$, and a polytope $C \subset \mathcal{X}$, we have $f([\mathcal{A}^+]) \cap f([\mathcal{A}^-]) \neq \emptyset$ for a splitting $C^\wedge = \mathcal{A}^+ \cup \mathcal{A}^-$, $\mathcal{A}^+ \cap \mathcal{A}^- = \emptyset$.

In addition, they also proved a generalization of Borsuk's theorem: Given a convex body $C \subset \mathcal{X}$ and a continuous map $f : \mathcal{X} \to \mathcal{X}_0$, $\dim \mathcal{X} > \dim \mathcal{X}_0$ as above, we have $f(C') = f(C'')$ for a pair of antipodal points $C', C'' \in \partial C$, where antipodal means that C' and C'' are contained in a pair of parallel hyperplanes supporting C.

For a continuous analogue of Tverberg's theorem above as well as various generalizations, see Eckhoff's article in [Gruber–Wills, 2.1].

1.4 Helly's Theorem

The third pillar of combinatorial convexity alluded to at the beginning of the previous section is the following:

Theorem 1.4.1 ([Helly]). *Let $\mathfrak{F}$ be a family of convex sets in $\mathcal{X}$. Assume that the intersection of any $n + 1$ members of $\mathfrak{F}$ is non-empty. Suppose that (i) $\mathfrak{F}$ is finite; or (ii) the members of $\mathfrak{F}$ are compact. Then $\bigcap \mathfrak{F} \neq \emptyset$.*

We first show that, in proving Helly's theorem, it suffices to assume the intersection of (i) and (ii), that is, we may assume that $\mathfrak{F}$ is a *finite* family of *compact* subsets.

To begin with, suppose that Helly's theorem holds under (i). Assuming now (ii) and applying Helly's theorem to the finite subsets of $\mathfrak{F}$ we obtain that $\mathfrak{F}$ is a family of compact sets with the *finite intersection property*, that is, every finite subfamily of $\mathfrak{F}$ has a non-empty intersection. Then, fixing (any) $\mathcal{F}_0 \in \mathfrak{F}$, the family $\mathfrak{F} \cap \mathcal{F}_0$ also has the finite intersection property. Any member $\mathcal{F} \cap \mathcal{F}_0$, $\mathcal{F} \in \mathfrak{F}$, of this family is *closed* in $\mathcal{F}_0$. Applying the usual definition of compactness to $\mathcal{F}_0$ and the complementary family of open sets $\mathcal{F}_0 \setminus \mathfrak{F}$, we obtain $\bigcap \mathfrak{F} \neq \emptyset$.

Thus, we reduced the proof of Helly's theorem to (i), that is, we may assume that $\mathfrak{F}$ is a *finite* family of convex sets with any $n + 1$ members intersecting non-trivially. We select a specific point in every intersection of $n + 1$ members of $\mathfrak{F}$. We denote the collection of these points by $\mathcal{A}$. The members of $\mathcal{A}$ are indexed by the $n + 1$ member subsets of $\mathfrak{F}$, and $\mathcal{A}$ consists of at most $\binom{|\mathfrak{F}|}{n+1}$ points. Letting $\mathfrak{F}' = \{[\mathcal{F} \cap \mathcal{A}] \mid \mathcal{F} \in \mathfrak{F}\}$, each $n + 1$ members of $\mathfrak{F}'$ intersect non-trivially and $\bigcap \mathfrak{F}' \subset \bigcap \mathfrak{F}$. Thus, it is enough to prove Helly's theorem for $\mathfrak{F}'$. Each member $[\mathcal{F} \cap \mathcal{A}] \in \mathfrak{F}'$, $\mathcal{F} \in \mathfrak{F}$ is a compact polyhedron (with possibly empty interior) so that (ii) along with (i) in Helly's theorem can be assumed.

These reductions indicate that the character of Helly's theorem is *combinatorial* rather than *topological*.

Proof à la *Helly*. We proceed by induction with respect to $n \geq 0$, the first step being obvious. Assume that the theorem is true in dimensions $< n$, and let $\mathfrak{F} \subset \mathfrak{K}$ be a finite family (of compact convex sets in $\mathcal{X}$), such that each $n + 1$ members of $\mathfrak{F}$ intersect non-trivially.

Assume, on the contrary, that $\bigcap \mathfrak{F} = \emptyset$. Replacing $\mathfrak{F}$ (if necessary) with a subfamily, we may assume that there exists $\mathcal{F}_0 \in \mathfrak{F}$ such that $\bigcap \mathfrak{F} = \emptyset$ but $\bigcap(\mathfrak{F} \setminus \{\mathcal{F}_0\}) = \mathcal{C} \neq \emptyset$. Since $\mathcal{F}_0$ and $\mathcal{C}$ are disjoint compact convex sets, by Corollary 1.2.2 (and the subsequent discussion), there exists a hyperplane $\mathcal{H} \subset \mathcal{X}$ that strictly separates $\mathcal{F}_0$ and $\mathcal{C}$.

Let $\mathcal{C}'$ be the intersection of any n members of $\mathfrak{F} \setminus \{\mathcal{F}_0\}$. (If $\mathfrak{F}$ consists of $\leq n+1$ members then the theorem is a tautology.) Clearly, $\mathcal{C} \subset \mathcal{C}'$ and, by assumption, $\mathcal{C}'$ intersects $\mathcal{F}_0$ (on the other side of $\mathcal{H}$). Convexity thus implies that $\mathcal{C}'$ intersects $\mathcal{H}$ in a compact convex set.

Summarizing, we obtain that $\mathfrak{F}' = \{\mathcal{F}' = \mathcal{F} \cap \mathcal{H} \mid \mathcal{F} \in \mathfrak{F} \setminus \{\mathcal{F}_0\}\}$ is a family of compact convex sets in $\mathcal{H}$ such that each n members of $\mathfrak{F}'$ intersect non-trivially. The induction hypothesis applies giving $\bigcap \mathfrak{F}' \neq \emptyset$. In particular, $\mathcal{C} \cap \mathcal{H}$ is non-empty, a contradiction.

Proof à la *Radon*. Assuming (i) (without (ii), actually) we use induction with respect to $|\mathfrak{F}| \geq n + 1$, the first step being a tautology.

For the general induction step, assume that the theorem is true for families with $< k$ members, where $k > n + 1$. Consider a family $\mathfrak{F}$ of k convex sets each $n + 1$ of which intersects non-trivially. By the induction hypothesis, for any $\mathcal{F} \in \mathfrak{F}$, we can select a point $p_\mathcal{F} \in \bigcap(\mathfrak{F} \setminus \{\mathcal{F}\})$. By Radon's theorem applied to the set $\{p_\mathcal{F} \mid \mathcal{F} \in \mathfrak{F}\}$, there is a partition $\mathfrak{F} = \mathfrak{F}^+ \cup \mathfrak{F}^-$, $\mathfrak{F}^+ \cap \mathfrak{F}^- = \emptyset$, such that the convex hulls $[\{p_\mathcal{F} \mid \mathcal{F} \in \mathfrak{F}^+\}]$ and $[\{p_\mathcal{F} \mid \mathcal{F} \in \mathfrak{F}^-\}]$ intersect non-trivially. Clearly, any point in this intersection is also in $\bigcap \mathfrak{F}$. Helly's theorem follows.

Remark. The first proof of Helly's theorem was published by [Radon] closely followed by [Helly, König 1]. Later proofs were given by [Rademacher–Schoenberg, Sandgren, Levi, Krasnosel'skiĭ]. In our treatment of Helly's theorem we followed [Danzer–Grünbaum–Klee]. This is the most comprehensive treatment of the subject. In addition to several other proofs, it contains many related problems, applications, and references. (See also some of the problems at the end of this chapter.) Many other books treat this subject, for example, [Eggleston 1, pp. 33–44], [Eckhoff], [Valentine, pp. 69–89], [Schneider 2, pp. 4–5], [Gruber, 3.2], [Gruber–Wills, 2.1].

Let $\mathcal{C} \in \mathfrak{B}$ and $O \in \mathrm{int}\,\mathcal{C}$. Given a boundary point $C \in \partial\mathcal{C}$, by Corollary 1.1.9, there exists a unique point $C^o \in \partial\mathcal{C}$ with O in the interior of the line segment $[C, C^o]$. The point C^o is called the *antipodal* of C with respect to O. (See Figure 1.4.1.)

The ratio $\Lambda(C, O) = d(C, O)/d(C^o, O)$ of lengths that O splits the chord $[C, C^o] \subset \mathcal{C}$ is called the *distortion ratio* of C (with respect to O). (By definition, a chord is a non-trivial intersection of a line with a convex set. We will study chords in Section 2.2.) Clearly, $\Lambda(C^o, O) = 1/\Lambda(C, O)$.

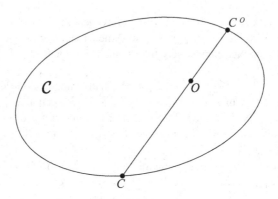

Fig. 1.4.1

Corollary 1.4.2 ([Minkowski 1, Radon]). *Let* $C \in \mathfrak{B}$. *Then there exists a point* $O \in \text{int}\,C$ *such that, for all* $C \in \partial C$, *we have*

$$\frac{1}{n} \leq \Lambda(C, O) \leq n.$$

Proof. Recall the similarity in (1.1.4). For $A \in C$, we define

$$C_A = S_{n/(n+1),A}(C) = \frac{1}{n+1}A + \frac{n}{n+1}C.$$

The family $\mathfrak{F} = \{C_A \mid A \in C\}$ consists of compact convex sets. In addition, given $\{A_1, \ldots, A_{n+1}\} \subset C$, the center of mass

$$A_0 = \frac{1}{n+1}\sum_{i=1}^{n+1} A_i \in C$$

belongs to $\bigcap_{j=1}^{n+1} C_{A_j}$. Indeed, for $j = 1, \ldots, n+1$, we have

$$A_0 = \frac{1}{n+1}A_j + \frac{n}{n+1}\sum_{1 \leq i \neq j \leq n+1} \frac{1}{n}A_i \in \frac{1}{n+1}A_j + \frac{n}{n+1}C = C_{A_j}.$$

Helly's theorem (with (ii)) applies, and we obtain that the family $\mathfrak{F}$ has non-empty intersection.

Let $O \in \bigcap \mathfrak{F} = \bigcap_{A \in C} C_A$. Clearly, $O \in \text{int}\,C$ since, for $A \in \text{int}\,C$, we have $C_A \subset \text{int}\,C$.

Finally, given $C \in \partial C$, the relation $O \in C_C$ means that $O = 1/(n+1)C + n/(n+1)A$ for some $A \in C$ (depending on C). Clearly, $O \in [C, A]$ so that $A \in [O, C^o]$, and we have

$$\Lambda(C, O) = \frac{d(C, O)}{d(C^o, O)} \leq \frac{d(C, O)}{d(A, O)} = n.$$

The corollary follows.

Remark 1. Corollary 1.4.2 was first proved by [Minkowski 1] for $n = 2, 3$ and by [Radon] in the general case. They actually proved that the *centroid* has this property. (A simple proof due to [Bonnesen-Fenchel, 34] (also quoted in [Valentine, Proposition 12.5, p. 190]) is outlined in Problem 11 at the end of Chapter 3.) The proof given here is due to [Yaglom–Boltyanskiĭ]; see also [Danzer–Grünbaum–Klee, p. 246].

Remark 2. In Corollary 1.4.2, the family $\mathfrak{F}$ is parametrized by the convex body C. We can restrict this parametrization to the boundary ∂C, that is, define $\mathfrak{F} = \{C_C |$ $C \in \partial C\}$, and the proof of the corollary still goes through.

Indeed, the only step that needs further elaboration is that $O \in \bigcap_{C \in \partial C} C_C$ implies $O \in \operatorname{int} C$. Assume, on the contrary, that $O \in \partial C$. Let $\mathcal{H}$ be a hyperplane supporting C at O, and $\mathcal{H}'$ a hyperplane parallel to $\mathcal{H}$ and also supporting C. (See Corollary 1.2.3 and the discussion therein.) Choose $C' \in \partial C \cap \mathcal{H}'$. Denote by $\mathcal{Q} \supset C$ the closed region bounded by the parallel hyperplanes $\mathcal{H}$ and $\mathcal{H}'$. Since $O \in \mathcal{H}$ and $C' \in \mathcal{H}'$, we have $O \notin S_{n/(n+1),C'}(\mathcal{Q})$. Thus, $O \notin S_{n/(n+1),C'}(C) = C_{C'}$. This is a contradiction since $C' \in \partial C$.

In Section 2.3 we will gain a further geometric insight into the proof of Corollary 1.4.2 above, including the seemingly ad hoc choice of the ratio $n/(n + 1)$, through Hammer's decomposition of a convex body.

Helly's theorem is extremely rich in applications. (See [Danzer–Grünbaum–Klee] and [Gruber–Wills, 2.1].) For our further developments, we need only one immediate application. A more complex refinement is deferred to Section 1.6.

Corollary 1.4.3 ([Vincensini, Klee 2]). *Let $\mathfrak{F}$ be a family of convex sets in $\mathcal{X}$. Assume that the intersection of any $n + 1$ members of $\mathfrak{F}$ contains the translate of a given convex set $\mathcal{K} \subset \mathcal{X}$. Suppose that (i) $\mathfrak{F}$ is finite; or (ii) $\mathcal{K}$ and the members of $\mathfrak{F}$ are compact; or (iii) $\mathcal{K}$ is open and the members of $\mathfrak{F}$ are bounded. Then $\bigcap \mathfrak{F}$ contains a translate of $\mathcal{K}$.*

Proof. For $\mathcal{F} \in \mathfrak{F}$, we define

$$\tilde{\mathcal{F}} = \{X \in \mathcal{X} \,|\, (\mathcal{K} + X) \subset \mathcal{F}\}.$$

Let $\tilde{\mathfrak{F}} = \{\tilde{\mathcal{F}} \,|\, \mathcal{F} \in \mathfrak{F}\}$. Clearly, $\tilde{\mathfrak{F}}$ consist of non-empty and convex subsets. Under (i), $\tilde{\mathfrak{F}}$ is finite, and, under (ii) or (iii), the members of $\tilde{\mathfrak{F}}$ are compact. In addition, by hypothesis, any $n + 1$ members of $\tilde{\mathfrak{F}}$ have non-empty intersection. Thus, by Helly's theorem, there is a point $X \in \bigcap \tilde{\mathfrak{F}}$. By definition, $\mathcal{K} + X$ is then contained in every member of $\mathfrak{F}$. The corollary follows.

Remark. In Corollary 1.4.3 the word "contains" can be replaced by "intersects" or "is contained in" (with corresponding modifications in the proof).

1.5   The Circumradius and Inradius in Euclidean Space

In this section we discuss classical geometric inequalities among the basic metric invariants of a convex body $C \in \mathfrak{B}_{\mathcal{X}}$ in a *Euclidean* space $\mathcal{X}$ of dimension n. The four invariants in question are the *circumradius* $R = R_C$, the *inradius* $r = r_C$, the *diameter* $D = D_C$, and the *minimal width* $d = d_C$. (Whenever convenient, we will suppress the subscript.) They are defined as follows:

$$R_C = \inf\{s > 0 \mid C \subset \bar{B}_s(X) \text{ for some } X \in \mathcal{X}\}$$

$$r_C = \sup\{s > 0 \mid C \supset \bar{B}_s(X) \text{ for some } X \in \mathcal{X}\}$$

$$D_C = \sup\{d(C, C') \mid C, C' \in C\}$$

$$d_C = \inf\{d(\mathcal{H}, \mathcal{H}') \mid \mathcal{H}, \mathcal{H}' \subset \mathcal{X} \text{ are parallel supporting hyperplanes of } C\}.$$

Since C is a convex body, the infima and suprema are clearly attained.

We now discuss these invariants along with some of their geometric properties.

We first claim that the circumradius $R = R_C$ is attained by a *unique* closed metric ball $\bar{B}_R(O) \supset C$, the *circumball* of C. (The corresponding sphere $S_R(O) = \partial B_R(O)$ is called the *circumsphere*, and the center O, the *circumcenter* of C.)

To give a quick proof of this, for $s > R$, we let $C_s = \{X \in \mathcal{X} \mid C \subset \bar{B}_s(X)\}$. Clearly, $\{C_s\}_{s>R}$ is a monotonic family of non-empty, compact subsets of $\mathcal{X}$. Thus, $C_R = \bigcap_{s>R} C_s$ is non-empty and compact. For $O \in C_R$, we have $C \subset \bar{B}_R(O)$, and hence the existence of a circumball follows. For unicity, if $O_1, O_2 \in C_R$, $O_1 \neq O_2$, then the intersection $\bar{B}_R(O_1) \cap \bar{B}_R(O_2) \supset C$ would be contained in a closed metric ball with center at $(O_1 + O_2)/2$ and radius $(R^2 - d(O_1, O_2)^2/4)^{1/2} < R$, a contradiction. Thus $C_R = \{O\}$ is a singleton, and unicity of the circumball also follows.

The circumsphere has the following geometric property:

$$O \in [C \cap S_R(O)]. \tag{1.5.1}$$

To show this, for simplicity, we may assume that O is the origin 0 (by performing a suitable translation). We first claim that, for each unit vector $N \in S$, there is $A \in C \cap S_R$ such that $\langle A, N \rangle \leq 0$.

Indeed, unicity of the circumsphere S_R implies that, for each $k \geq 1$, there exists $A_k \in C$ such that $(d(A_k, 0) \leq)R < d(A_k, N/k)$. In particular, from the triangle $[A_k, 0, N/k]$, we obtain $\langle A_k, N \rangle < 0$, $k \geq 1$. Now, an accumulation point $A \in C$ of the sequence $\{A_k\}_{k\geq 1} \subset C$ satisfies $d(A, 0) = R$ and $\langle A, N \rangle \leq 0$. The claim follows. Applying this to all pairs $\pm N \in S$, we conclude that the origin cannot be strictly separated from $C \cap S_R$. (See the discussion after Corollary 1.2.2.) Now (1.5.1) follows.

As a byproduct, we obtain $O \in C$, in particular, we have $R_C < D_C$ *with strict inequality*. (Note that the circumcenter O can be on the boundary of C as the case of the half-disk shows.)

Remark. The Euclidean structure of $\mathcal{X}$ is essential to guarantee the unicity of the circumball and (1.5.1). For example, in Minkowski space a circumball exists but it is *not unique*, and (1.5.1) may not hold.

Finally, note that the discussion above holds for *any* compact subset $\mathcal{A} \subset \mathcal{C}$ by setting $\mathcal{C} = [\mathcal{A}]$ (via Corollary 1.3.2).

The fact that the inradius $r = r_\mathcal{C}$ is attained as the radius of a closed metric ball in $\mathcal{C}$, an *inball* of $\mathcal{C}$, is a textbook application of Blaschke's selection theorem (Section 1.1/B) (by considering a minimizing sequence for the supremum defining $r_\mathcal{C}$). Note also that, as the example of a circular cylinder shows, one cannot expect the inball to be unique.

The supremum defining the diameter $D_\mathcal{C}$ of $\mathcal{C}$ is clearly attained (at boundary points of $\mathcal{C}$). A corresponding chord $[C, C'] \subset \mathcal{C}$ with $C, C' \in \partial \mathcal{C}$ and $d(C, C') = D_\mathcal{C}$ is called a *metric diameter*.

A metric diameter $[C, C']$ is a *double normal* of $\mathcal{C}$ in the sense that the (parallel) hyperplanes $\mathcal{H} \ni C$ and $\mathcal{H}' \ni C'$ orthogonal to $[C, C']$ both support $\mathcal{C}$. In addition, we also have $\mathcal{C} \cap \mathcal{H} = \{C\}$ and $\mathcal{C} \cap \mathcal{H}' = \{C'\}$, that is, C and C' are exposed points of $\mathcal{C}$ (Section 2.1). (These statements are clear: If $C'' \in \mathcal{C} \cap \mathcal{H}'$, $C'' \neq C'$, then $d(C'', C) > d(C', C)$, contradicting to maximality of $[C, C']$.)

As the example of a generic ellipsoid shows, a double normal is not necessarily a metric diameter. (See Problem 11.)

Remark. In 1960, V. Klee conjectured that a convex body $\mathcal{C}$ has at least n double normals; see [Klee 3]. This was solved affirmatively by [Kuiper] in 1964. (For a brief history of this problem and related problems, see the survey article of [Soltan, 2.3].)

The infimum defining the minimal width $d_\mathcal{C}$ is also attained at a (not necessarily unique) pair $\mathcal{H}, \mathcal{H}' \subset \mathcal{X}$ of parallel hyperplanes supporting $\mathcal{C}$ and having minimal distance $d_\mathcal{C} = d(\mathcal{H}, \mathcal{H}')$.

An important property of this minimal configuration is the following: Under the orthogonal projection $\pi : \mathcal{X} \to \mathcal{H}$ (onto $\mathcal{H}$) the image of $\mathcal{C} \cap \mathcal{H}'$ intersects $\mathcal{C} \cap \mathcal{H}$. Consequently, $\mathcal{H}$ and $\mathcal{H}'$ are connected by a double normal of $\mathcal{C}$.

To prove this, assume that $\pi(\mathcal{C} \cap \mathcal{H}')$ is *disjoint* from $\mathcal{C} \cap \mathcal{H}$. Being compact convex subsets of $\mathcal{H}$, these can be strictly separated by a hyperplane $\mathcal{E}$ in $\mathcal{H}$ (Section 1.2). Let $\mathcal{H}'' = \pi^{-1}(\mathcal{E})$, an extension of $\mathcal{E}$ to a hyperplane of $\mathcal{X}$. Then $\mathcal{E}' = \mathcal{H}' \cap \mathcal{H}''$ is a hyperplane in $\mathcal{H}'$. We can now simultaneously rotate $\mathcal{H}$ about $\mathcal{E}$ and $\mathcal{H}'$ about $\mathcal{E}'$ to a new configuration of parallel hyperplanes neither of which intersects $\mathcal{C}$. Since their distance in the new position is less than the original $d_\mathcal{C}$, this is a contradiction. (For this proof, see [Eggleston 1, 4.5]. For a different proof, see [Soltan, 2.2].)

Returning to the main line, as a direct consequence of the definitions, the four invariants are connected through the inequalities

$$2r_\mathcal{C} \leq d_\mathcal{C} \leq D_\mathcal{C} \leq 2R_\mathcal{C}.$$

The middle inequality can significantly be improved; in fact, we claim that the *maximal* width is equal to the diameter:

$$D_C = \sup\{d(\mathcal{H}, \mathcal{H}') \mid \mathcal{H}, \mathcal{H}' \subset \mathcal{X} \text{ are parallel supporting hyperplanes of } C\}. \tag{1.5.2}$$

To show this claim, for a moment, let D_C' denote the right-hand side of (1.5.2). If $\mathcal{H}, \mathcal{H}' \subset C$ are parallel supporting hyperplanes of C then, for $C \in C \cap \mathcal{H}$ and $C' \in C \cap \mathcal{H}'$, we have $D_C \geq d(C, C') \geq d(\mathcal{H}, \mathcal{H}')$. Taking the supremum in the parallel pair $\mathcal{H}, \mathcal{H}'$, we obtain $D_C \geq D_C'$.

Conversely, letting $C, C' \in \partial C$ realize the diameter $D_C = d(C, C')$, then, by the discussion above, $[C, C']$ is a metric diameter. By definition, the pair of parallel hyperplanes $\mathcal{H} \ni C$ and $\mathcal{H}' \ni C'$ orthogonal to $[C, C']$ are supporting C so that $D_C = d(C, C') = d(\mathcal{H}, \mathcal{H}') \leq D_C'$. The claim follows.

Remark 1. If $\mathcal{H}, \mathcal{H}' \subset \mathcal{X}$ are parallel supporting hyperplanes of a convex body $C \in \mathfrak{B}$ then their *distance* can be expressed using the support function h_C (Section 1.2):

$$d(\mathcal{H}, \mathcal{H}') = h_C(N) + h_C(-N),$$

where $N \in \mathcal{S}$ is orthogonal to the pair $\mathcal{H}, \mathcal{H}'$. It follows that we have

$$d_C = \inf\{h_C(N) + h_C(-N) \mid N \in \mathcal{S}\}.$$

By what we just proved, we also have

$$D_C = \sup\{h_C(N) + h_C(-N) \mid N \in \mathcal{S}\}.$$

Remark 2. In addition to the inequalities above, we also have

$$r_C + R_C \leq D_C, \quad C \in \mathfrak{B}.$$

(See Problem 15 at the end of Chapter 2.)

A quick case-by-case check shows that, among the twelve possible ratios of our four metric invariants, there are only two left that may have universal upper bounds (depending only on $n = \dim C$); the "dual" pair R_C/D_C and d_C/r_C, $C \in \mathfrak{B}$.

In *Euclidean* setting, the universal sharp upper bounds for these ratios are due to Jung and Steinhagen, respectively. In our treatment of these classical results we follow [Berger, 11.5.8] and [Eggleston 1, 6.3].

Theorem 1.5.1 ([Jung]). *Let $\mathcal{X}$ be Euclidean and $C \in \mathfrak{B}_{\mathcal{X}}$. We have*

$$\frac{R_C}{D_C} \leq \sqrt{\frac{n}{2(n+1)}}. \tag{1.5.3}$$

Proof. By (1.5.1) and Carathéodory's theorem (Section 1.3), the circumcenter O is in the convex hull of $n + 1$ (or less) points $A_0, \ldots, A_n \in C \cap S_R(O)$. As usual, we assume that O is at the origin and write $\sum_{i=0}^{n} \lambda_i A_i = 0$, $\sum_{i=0}^{n} \lambda_i = 1$, $\{\lambda_0, \ldots, \lambda_n\} \subset [0, 1]$.

For fixed $i = 0, \ldots, n$, we estimate

$$(1 - \lambda_i) D_C^2 \geq \sum_{j=0}^{n} \lambda_j d(A_i, A_j)^2 = \sum_{j=0}^{n} \lambda_j(|A_i|^2 + |A_j|^2 - 2\langle A_i, A_j \rangle)$$

$$\geq \sum_{j=0}^{n} \lambda_j(2R_C^2 - 2\langle A_i, A_j \rangle) = 2R_C^2,$$

where, in the first sum, the ith term is zero, and $1 - \lambda_i = \sum_{j=0; j \neq i}^{n} \lambda_j$. Summing over $i = 0, \ldots, n$, we obtain $nD_C^2 \geq 2(n + 1)R_C^2$. The estimate in (1.5.3) follows.

Remark 1. Jung's theorem holds for any compact subset $\mathcal{A} \subset \mathcal{X}$ by setting $C = [\mathcal{A}]$ (in the affine span $\langle \mathcal{A} \rangle$ as $\mathcal{X}$).

Remark 2. A quick check of the proof shows that equality holds in all the estimates (and hence also in (1.5.3)) for a *regular* simplex.

The upper bound in (1.5.3) is attained by many other convex bodies, however. For example, equality holds in (1.5.3) for *any* convex body *between a regular simplex and any of its completions*. (See Section 2.5 for the relevant discussion.)

As a partial converse, a theorem of [Melzak] asserts that if, for a convex body $C \in \mathfrak{B}$ *of constant width* d, equality holds in (1.5.3) then C contains a regular simplex Δ of diameter d, therefore C is a completion of Δ.

Remark 3. The analogue of Jung's theorem in Minkowski space, due to Bohnenblust, will be treated in Section 3.9.

Theorem 1.5.2 ([Steinhagen]). *Let $\mathcal{X}$ be Euclidean and $C \in \mathcal{B}_\mathcal{X}$. We have*

$$\frac{d_C}{r_C} \leq \begin{cases} 2\sqrt{n} & \text{if } n \text{ is odd} \\ \frac{2(n+1)}{\sqrt{n+2}} & \text{if } n \text{ is even.} \end{cases} \tag{1.5.4}$$

Proof. Fix an insphere $S_r(O) \subset C$, $r = r_C$. Due to the maximality of r, we have $O \in [\partial C \cap S_r(O)]$. By Carathéodory's theorem (Section 1.3), there exists an affinely independent set $\{C_1, \ldots, C_{k+1}\} \subset \partial C \cap S_r(O)$, $1 \leq k \leq n$, such that O is in the (relative) interior of the k-simplex $[C_1, \ldots, C_{k+1}]$.

As an initial step in proving (1.5.4), we claim that, without loss of generality, we may assume $k = n$.

To show this, we introduce the following notation: For $C \in S_r(O)$, we let $\mathcal{H}_C \subset \mathcal{X}$ denote the hyperplane tangent to $S_r(O)$ at C. This hyperplane bounds a unique closed half-space $\mathcal{G}_C$ which contains $S_r(O)$. Note that if $C \in \partial C \cap S_r(O)$ then $\mathcal{H}_C$ supports C (by convexity), and consequently $C \subset \mathcal{G}_C$.

Now, assume that $k < n$. Then, for every $\epsilon > 0$ we can find an affinely independent set $\{C'_1, \ldots, C'_{n+1}\} \subset S_r(O)$ (near $\{C_1, \ldots, C_{k+1}\}$) such that O is in the interior of the n-simplex $[C'_1, \ldots, C'_{n+1}]$, and, for the minimal width $d_{C'}$ of the truncated convex body $C' = C \cap \bigcap_{i=1}^{n+1} \mathcal{G}_{C'_i}$, we have $d_C - \epsilon \leq d_{C'} \leq d_{C'}$. Since $\{C'_1, \ldots, C'_{n+1}\} \subset \partial C'$ and $S_r(O)$ is an insphere for C' as well, once (1.5.4) is proved for C', we then let $\epsilon \to 0$ to obtain (1.5.4) in general.

Summarizing (and reverting to the earlier notation), we may assume that there is an affinely independent set $\{C_1, \ldots, C_{n+1}\} \subset \partial C \cap S_r(O)$ such that O is in the interior of the n-simplex $[C_1, \ldots, C_{n+1}]$.

Let $\mathcal{T} = \bigcap_{i=1}^{n+1} \mathcal{G}_{C_i}$ be the n-simplex bounded by the hyperplanes $\mathcal{H}_{C_i}$ tangent to $S_r(O)$ at the points C_i, $i = 1, \ldots, n + 1$. Since $C_i \in \partial C \cap S_r(O)$, as noted above, the hyperplanes $\mathcal{H}_i$ support C, and we have $C \subset \mathcal{T}$. In particular, we have $d_{\mathcal{T}} \geq d_C$. Since, by construction, $r_{\mathcal{T}} = r_C$, it is enough to prove (1.5.4) for $\mathcal{T}$.

Let $\{B_1, \ldots, B_{n+1}\}$ denote the vertices of $\mathcal{T}$ with B_i, $i = 1, \ldots, n + 1$, opposite to the ith face $\mathcal{T}_i = \mathcal{T} \cap \mathcal{H}_{C_i}$ of $\mathcal{T}$. Finally, for $i = 1, \ldots, n + 1$, let $V_i \in \mathcal{X}$ be the vector orthogonal to $\mathcal{T}_i$, pointing outward from $\mathcal{T}$, and having norm equal to the $(n-1)$-dimensional volume $\mathrm{vol}_{n-1}(\mathcal{T}_i)$.

We first claim that

$$\sum_{i=1}^{n+1} V_i = 0. \tag{1.5.5}$$

Indeed, if $N \in S$ is any unit vector then $\left\langle \sum_{i=1}^{n+1} V_i, N \right\rangle$ is equal to the sum of signed volumes of the faces $\mathcal{T}_i \subset \mathcal{T}$, $i = 1, \ldots, n + 1$, projected to the line $\mathbb{R} \cdot N$. For any polytope this sum vanishes. We obtain that $\left\langle \sum_{i=1}^{n+1} V_i, N \right\rangle = 0$. Since $N \in S$ was arbitrary, (1.5.5) follows.

Now, taking norms, we also have

$$\sum_{i=1}^{n+1} |V_i|^2 + 2 \sum_{1 \leq i < j \leq n+1} \langle V_i, V_j \rangle = 0. \tag{1.5.6}$$

Let $\mathcal{I} = \{1, \ldots, n + 1\}$. For $k = 1, \ldots, n$, we let $\mathfrak{I}_k$, denote the family of k-element subsets of $\mathcal{I}$. (Only proper subsets of $\mathcal{I}$ will be used here.) We have $|\mathfrak{I}_k| = \binom{n+1}{k}$.

For $\mathcal{A} \in \mathfrak{I}_k$, $k = 1, \ldots, n$, we write $V(\mathcal{A}) = \sum_{i \in \mathcal{A}} V_i$. In particular, $V(\{V_i\}) = V_i$, $i = 1, \ldots, n + 1$. By (1.5.5), we have

$$V(\mathcal{A}^c) = -V(\mathcal{A}), \quad \mathcal{A} \in \mathfrak{I}_k, \ k = 1, \ldots, n, \tag{1.5.7}$$

where $\mathcal{A}^c = \mathcal{I} \setminus \mathcal{A}$ is the complement of $\mathcal{A}$.

Now, fix $A \in \mathfrak{I}_k$, for some $k = 1, \ldots, n$. By definition $V(A)$ is orthogonal to the affine span $\langle \{B_i\}_{i \in A^c} \rangle$ (since V_i is orthogonal to the affine span $\langle B_1, \ldots, \widehat{B_i}, \ldots, B_{n+1} \rangle \supset \mathcal{T}_i, i \in A$). By (1.5.7), $V(A)$ is also orthogonal to $\langle \{B_i\}_{i \in A} \rangle$.

Let $\mathcal{H}_A$ and $\mathcal{H}_{A^c}$ be the (unique) parallel pair of hyperplanes supporting the simplex $\mathcal{T}$ and containing its respective disjoint faces $[\{B_i\}_{i \in A}] \subset \mathcal{H}_A$ and $[\{B_i\}_{i \in A^c}] \subset \mathcal{H}_{A^c}$. By what we just concluded, $V(A)$ is orthogonal to this pair. Let $d(A) = d(\mathcal{H}_A, \mathcal{H}_{A^c})(\geq d_{\mathcal{T}})$.

In addition, with $N = V(A)/|V(A)|$, the norm $|V(A)| = \langle V(A), N \rangle = \langle \sum_{i \in A} V_i, N \rangle$, is the sum over $i \in A$ of the (signed) volumes of the faces $\mathcal{T}_i$ projected to $\mathbb{R} \cdot V(A)$. For each $i \in A$, the projection factor (the cosine of the respective angle between V_i and V_A) multiplied by $d(A)$ is equal to the length of the altitude of the sub-simplex $[\mathcal{T}_i, C] \subset \mathcal{T}$ over the base $\mathcal{T}_i$, where the extra vertex C is the endpoint in $\mathcal{H}_A$ of the unique double normal of $\mathcal{T}$ connecting $\mathcal{H}_A$ and $\mathcal{H}_{A^c}$. These sub-simplices form a subdivision of $\mathcal{T}$, so that we have

$$\operatorname{vol} \mathcal{T} = \frac{1}{n} |V(A)| d(A). \tag{1.5.8}$$

On the other hand, another subdivision of $\mathcal{T}$ is given by the simplices $[\mathcal{T}_i, O]$, $i = 1, \ldots, n+1$, with the extra vertex O, the center of the insphere $\mathcal{S}_r(O)$. We thus have

$$\operatorname{vol} \mathcal{T} = \frac{r}{n} \sum_{i=1}^{n+1} |V_i|. \tag{1.5.9}$$

Combining (1.5.8)–(1.5.9), we obtain

$$d_{\mathcal{T}} \leq \min_{1 \leq k \leq n} \min_{A \in \mathfrak{I}_k} d(A) = r \min_{1 \leq k \leq n} \min_{A \in \mathfrak{I}_k} \frac{\sum_{i=1}^{n+1} |V_i|}{|V(A)|} = \min_{1 \leq k \leq n} \frac{\sum_{i=1}^{n+1} |V_i|}{\max_{A \in \mathfrak{I}_k} |V(A)|}. \tag{1.5.10}$$

Since $|\mathfrak{I}_k| = \binom{n+1}{k}$, we have

$$\left(\max_{A \in \mathfrak{I}_k} |V(A)| \right)^2 \geq \frac{1}{\binom{n+1}{k}} \sum_{A \in \mathfrak{I}_k} |V(A)|^2. \tag{1.5.11}$$

We estimate

$$\sum_{A \in \mathfrak{I}_k} |V(A)|^2 = \sum_{A \in \mathfrak{I}_k} \left| \sum_{i \in A} V_i \right|^2 = \sum_{A \in \mathfrak{I}_k} \left(\sum_{i \in A} |V_i|^2 + 2 \sum_{i < j, i,j \in A} \langle V_i, V_j \rangle \right)$$

$$= \binom{n}{k-1} \sum_{i=1}^{n+1} |V_i|^2 + 2\binom{n-1}{k-2} \sum_{1 \le i < j \le n+1} \langle V_i, V_j \rangle \qquad (1.5.12)$$

$$= \binom{n-1}{k-1} \sum_{i=1}^{n+1} |V_i|^2,$$

where in the last equality we used (1.5.6).

We also have

$$\left(\sum_{i=1}^{n+1} |V_i| \right)^2 = \sum_{i=1}^{n+1} |V_i|^2 + 2 \sum_{1 \le i < j \le n+1} |V_i|\,|V_j| \le (n+1) \sum_{i=1}^{n+1} |V_i|^2, \qquad (1.5.13)$$

where the last inequality can be established by a simple induction.
Combining (1.5.11)–(1.5.13), we obtain

$$\left(\max_{\mathcal{A} \in \mathfrak{I}_k} |V(\mathcal{A})| \right)^2 \ge \frac{\binom{n-1}{k-1}}{\binom{n+1}{k}} \sum_{i=1}^{n+1} |V_i|^2$$

$$\ge \frac{1}{n+1} \frac{\binom{n-1}{k-1}}{\binom{n+1}{k}} \left(\sum_{i=1}^{n+1} |V_i| \right)^2$$

$$= \frac{k(n-k+1)}{n(n+1)^2} \left(\sum_{i=1}^{n+1} |V_i| \right)^2.$$

Using this in (1.5.10), we arrive at the estimate

$$d_T \le r \min_{1 \le k \le n} \frac{(n+1)\sqrt{n}}{\sqrt{k(n-k+1)}}.$$

Finally, the minimum on the right-hand side is attained for $k = (n+1)/2$ if n is odd, and for $k = n/2$ if n is even. The estimate in (1.5.4) follows.

Remark. The analogue of Steinhagen's theorem in Minkowski space will be treated in Section 3.9.

1.6 A Helly-Type Theorem of Klee

The natural domain of the Hausdorff-distance is $\mathfrak{C}_{\mathcal{X}}$, the set of all compact subsets of $\mathcal{X}$. For *unbounded* (closed) subsets this metric does not reflect the intuitive concept of limit. Indeed, a sequence $\{\bar{\mathcal{B}}_{r_k}\}_{k \ge 1}$ of nested closed metric balls with

$\lim_{k\to\infty} r_k = \infty$ can be thought of having the entire space $\mathcal{X}$ as its limit; however, $\lim_{k\to\infty} d_H(\bar{\mathcal{B}}_{r_k}, \mathcal{X}) = 0$ cannot hold since $\mathcal{X} \subset (\bar{\mathcal{B}}_{r_k})_r = \bar{\mathcal{B}}_{r_k+r}$ clearly fails for any $r > 0$ and $k \geq 1$ (while $\bar{\mathcal{B}}_{r_k} \subset \overline{\mathcal{X}_r} = \mathcal{X}$ is obviously valid).

Klee's theorem to be discussed in this section requires a concept of convergence for *unbounded* subsets in $\mathcal{X}$. This has been initiated by Painlevé and developed by Kuratowski. We begin with the definition and basic properties of this *Painlevé–Kuratowski* convergence of sequences of subsets in a topological space.

Although this concept of convergence can be stated in much more general setting (for topological spaces satisfying the second axiom of countability), for our purposes it will be sufficient to assume that the ambient space is *separable metric*.

For a sequence $\{\mathcal{A}_k\}_{k\geq 1}$ of sets in $\mathcal{X}$, we define

$$\liminf_{k\to\infty} \mathcal{A}_k = \{X \in \mathcal{X} \mid \limsup_{k\to\infty} d(X, \mathcal{A}_k) = 0\}$$

$$= \{X \in \mathcal{X} \mid \exists X_k \in \mathcal{A}_k, k \geq 1, \lim_{k\to\infty} X_k = X\},$$

$$\limsup_{k\to\infty} \mathcal{A}_k = \{X \in \mathcal{X} \mid \liminf_{k\to\infty} d(X, \mathcal{A}_k) = 0\}$$

$$= \{X \in \mathcal{X} \mid \exists X_{k_l} \in \mathcal{A}_{k_l}, k_{l+1} > k_l, l \geq 1, \lim_{l\to\infty} X_{k_l} = X\}.$$

Equivalently, $X \in \liminf_{k\to\infty} \mathcal{A}_k$ if and only if for every $\epsilon > 0$ there exists $k_0 \geq 1$ such that $\mathcal{B}_\epsilon(X) \cap \mathcal{A}_k \neq \emptyset$, $k \geq k_0$. In a similar vein, $X \in \limsup_{k\to\infty} \mathcal{A}_k$ if and only if for every $\epsilon > 0$ we have $\mathcal{B}_\epsilon(X) \cap \mathcal{A}_k \neq \emptyset$ for infinitely many $k \geq 1$.

Remark. As noted by [Kuratowski, Section 25, VIII, p. 156, footnote #2], the concepts of limit inferior and limit superior are due to [Painlevé, p. 1156].

It follows that the limit inferior and limit superior are both *closed* subsets. We obviously have

$$\liminf_{k\to\infty} \mathcal{A}_k \subset \limsup_{k\to\infty} \mathcal{A}_k.$$

The limit superior has the following useful explicit formula

$$\limsup_{k\to\infty} \mathcal{A}_k = \bigcap_{k\geq 1} \overline{\bigcup_{l\geq k} \mathcal{A}_l} \left(\subset \overline{\bigcup_{l\geq 1} \mathcal{A}_l} \right).$$

Remark. Note that

$$\left(\bigcap_{l\geq 1} \mathcal{A}_l \subset \right) \bigcup_{k\geq 1} \bigcap_{l\geq k} \mathcal{A}_l \subset \liminf_{k\to\infty} \mathcal{A}_k,$$

with inclusion only. Referring to the formula above for the limit superior Kuratowski noted *"on ne connaît pas de formule analogue"* for the limit inferior; see [Kuratowski, Section 25, IV, p. 154].

On the other hand, one may note that the families of subsets (of $\mathbb{N} = \{1, 2, \ldots\}$):

$$\mathfrak{N}_\infty = \{\mathcal{N} \subset \mathbb{N} \mid |\mathbb{N} \setminus \mathcal{N}| < \infty\}$$

$$\mathfrak{N}_\infty^\# = \{\mathcal{N} \subset \mathbb{N} \mid |\mathbb{N} \setminus \mathcal{N}| = \infty\}$$

are "dual" in the sense

$$\mathfrak{N}_\infty^\# = \{\mathcal{N} \subset \mathbb{N} \mid \mathcal{N} \cap \mathcal{N}' \neq \emptyset, \ \mathcal{N}' \subset \mathfrak{N}_\infty\},$$

$$\mathfrak{N}_\infty = \{\mathcal{N} \subset \mathbb{N} \mid \mathcal{N} \cap \mathcal{N}' \neq \emptyset, \ \mathcal{N}' \subset \mathfrak{N}_\infty^\#\}.$$

($\mathfrak{N}_\infty$ defines the cofinite topology on $\mathbb{N}$.) With these, we have

$$\limsup_{k \to \infty} \mathcal{A}_k = \bigcap_{\mathcal{N} \in \mathfrak{N}_\infty} \overline{\bigcup_{k \in \mathcal{N}} \mathcal{A}_k}$$

$$\liminf_{k \to \infty} \mathcal{A}_k = \bigcap_{\mathcal{N} \in \mathfrak{N}_\infty^\#} \overline{\bigcup_{k \in \mathcal{N}} \mathcal{A}_k}.$$

For a thorough treatise on set convergence, see [Rockafellar–Wets, IV].

For any subsequence $\{\mathcal{A}_{k_l}\}_{l \geq 1}$ of $\{\mathcal{A}_k\}_{k \geq 1}$, we clearly have

$$\liminf_{k \to \infty} \mathcal{A}_k \subset \liminf_{l \to \infty} \mathcal{A}_{k_l} \subset \limsup_{l \to \infty} \mathcal{A}_{k_l} \subset \limsup_{k \to \infty} \mathcal{A}_k. \tag{1.6.1}$$

Finally, for two sequences $\{\mathcal{A}_k\}_{k \geq 1}$ and $\{\mathcal{B}_k\}_{k \geq 1}$, we have

$$\limsup_{k \to \infty} (\mathcal{A}_k \cap \mathcal{B}_k) \subset \limsup_{k \to \infty} \mathcal{A}_k \cap \limsup_{k \to \infty} \mathcal{B}_k. \tag{1.6.2}$$

We say that the sequence $\{\mathcal{A}_k\}_{k \geq 1}$ *converges* to $\mathcal{A}$, written as $\lim_{k \to \infty} \mathcal{A}_k = \mathcal{A}$, if

$$\liminf_{k \to \infty} \mathcal{A}_k = \limsup_{k \to \infty} \mathcal{A}_k = \mathcal{A}.$$

Remark. Let $\{\mathcal{C}_k\}_{k \geq 1} \subset \mathfrak{C}$ be a sequence of *compact* subsets in the Euclidean space $\mathcal{X}$. A quick comparison of (1)–(2) in Theorem 1.1.11 with the definitions of limit inferior and limit superior shows that this sequence converges in the Hausdorff metric if and only if it converges in the Painlevé–Kuratowski sense above.

More generally, in a separable metric space in which closed and bounded sets are compact (such as the Euclidean space $\mathcal{X}$) the Painlevé–Kuratowski convergence of sequences of sets *coincides* with the convergence in the *bounded-Hausdorff* topology; or equivalently, $\lim_{k\to\infty} A_k = A$ if and only if $\lim_{k\to\infty} d(.,A_k) = d(.,A)$ *uniformly* on *bounded* subsets.

We will repeatedly use *sequential compactness* (the Bolzano–Weierstrass property) of the Painlevé–Kuratowski convergence: Any sequence $\{A_k\}_{k\geq 1}$ of subsets *subconverges*, that is, contains a convergent subsequence.

Remark. For a classical proof in separable metric spaces (using second countability), see [Kuratowski, Section 25, VIII, pp. 156–157] (and also [Aubin–Frankowska, p. 23]).

Note that the Bolzano–Weierstrass property for the Painlevé–Kuratowski convergence was first observed by S. Mrówka (for nets) and [Sierpiński] (for sequences). Emphasizing that this is the property of the respective space of all closed subsets, it is usually stated for sequences of *closed* sets. Since $d(X,A) = d(X,\bar{A})$, $X \in \mathcal{X}$, $A \subset \mathcal{X}$, for any sequence $\{A_k\}_{k\geq 1}$, we have $\liminf_{k\to\infty} A_k = \liminf_{k\to\infty} \bar{A}_k$ and $\limsup_{k\to\infty} A_k = \limsup_{k\to\infty} \bar{A}_k$, and hence this restriction is irrelevant.

We now return to our Euclidean space $\mathcal{X}$. A family $\mathfrak{I}$ of *open* subsets of $\mathcal{X}$ is called *interior-complete* if for every convergent sequence $\{\mathcal{I}_k\}_{k\geq 1} \subset \mathfrak{I}$ with $\lim_{k\to\infty} \mathcal{I}_k = \mathcal{I}$ we have $\text{int}\,\mathcal{I} \in \mathfrak{I}$.

Theorem 1.6.1 ([Klee 2]). *Let $\mathfrak{I}$ be an interior-complete family of convex subsets of $\mathcal{X}$. Assume that there exists $\mathcal{I} \in \mathfrak{I}$, $\mathcal{I} \neq \mathcal{X}$ such that whenever $\{\mathcal{I}_1,\ldots,\mathcal{I}_n\} \subset \mathfrak{I}$ then $\mathcal{I}_1 \cap \ldots \cap \mathcal{I}_n \cap \mathcal{I}$ contains points arbitrarily far from $\partial\mathcal{I}$. Then $\text{int}\,(\bigcap \mathfrak{I}) \neq \emptyset$.*

The proof of Klee's theorem is technical and will be preceded by two lemmas. We follow the original paper of [Klee 2] with some modifications.

Lemma 1.6.2. *Let $\{C_k\}_{k\geq 1}$ be a sequence of convex sets in $\mathcal{X}$ with $\lim_{k\to\infty} C_k = C$. Assume that $O \in \text{int}\,C$. Then $O \in \text{int}\,\bigcap_{k\geq k_0} C_k$ for some $k_0 \geq 1$.*

Proof. We may assume that $O = 0$, the origin, and that $\{C_k\}_{k\geq 1}$ is *uniformly bounded*, that is, the sequence is contained in a ball $\mathcal{B}_R$ of radius $R > 0$ (centered at 0).

We first claim that there exists $k_0 \geq 1$ such that $0 \in \text{int}\,C_k$, $k \geq k_0$. Otherwise, by convexity (Corollary 1.2.3), for a subsequence $\{C_{k_l}\}_{l\geq 1}$, we would have $C_{k_l} \subset \mathcal{G}_l$, $l \geq 1$, where $\mathcal{G}_l$ is a closed half-space with boundary hyperplane containing 0. Since the sequence $\{\mathcal{G}_l\}_{l\geq 1}$ subconverges to a closed half-space $\mathcal{G}$ with boundary hyperplane once again containing 0, we would have $C = \liminf_{k\to\infty} C_k \subset \liminf_{l\to\infty} C_{k_l} \subset \liminf_{l\to\infty} \mathcal{G}_l \subset \mathcal{G}$. Thus 0 cannot be an interior point of C, a contradiction. The claim follows.

Now, for $k \geq k_0$, let $r_k > 0$ be the radius of the *largest* open metric ball with center at 0 which is contained in C_k. The lemma asserts that $\liminf_{k\to\infty} r_k > 0$.

Assuming the contrary, let $\{C_{k_l}\}_{l\geq 1}$ be a subsequence such that $\lim_{l\to\infty} r_{k_l} = 0$. By convexity (and the extremal property of the radius), there exists a unit vector

$V_{k_l} \in \mathcal{S}$ such that $r_{k_l} = \sup\{\langle V_{k_l}, X \rangle \mid X \in \mathcal{C}_{k_l}\}$. Passing to a subsequence if necessary, we may assume that $\lim_{l \to \infty} V_{k_l} = V$. Setting $\epsilon_{k_l} = |V - V_{k_l}|$, we have

$$\langle V, X \rangle = \langle V - V_{k_l}, X \rangle + \langle V_{k_l}, X \rangle \le |V - V_{k_l}||X| + r_{k_l} \le \epsilon_{k_l} R + r_{k_l},$$

or equivalently

$$\mathcal{C}_{k_l} \subset \{X \in \mathcal{X} \mid \langle V, X \rangle \le \epsilon_{k_l} R + r_{k_l}\}.$$

We now have

$$\mathcal{C} = \liminf_{k \to \infty} \mathcal{C}_k \subset \liminf_{l \to \infty} \mathcal{C}_{k_l} \subset \{X \in \mathcal{X} \mid \langle V, X \rangle \le 0\}.$$

This contradicts to $0 \in \mathrm{int}\,\mathcal{C}$. The lemma follows.

Lemma 1.6.3. *Let $\mathfrak{I}$ be an interior-complete family of convex sets in $\mathcal{X}$. Assume that any $n + 1$ members of $\mathfrak{I}$ intersect non-trivially. Then $\bigcap \mathfrak{I}$ has a non-empty interior.*

Proof. We will make use of Corollary 1.4.3(iii) of Helly's theorem. We claim that there exists $k \ge 1$ such that any $n + 1$ members of $\mathfrak{I}$ contain a metric ball $\mathcal{B}_{1/k}(X)$ within $\mathcal{B}_k$ (with X depending on the members).

Assuming this, we can apply this corollary to $\mathcal{K} = \mathcal{B}_{1/k}$ and $\mathfrak{I}_k = \{\mathcal{I} \cap \mathcal{B}_k \mid \mathcal{I} \in \mathfrak{I}\}$, and obtain that a translate of $\mathcal{B}_{1/k}$ is contained in $\bigcap \mathfrak{I}_k$. Therefore $\mathrm{int}\,\bigcap \mathfrak{I}_k \ne \emptyset$, in particular, $\mathrm{int}\,\bigcap \mathfrak{I} \ne \emptyset$. Our lemma then follows.

It remains to prove the claim. We argue by contradiction, and assume that, for any $k \ge 1$, there exists $\{\mathcal{I}_k^1, \ldots, \mathcal{I}_k^{n+1}\} \subset \mathfrak{I}$ such that $\mathcal{B}_k \cap \mathcal{I}_k^1 \cap \ldots \cap \mathcal{I}_k^{n+1}$ contains no $1/k$-sphere. By sequential compactness of the Painlevé–Kuratowski convergence, we may assume that $\lim_{k \to \infty} \mathcal{I}_k^i = \mathcal{I}_0^i$ exist for all $i = 1, \ldots, n + 1$. Being limits, $\mathcal{I}_0^i, i = 1, \ldots, n + 1$, are closed sets. On the other hand, interior-completeness gives $\mathrm{int}\,\mathcal{I}_0^i \in \mathfrak{I}, i = 1, \ldots, n + 1$, and therefore, by assumption, $\cap_{i=1}^{n+1} \mathrm{int}\,\mathcal{I}_0^i \ne \emptyset$.

Letting $\mathcal{D}_k = \mathcal{I}_k^1 \cap \ldots \cap \mathcal{I}_k^{n+1}, k \ge 0$, we see that $\mathcal{D}_0$ is a closed set with $\mathrm{int}\,\mathcal{D}_0 \ne \emptyset$.

We now claim that there is a convergent subsequence $\{\mathcal{D}_{k_l}\}_{l \ge 1}$ such that

$$\lim_{l \to \infty} \mathcal{D}_{k_l} = \mathcal{D}_0. \tag{1.6.3}$$

First of all, for any subsequence $\{\mathcal{D}_{k_l}\}_{l \ge 1}$ (convergent or not), by (1.6.1) and (1.6.2), we have

$$\limsup_{l \to \infty} \mathcal{D}_{k_l} \subset \limsup_{k \to \infty} \mathcal{D}_k = \limsup_{k \to \infty} (\mathcal{I}_k^1 \cap \ldots \cap \mathcal{I}_k^{n+1})$$

$$\subset \limsup_{k \to \infty} \mathcal{I}_k^1 \cap \ldots \cap \limsup_{k \to \infty} \mathcal{I}_k^{n+1} \tag{1.6.4}$$

$$= \mathcal{I}_0^1 \cap \ldots \cap \mathcal{I}_0^{n+1} = \mathcal{D}_0.$$

For the opposite inclusion, we exhibit a subsequence $\{\mathcal{D}_{k_l}\}_{l\geq 1}$ such that $\text{int }\mathcal{D}_0 \subset \liminf_{l\to\infty} \mathcal{D}_{k_l}$. Since the limit inferior is a closed set, using (1.6.4), this will give (1.6.3).

Given $X \in \text{int }\mathcal{D}_0$, we have $X \in \text{int }\mathcal{I}_0^i$ for all $i = 1,\ldots,n+1$. Since $\mathcal{I}_0^i = \lim_{k\to\infty}\mathcal{I}_k^i$, Lemma 1.6.2 applies giving $X \in \text{int}\left(\bigcap_{k\geq k_0(X,i)}\mathcal{I}_k^i\right)$, for some $k_0(X,i) \geq 1$ (depending on X and $i = 1,\ldots,n+1$). Letting $k_0(X) = \max_{1\leq i\leq n+1} k_0(X,i)$, we obtain $X \in \text{int}\left(\bigcap_{k\geq k_0(X)}\mathcal{D}_k\right)$.

We now appeal to separability and apply Lindelöf's theorem to obtain a countable set arranged in a sequence $\{X_l\}_{l\geq 1} \subset \text{int }\mathcal{D}_0$ such that

$$\text{int }\mathcal{D}_0 \subset \bigcup_{l\geq 1} \text{int} \bigcap_{k\geq k_0(X_l)} \mathcal{D}_k. \tag{1.6.5}$$

Letting $k_l = k_0(X_l)$, $l \geq 1$, we consider the subsequence $\{\mathcal{D}_{k_l}\}_{l\geq 1}$ of $\{\mathcal{D}_k\}_{k\geq 1}$. By (1.6.5) and the definition of the limit inferior, we have

$$\text{int }\mathcal{D}_0 \subset \bigcup_{l\geq 1}\bigcap_{k\geq k_l}\mathcal{D}_k \subset \bigcup_{l\geq 1}\bigcap_{j\geq l}\mathcal{D}_{k_j} \subset \liminf_{l\to\infty}\mathcal{D}_{k_l}.$$

We obtain (1.6.3).

Finally, for $X \in \text{int }\mathcal{D}_0$, by Lemma 1.6.2 this time applied to the sequence $\{\mathcal{D}_{k_l}\}_{l\geq 1}$, we have $\mathcal{B}_\epsilon(X) \subset \bigcap_{l\geq l_0}\mathcal{D}_{k_l}$ for some $\epsilon > 0$ and $l_0 \geq 1$. This shows that, for $k_l \geq \max(1/\epsilon, 2d(X,0))$, $l \geq l_0$, the set $\mathcal{B}_{k_l} \cap \mathcal{D}_{k_l}$ contains a $1/k_l$-sphere, a contradiction. The lemma follows.

Proof of Theorem 1.6.1. We proceed by induction with respect to $n = \dim \mathcal{X} \geq 1$. For $n = 1$, the assumptions imply that $\mathcal{I}$ is an open half-infinite interval which we may assume to be $(0,\infty)$. Moreover, any other member $\mathcal{I}' \in \mathfrak{J}$, $\mathcal{I}' \neq \mathcal{X}$, must be of the form (a,∞), where $a \in \mathbb{R}$. The statement of the lemma is now equivalent to finiteness of $\sup\{a \mid (a,\infty) \in \mathfrak{J}\}$. If the supremum is infinite then we can select a monotonic sequence $\{(a_k,\infty)\}_{k\geq 1} \subset \mathfrak{J}$ with $\lim_{k\to\infty} a_k = \infty$. Then

$$\lim_{k\to\infty}(a_k,\infty) = \overline{\limsup_{k\to\infty}(a_k,\infty)} = \bigcap_{k\geq 1}\bigcup_{l\geq k}(a_l,\infty) = \bigcap_{k\geq 1}[a_k,\infty) = \emptyset,$$

contradicting to interior-completeness.

For the general induction step, we assume that the statement holds in dimensions $< n$ ($n > 1$) and prove it for $n = \dim \mathcal{X}$. In view of Lemma 1.6.3, we need to show that, for $\{\mathcal{I}_0,\ldots,\mathcal{I}_n\} \subset \mathfrak{J}$, we have $\bigcap_{0\leq i\leq n}\mathcal{I}_i \neq \emptyset$.

Assuming the contrary, there exists $\{\mathcal{I}_0,\ldots,\mathcal{I}_n\} \subset \mathfrak{J}$ such that the open convex sets $\mathcal{I}_0$ and $\mathcal{C} = \bigcap_{1\leq i\leq n}\mathcal{I}_i$ are disjoint. Let $\mathcal{H} \subset \mathcal{X}$ be a hyperplane (strictly) separating $\mathcal{I}_0$ and $\mathcal{C}$.

Consider now $\mathcal{I} \in \mathfrak{J}$ as in Theorem 1.6.1. We first claim that $\mathcal{H} \not\subset \mathcal{I}$. Indeed, if $\mathcal{H} \subset \mathcal{I}$ then, by convexity of $\mathcal{I}$, its boundary $\partial \mathcal{I}$ must be comprised by one or two parallel translates of $\mathcal{H}$ (Corollary 1.2.4). This, however, cannot happen since, by assumption, both $\mathcal{I} \cap \mathcal{I}_0$ and $\mathcal{I} \cap \mathcal{C}$ contain points arbitrarily far from $\partial \mathcal{I}$.

We obtain that $\mathcal{I}' = \mathcal{I} \cap \mathcal{H}$ is a non-empty open convex proper subset of $\mathcal{H}$. Given $k = 1 \ldots, n$, by assumption, $\bigcap_{1 \le i \ne k \le n} \mathcal{I}_i$ must intersect $\mathcal{I}_0$. Since $\bigcap_{1 \le i \ne k \le n} \mathcal{I}_i$ is convex, contains $\mathcal{C}$, and intersects $\mathcal{I}_0$ on the other side of $\mathcal{H}$, we obtain that $\bigcap_{1 \le i \ne k \le n} \mathcal{I}_i \cap \mathcal{H}$ is non-empty. In particular, since $n \ge 2$, the intersections $\mathcal{I}'_i = \mathcal{I}_i \cap \mathcal{H}$, $i = 1, \ldots, n$, themselves are non-empty. In addition, it also follows from the assumption that $\bigcap_{1 \le i \ne k \le n} \mathcal{I}'_i \cap \mathcal{I}'$ contains points arbitrarily far from the relative boundary $\partial \mathcal{I}'$ in $\mathcal{H}$. Now the induction hypothesis applies to $\mathfrak{J}' = \{\mathcal{I}'_1, \ldots, \mathcal{I}'_n, \mathcal{I}'\}$ as $\dim \mathcal{H} = n - 1$. We obtain that $\bigcap_{1 \le i \le n} \mathcal{I}'_i \ne \emptyset$ contradicting to $\mathcal{C} \cap \mathcal{H} = \emptyset$. Klee's theorem follows.

Remark. The proof of the general induction step follows the original proof of Helly's as well as the one by [König 1]; see [Danzer–Grünbaum–Klee].

Corollary 1.6.4. *Let $\mathcal{K} \subset \mathcal{X}$ be an affine subspace of dimension $k < n$ and $\mathfrak{J}$ an interior-complete family of convex subsets of $\mathcal{X}$ such that each element of $\mathfrak{J}$ contains a translate of $\mathcal{K}$. Assume that there exists $\mathcal{I} \in \mathfrak{J}$, $\mathcal{I} \ne \mathcal{X}$, such that whenever $\{\mathcal{I}_1, \ldots, \mathcal{I}_{n-k}\} \subset \mathfrak{J}$ then $\mathcal{I}_1 \cap \ldots \cap \mathcal{I}_{n-k} \cap \mathcal{I}$ contains points arbitrarily far from $\partial \mathcal{I}$. Then $\operatorname{int}(\bigcap \mathfrak{J})$ contains a translate of $\mathcal{K}$.*

Proof. Let $\mathcal{K}' \subset \mathcal{X}$ be an affine subspace complementary to $\mathcal{K}$. Let $\mathfrak{J}_{\mathcal{K}'} = \{\mathcal{I} \cap \mathcal{K}' \mid \mathcal{I} \in \mathfrak{J}\}$ and apply Theorem 1.6.1 to $\mathfrak{J}_{\mathcal{K}'}$.

Exercises and Further Problems

1.* Show that a convex set $\mathcal{C} \subset \mathcal{X}$ is closed if and only if its intersection with any line is a closed (possibly infinite) line segment.
2. Show that $[\mathcal{A} + \mathcal{A}'] = [\mathcal{A}] + [\mathcal{A}']$, $\mathcal{A}, \mathcal{A}' \subset \mathcal{X}$.
3. Show that if $\mathcal{A} \subset \mathcal{X}$ is open then so is $[\mathcal{A}] \subset \mathcal{X}$.
4. Show that taking the convex hull $\mathcal{A} \mapsto [\mathcal{A}]$, $\mathcal{A} \in \mathfrak{C}$, is a continuous projection $\mathfrak{C} \to \mathfrak{K}$; in fact, a Lipschitz map (with Lipschitz constant one): $d_H([\mathcal{A}], [\mathcal{A}']) \le d_H(\mathcal{A}, \mathcal{A}')$, $\mathcal{A}, \mathcal{A}' \in \mathfrak{C}$.
5.* For $\mathcal{C}_1, \mathcal{C}_2, \mathcal{C}'_1, \mathcal{C}'_2 \in \mathfrak{K}$, derive the following inequalities:

$$d_H(\mathcal{C}_1 + \mathcal{C}_2, \mathcal{C}'_1 + \mathcal{C}'_2) \le d_H(\mathcal{C}_1, \mathcal{C}'_1) + d_H(\mathcal{C}_2, \mathcal{C}'_2)$$

$$d_H(\mathcal{C}_1 \cup \mathcal{C}_2, \mathcal{C}'_1 \cup \mathcal{C}'_2) \le \max\left(d_H(\mathcal{C}_1, \mathcal{C}'_1), d_H(\mathcal{C}_2, \mathcal{C}'_2)\right).$$

6. Derive positivity of the symmetric difference function: For $\mathcal{C}, \mathcal{C}' \in \mathfrak{B}$, we have $d_\Delta(\mathcal{C}, \mathcal{C}') = 0$ if and only if $\mathcal{C} = \mathcal{C}'$.

7.* Let $C \subset \mathcal{X}$ be convex and $X \in \mathcal{X}$. Show that there is *at most one* point $C \in \mathcal{C}$ such that $d(X, C) = d(X, \mathcal{C})$.

8. Another construction of a supporting hyperplane to a closed convex set $C \subset \mathcal{X}$ with no control over the point of support is as follows. Given $X \notin C$ there is a point $C \in \mathcal{C}$ such that $d(X, C) = d(X, \mathcal{C})$. (By Problem 7, C is unique. C is called the *metric projection* of X to $\mathcal{C}$.) Show that the hyperplane $\mathcal{H}$ through C with normal vector $N = X - C$ supports $\mathcal{C}$ at C.

9. Let $C \subset \mathcal{X}$ be a planar convex body. Show that the set $C^{\wedge} \subset \partial\mathcal{C}$ of extremal points is closed.

10.* Show that the set of vertices (Section 1.2) of a convex set is at most countable.

11. Show that a (scalene) ellipsoid

$$\left\{ (x_1, \ldots, x_n) \in \mathbb{R}^n \,\middle|\, \sum_{i=1}^{n} \frac{x_i^2}{a_i^2} \leq 1 \right\},$$

with $a_1 > a_2 \ldots > a_n > 0$, has n double normals and one metric diameter.

12. Show that $D_{\mathcal{C}+\mathcal{C}'} \leq D_{\mathcal{C}} + D_{\mathcal{C}'}$ and $d_{\mathcal{C}+\mathcal{C}'} \geq d_{\mathcal{C}} + d_{\mathcal{C}'}, \mathcal{C}, \mathcal{C}' \in \mathfrak{K}$.

13.* Let $C \subset \mathcal{X}$ be a convex set and $\mathfrak{F}$ a finite family of subsets of $\mathcal{X}$ covering $\mathcal{C}$ such that, for every $\mathcal{E} \in \mathfrak{F}$, the intersection $C \setminus \mathcal{E}$ is convex. Use Helly's theorem (Section 1.4) to prove that there is a subfamily in $\mathfrak{F}$ consisting of $\leq n + 1$ members that still covers $\mathcal{C}$.

14.* Let $\mathcal{A}$ and $\mathcal{B}$ be two finite subsets of $\mathcal{X}$. Show that $\mathcal{A}$ and $\mathcal{B}$ can be strictly separated if and only if, for every subset $\mathcal{C} \subset \mathcal{A} \cup \mathcal{B}$ of $\leq n + 2$ points, the intersections $C \cap \mathcal{A}$ and $C \cap \mathcal{B}$ can be strictly separated.

15.* Verify that the convex set

$$\mathcal{C}_n = \left\{ (x_1, \ldots, x_n) \in \mathbb{R}^n \,\middle|\, \sum_{i=1}^{n} |x_i| \leq 1 \right\}$$

is a convex polytope with the $2n$ vertices $(0, \ldots, 0, \pm 1, 0 \ldots, 0) \in \mathbb{R}^n$. $\mathcal{C}_n$ is called the *cross-polytope*. Show that the origin $0 \in \mathrm{int}\,\mathcal{C}$ is not in the *interior* of the convex hull of a set of less than $2n$ vertices.

  Prove the following theorem of [Steinitz]: Let $\mathcal{A} \subset \mathcal{X}$ and $O \in \mathrm{int}\,[\mathcal{A}]$. Then there exists a subset $\mathcal{A}_0 \subset \mathcal{A}$ consisting of *at most* $2n$ points such that $O \in \mathrm{int}\,[\mathcal{A}_0]$.

16. Show that if $\{\mathcal{A}_k\}_{k \geq 1}$ is a sequence of convex sets then $\bigcup_{k \geq 1} \left(\bigcap_{l \geq k} \mathcal{A}_l \right)$ is convex.

17.* Show that the diameter map $D : \mathfrak{K} \to \mathbb{R}$, associating to a compact subset $\mathcal{A} \subset \mathcal{X}$ its diameter $D_{\mathcal{A}}$, is Lipschitz with constant 2.

18. Let $\{\bar{B}_{r_k}(O_k)\}_{k \geq 1} \subset \mathfrak{C}_{\mathcal{X}}$ be a sequence of closed metric balls such that $\lim_{k \to \infty} O_k = O$ and $\lim_{k \to \infty} r_k = \infty$. Show that the Painlevé–Kuratowski limit of this sequence is the whole space $\mathcal{X}$.

19.* Let $C, C' \subset \mathcal{X}$ be closed subsets. Consider the alternating sequence $\{C, C', C, C', \ldots\}$. Show that the limit inferior is $C \cap C'$ while the limit superior is $C \cup C'$.

20.* In this exercise we derive several elementary properties of convex functions (of one variable). Let $f : I \to \mathbb{R}$ be a convex function defined on an interval $I \subset \mathbb{R}$.

(a) Show that $f : I \to \mathbb{R}$ is bounded and Lipschitz continuous on any closed subinterval of $\operatorname{int} I$. (Thus f is continuous on $\operatorname{int} I$; in fact, absolutely continuous on every closed subinterval of $\operatorname{int} I$. Consequently, f almost everywhere differentiable on I.)

(b) Let $x_1, x_2, x_3, x_4 \in \operatorname{int} I$ be consecutive points: $x_1 < x_2 < x_3 < x_4$. Show that

$$\frac{f(x_2) - f(x_1)}{x_2 - x_1} \leq \frac{f(x_3) - f(x_1)}{x_3 - x_1} \leq \frac{f(x_3) - f(x_2)}{x_3 - x_2}$$

$$\leq \frac{f(x_4) - f(x_2)}{x_4 - x_2} \leq \frac{f(x_4) - f(x_3)}{x_4 - x_3}.$$

(What is the geometric meaning of these inequalities?) Use these to conclude that the left- and right-derivatives

$$f'_-(a) = \lim_{x \to a^-} \frac{f(x) - f(a)}{x - a} \quad \text{and} \quad f'_+(a) = \lim_{x \to a^+} \frac{f(x) - f(a)}{x - a}$$

both exist for $a \in \operatorname{int} I$, and they are increasing functions on $\operatorname{int} I$.

(c) Show that

$$\lim_{x \to a^\pm} f'_\pm(x) = f'_\pm(a) \quad \text{and} \quad \lim_{x \to a^\mp} f'_\pm(x) = f'_\mp(a), \quad a \in \operatorname{int} I.$$

21. (a) Show that the volume functional (Lebesgue measure) is upper semi-continuous on $\mathcal{C}$ with respect to the Hausdorff distance: If $\lim_{k \to \infty} C_k = C$ in $\mathcal{C}$ with respect to d_H then $\lim \sup_{k \to \infty} \operatorname{vol} C_k \leq \operatorname{vol} C$.

Chapter 2
Affine Diameters and the Critical Set

2.1 The Distortion Ratio

Let $\mathcal{X}$ be a Euclidean space of dimension n. Recall that $\mathfrak{B} = \mathfrak{B}_{\mathcal{X}}$ denotes the space of all convex bodies in $\mathcal{X}$ equipped with the Hausdorff metric d_H. In Section 1.4, in an application of Helly's theorem, we briefly encountered the concept of *distortion ratio* (Corollary 1.4.2). In the present section we will make a more elaborate analysis of this concept. Throughout, $\mathcal{C} \in \mathfrak{B}$ will denote a convex body in $\mathcal{X}$. The distortion ratio can be defined with respect to interior and exterior points of $\mathcal{C}$. We split our treatment accordingly.

A. Interior Points. Let $O \in \operatorname{int} \mathcal{C}$. For $C \in \partial \mathcal{C}$, let $\Lambda(C, O)$ denote the ratio into which O divides the chord in $\mathcal{C}$ passing through C and O with other endpoint $C^o \in \partial \mathcal{C}$:

$$\Lambda(C, O) = \frac{d(C, O)}{d(C^o, O)}. \tag{2.1.1}$$

(See Figure 1.4.1 in Section 1.4.) We call C^o the *antipodal* of C (with respect to O). (For the existence and uniqueness of the antipodal, see Corollary 1.1.9. In this section we will repeatedly use this corollary in various settings without making explicit references.) This defines the *distortion ratio* $\Lambda = \Lambda_{\mathcal{C}} : \partial \mathcal{C} \times \operatorname{int} \mathcal{C} \to \mathbb{R}$. (The dependence on $\mathcal{C}$ will be indicated by subscript if necessary.)

Clearly, $(C^o)^o = C$, so that the antipodal map $C \mapsto C^o$, $C \in \partial \mathcal{C}$, is an involution of $\partial \mathcal{C}$, and we have $\Lambda(C^o, O) = 1/\Lambda(C, O)$, $C \in \partial \mathcal{C}$. More specifically, we have

$$C - O = -\Lambda(C, O)(C^o - O), \quad C \in \partial \mathcal{C}. \tag{2.1.2}$$

Proposition 2.1.1. *(a) For fixed $O \in \operatorname{int} \mathcal{C}$, the function $\Lambda(., O) : \partial \mathcal{C} \to \mathbb{R}$ is continuous; (b) the family of functions $\{\Lambda(C, .)\}_{C \in \partial \mathcal{C}}$ is equicontinuous; (c) $\Lambda : \partial \mathcal{C} \times \operatorname{int} \mathcal{C} \to \mathbb{R}$ is continuous.*

© Springer International Publishing Switzerland 2015
G. Toth, *Measures of Symmetry for Convex Sets and Stability*,
Universitext, DOI 10.1007/978-3-319-23733-6_2

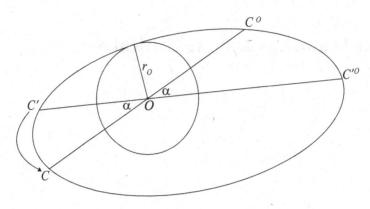

Fig. 2.1.1

Proof. Albeit fairly simple, we give detailed and elementary proofs.

(a) Let $C' \to C$ in $\partial \mathcal{C}$. We need to show that $\Lambda(C', O) \to \Lambda(C, O)$.
    The triangle $[C, O, C']$ gives

$$d(C', C)^2 = (d(C', O) - d(C, O))^2 + 4d(C', O)d(C, O) \sin^2(\alpha/2),$$

where $\alpha = \angle COC'$ (see Figure 2.1.1).
    As $C' \to C$, this implies that $d(C', O) \to d(C, O)$, and

$$0 \le r_O^2 \sin^2(\alpha/2) \le d(C', O)d(C, O) \sin^2(\alpha/2) \to 0,$$

where $r_O = d(O, \partial \mathcal{C}) > 0$ is the radius of the largest metric ball with center O which can be inscribed into $\mathcal{C}$. We thus have $\alpha \to 0$.
    Next, let C^o and C'^o be the antipodals of C and C' with respect to O, respectively. We have $\alpha = \angle C^o O C'^o$. Let $\pi_O : \mathcal{S}(O) \to \partial \mathcal{C}$ be the radial projection of the unit sphere $\mathcal{S}(O)$ centered at O to $\partial \mathcal{C}$. By Proposition 1.1.10, π_O is a homeomorphism.
    Since

$$d(\pi_O^{-1}(C'^o), \pi_O^{-1}(C^o)) = 2\sin(\alpha/2) \to 0,$$

we conclude that $C'^o \to C^o$ as $C' \to C$. Taking ratios, (a) follows.

(b) Let $O' \to O$ in $\operatorname{int} \mathcal{C}$, $C \in \partial \mathcal{C}$ a variable point, and C^o and C'^o the antipodals of C with respect to O and O', respectively. Let $\alpha = \angle OCO'$, $\beta = \angle OC'^o C$, and $\gamma = \angle C^o O C'^o$. The dependence of these angles on C, O, and O' will be indicated if necessary. (See Figure 2.1.2.)

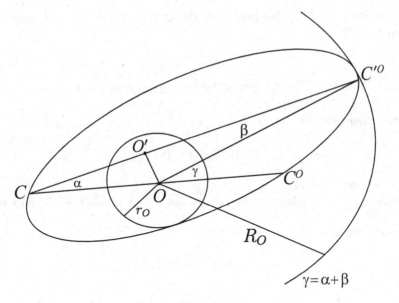

Fig. 2.1.2

Since γ is an exterior angle of the triangle $[O, C, C'^o]$, we have $\gamma = \alpha + \beta$. Let r_O be as in (a) and R_O the radius of the smallest metric ball with center at O that can be circumscribed about $\mathcal{C}$. We have $R_O \geq r_O > 0$.

The triangle $[O, O', C]$ gives

$$d(O', O)^2 = (d(C, O') - d(C, O))^2 + 4d(C, O')d(C, O)\sin^2(\alpha/2).$$

In particular, we obtain

$$d(O', O)^2 \geq 4d(C, O')d(C, O)\sin^2(\alpha/2).$$

Let $r \in (0, r_O/2)$ and assume that $d(O', O) < r$. Using the triangle inequality, we have $d(C, O') \geq r_O - r > r_O/2$, and the inequality above reduces to

$$d(O', O) \geq \sqrt{2}r_O \sin(\alpha/2).$$

We obtain that $\alpha = \alpha(C, O, O') \to 0$ as $O' \to O$ uniformly in $C \in \partial \mathcal{C}$.

The triangle $[C, O, C'^o]$ gives

$$\sin \beta = \frac{d(C, O)}{d(C'^o, O)} \sin \alpha \leq \frac{R_O}{r_O} \sin \alpha.$$

Hence $\beta = \beta(C, O, O') \to 0$ as $O' \to O$ uniformly in $C \in \partial \mathcal{C}$.

Since $\gamma = \alpha + \beta$, we have $\gamma = \gamma(C, O, O') \to 0$ as $O' \to O$ uniformly in $C \in \partial\mathcal{C}$.

As in (a)

$$d(\pi_O^{-1}(C'^o), \pi_O^{-1}(C^o)) = 2\sin(\gamma/2),$$

and we obtain that

$$\pi_O^{-1}(C'^o) \to \pi_O^{-1}(C^o)$$

as $O' \to O$ uniformly in $C \in \partial\mathcal{C}$.

Once again, π_O is a homeomorphism, so that $C'^o \to C^o$ as $O' \to O$ uniformly in $C \in \partial\mathcal{C}$.

Finally, we have

$$|d(C, O') - d(C, O)| \le d(O', O),$$

and

$$|d(C'^o, O') - d(C^o, O)| \le |d(C'^o, O') - d(C'^o, O)| + |d(C'^o, O) - d(C^o, O)|$$
$$\le d(O', O) + d(C'^o, C^o),$$

and both converge to zero uniformly in $C \in \partial\mathcal{C}$ as $O' \to O$.

Combining these, we obtain that $|\Lambda(C, O) - \Lambda(C, O')|$ converges to zero uniformly in $C \in \partial\mathcal{C}$ as $O' \to O$. (b) follows.

Finally, (a) and (b) imply (c). The proposition follows.

We define the *maximum distortion* $\mathfrak{m} = \mathfrak{m}_\mathcal{C} : \mathrm{int}\,\mathcal{C} \to \mathbb{R}$ as

$$\mathfrak{m}(O) = \max_{C \in \partial\mathcal{C}} \Lambda(C, O), \quad O \in \mathrm{int}\,\mathcal{C}. \tag{2.1.3}$$

(Once again the dependence on $\mathcal{C}$ will be indicated by subscript when necessary.)

Since $\Lambda(C^o, .) = 1/\Lambda(C, .)$, $C \in \partial\mathcal{C}$, we have $\mathfrak{m} \ge 1$. Moreover, if $\mathfrak{m}(O) = 1$ for some $O \in \mathrm{int}\,\mathcal{C}$ then $\Lambda(., O) = 1$ identically on $\partial\mathcal{C}$, and, consequently, $\mathcal{C}$ is *centrally symmetric* with respect to O.

As we will discuss below, $\mathfrak{m}$ does not have an upper bound on $\mathrm{int}\,\mathcal{C}$ (Lemma 2.1.6).

Theorem 2.1.2. *For $C \in \partial\mathcal{C}$, the function $1/(\Lambda(C, .) + 1)$ is concave on $\mathrm{int}\,\mathcal{C}$. The function $1/(\mathfrak{m} + 1)$ is continuous and concave on $\mathrm{int}\,\mathcal{C}$, and extends to $\mathcal{C}$ as a continuous concave function by setting it equal to zero on $\partial\mathcal{C}$.*

The proof of this theorem will be broken up into several steps below. We first point out the following consequence:

Corollary 2.1.3. *The function* $\mathfrak{m} : \operatorname{int} C \to \mathbb{R}$ *is convex.*

This corollary is a simple consequence of Theorem 2.1.2 along with the general fact that the composition of a *concave* function (such as $1/(\mathfrak{m}+1)$) with a *convex and decreasing* function (such as $f(x) = 1/x - 1, x > 0$) is convex.

Remark. In Chapter 4 we will make an extensive use the function $1/(\mathfrak{m}+1)$, and its concavity (asserted in Theorem 2.1.2) will play a principal role. In contrast, as the next example shows, the reciprocal $1/\mathfrak{m}$ is, in general, *not concave*.

Example 2.1.4. Consider the closed unit ball $\bar{B} \subset \mathbb{R}^n$. As simple computation shows

$$\mathfrak{m}_{\bar{B}}(O) = \frac{1+|O|}{1-|O|}, \quad O \in \mathcal{B}.$$

Clearly, $\mathfrak{m}_{\bar{B}}$ is convex but $1/\mathfrak{m}_{\bar{B}}$ is not concave (albeit quasi-convex, that is, the level-sets are convex; in fact, the level-sets are the concentric closed metric balls $\bar{B}_r$, $0 \le r \le 1$).

On the other hand, as in Theorem 2.1.2, the ratio $1/(\mathfrak{m}_{\bar{B}}+1)$ is concave, since

$$\frac{1}{\mathfrak{m}_{\bar{B}}(O)+1} = \frac{1-|O|}{2}, \quad O \in \mathcal{B}.$$

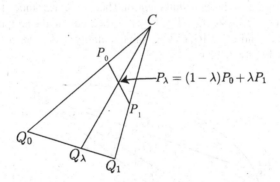

Fig. 2.1.3

We now begin the proof of Theorem 2.1.2 with the following:

Lemma 2.1.5. *Let two line segments* $[P_0, P_1]$ *and* $[Q_0, Q_1]$ *in* $\mathcal{X}$ *be related by a perspectivity centered at* C. *(See Figure 2.1.3.) Let* $P_\lambda = (1-\lambda)P_0 + \lambda P_1$, $0 \le \lambda \le 1$, *and denote by* Q_λ *the corresponding point in* $[Q_0, Q_1]$. *Setting* $\rho_\lambda = d(C, P_\lambda)/d(C, Q_\lambda), 0 \le \lambda \le 1$, *we have*

$$\rho_\lambda = (1-\lambda)\rho_0 + \lambda\rho_1. \tag{2.1.4}$$

Proof. By definition, we have

$$C - Q_\lambda = \frac{1}{\rho_\lambda}(P_\lambda - C).$$

Letting $Q_\lambda = (1 - \mu)Q_0 + \mu Q_1$ (with μ depending on λ), we compute

$$C - Q_\lambda = (1 - \mu)(C - Q_0) + \mu(C - Q_1)$$
$$= \frac{1 - \mu}{\rho_0}(P_0 - C) + \frac{\mu}{\rho_1}(P_1 - C)$$
$$= \frac{1 - \lambda}{\rho_\lambda}(P_0 - C) + \frac{\lambda}{\rho_\lambda}(P_1 - C).$$

Assuming (without loss of generality) that $P_0 - C$ and $P_1 - C$ are linearly independent, we obtain

$$\frac{1 - \mu}{\rho_0} = \frac{1 - \lambda}{\rho_\lambda}$$

$$\frac{\mu}{\rho_1} = \frac{\lambda}{\rho_\lambda}$$

Eliminating μ we arrive at (2.1.4).

Lemma 2.1.5 implies the first statement of Theorem 2.1.2. Indeed, let $O_0, O_1 \in$ int $\mathcal{C}$ and $O_\lambda = (1 - \lambda)O_0 + \lambda O_1, 0 \leq \lambda \leq 1$. Let C_λ^o denote the antipodal of C with respect to O_λ in $\mathcal{C}$, and $\bar{C}_\lambda^o \in [C, C_\lambda^o]$ the same quantity *with respect to the triangle* $[C, C_0^o, C_1^o]$. (See Figure 2.1.4.)

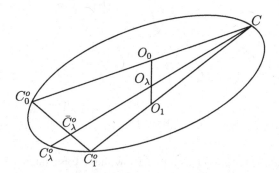

Fig. 2.1.4

Using the notation of Lemma 2.1.5 in our setting, we have

$$\rho_\lambda = \frac{d(C, O_\lambda)}{d(C, \bar{C}_\lambda^o)} \geq \frac{d(C, O_\lambda)}{d(C, C_\lambda^o)} = \frac{d(C, O_\lambda)}{d(C, O_\lambda) + d(C_\lambda^o, O_\lambda)} = 1 - \frac{1}{\Lambda(C, O_\lambda) + 1}$$

with equality for $\lambda = 0, 1$. By (2.1.4), we have

$$\frac{1}{\Lambda(C, O_\lambda) + 1} \geq \frac{1 - \lambda}{\Lambda(C, O_0) + 1} + \frac{\lambda}{\Lambda(C, O_1) + 1}.$$

Concavity of the function $1/(\Lambda(C, .) + 1)$ follows.

The family of concave functions $\{1/(\Lambda(C, .) + 1)\}_{C \in \partial C}$ is equicontinuous by part (b) of Proposition 2.1.1. Thus, the pointwise minimum $1/(m + 1)$ is continuous and concave. This is the second statement of Theorem 2.1.2.

The last statement follows from the following:

Lemma 2.1.6. *We have*

$$\lim_{d(O, \partial C) \to 0} m(O) = \infty. \tag{2.1.5}$$

Proof. Fix $O \in \text{int} C$ and $0 < \epsilon < r_0$. We will show that, for $d(O', \partial C) < \epsilon$, $O' \in \text{int} C$, we have

$$\frac{1}{\epsilon} \frac{r_O^2}{R_O} < m(O').$$

Indeed, let $B \in \partial C$ be on the extension of the ray emanating from O and passing through O', and choose $C' \in \partial C$ such that $d(C', O') < \epsilon$. We may assume that $C' \neq B$ since otherwise the proof is simpler. (See Figure 2.1.5.)

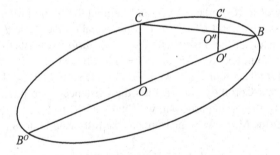

Fig. 2.1.5

Let $C \in \partial C$ be the unique point such that $C - O = \lambda(C' - O')$ for some $\lambda > 0$. Finally, let $O'' = [C', O'] \cap [B, C]$. The triangle $[O, B, C]$ gives

$$\frac{d(B, O')}{d(B, O)} = \frac{d(O'', O')}{d(C, O)} \leq \frac{d(C', O')}{d(C, O)}.$$

Estimating, we obtain

$$d(B, O') \le d(C', O')\frac{d(B, O)}{d(C, O)} \le d(C', O')\frac{R_O}{r_O} < \epsilon\frac{R_O}{r_O}.$$

Hence

$$\Lambda(B, O') = \frac{d(B, O')}{d(B^o, O')} \le \frac{d(B, O')}{d(B^o, O)} < \epsilon\frac{R_O}{r_O^2},$$

where the antipodal B^o of B is with respect to O (or O'). We conclude

$$m(O') = \max_{\partial C} \Lambda(., O') \ge \Lambda(B^o, O') = \frac{1}{\Lambda(B, O')} > \frac{1}{\epsilon}\frac{r_O^2}{R_O}.$$

The lemma follows.

A fundamental concept in convex geometry is the *Minkowski measure (of symmetry)* defined by

$$m^* = m_C^* = \inf_{O \in \text{int} C} m(O) = \inf_{O \in \text{int} C} \max_{C \in \partial C} \Lambda(C, O), \quad C \in \mathfrak{B}, \tag{2.1.6}$$

where $m = m_C : \text{int} C \to \mathbb{R}$ is the maximum distortion as in (2.1.3). When discussing a specific convex body, the subscript C will often be suppressed. On the other hand, in Chapter 3 we will consider m^* as function on $\mathfrak{B}$ and will write $m^*(C) = m_C^*, C \in \mathfrak{B}$.

By convexity of m (Corollary 2.1.3) and Lemma 2.1.6, the infimum is attained on a *compact convex* subset $C^* \subset \text{int} C$. This is called the *critical set* of C. The corresponding distortion ratio is called the *critical ratio*. We will study the critical set and critical ratio in Sections 2.3 and 2.4 and in Chapter 3.

As noted before Theorem 2.1.2, we have $m \ge 1$, so that $m^* \ge 1$. At the other extreme, recall that Corollary 1.4.2 asserts the existence of an interior point $O \in \text{int} C$ such that $m(O) \le n$. In particular, we obtain $m^* \le n$.

Summarizing, the Minkowski measure has the following bounds

$$1 \le m_C^* \le n, \quad C \in \mathfrak{B}. \tag{2.1.7}$$

For the lower bound, if $m^* = 1$ then, for any $O^* \in C^*$, we have $m(O^*) = m^* = 1$. Therefore $\Lambda(., O^*) = 1$ *identically* on ∂C. This means that C is *symmetric* with respect to O^* (and, consequently, O^* is unique, that is, the critical set C^* is a singleton). (See also the discussion before Theorem 2.1.2.)

For the upper bound in (2.1.7), we claim that $m^* = n$ if and only if C is a simplex. The "if" part is a simple computation, and it is the content of the next example. The "only if" part can also be shown directly, but will follow from the much stronger Theorem 3.2.4 ($\epsilon \to 0$), and also as a byproduct of the results in Section 4.1. (See Corollary 4.1.3.)

Remark. Some authors define the reciprocal $1/m_C^*$ as the *Minkowski measure*. In either case, it is usually denoted by as(C) or as$_\infty(C)$, where the notation indicates that it is also called the Minkowski measure of *asymmetry*.

Example 2.1.7. Let $\Delta = [C_0, \ldots, C_n] \in \mathfrak{B}$ be an *n*-simplex with vertices $C_0, \ldots, C_n$. We claim that $m_\Delta^* = n$. Given $O \in \text{int } \Delta$, we write $O = \sum_{i=0}^n \lambda_i C_i$, $\sum_{i=0}^n \lambda_i = 1$, $\{\lambda_0, \ldots, \lambda_n\} \subset [0, 1]$. Due to the simple geometry of the simplex, its distortion ratio $\Lambda(., O)$ attains its maximum at one of the vertices and $\Lambda(C_i, O) = (1 - \lambda_i)/\lambda_i$, $i = 0, \ldots, n$. Hence, we have

$$m_\Delta(O) = \max_{\partial \Delta} \Lambda(., O) = \max_{0 \le i \le n} \Lambda(C_i, O) = \max_{0 \le i \le n} \frac{1 - \lambda_i}{\lambda_i} = \frac{1}{\min_{0 \le i \le n} \lambda_i} - 1.$$

On the other hand, $\sum_{i=0}^n \lambda_i = 1$ implies that $\min_{0 \le i \le n} \lambda_i \le 1/(n + 1)$ and hence $m_\Delta(O) \ge n$. This lower bound is attained for $\lambda_0 = \ldots = \lambda_n = 1/(n + 1)$. Thus, we have $m_\Delta^* = \inf_{O \in \text{int } \Delta} m_\Delta(O) = n$.

We now return to the general setting, and let $C \in \mathfrak{B}$. By Theorem 2.1.2, for $0 < r \le 1$, the level-sets

$$C_r = \left\{ O \in C \,\middle|\, \frac{1}{m(O) + 1} \ge 1 - r \right\} \tag{2.1.8}$$

form a monotonic family of compact convex subsets of C with $C_1 = C$, and smallest non-empty level-set

$$C_{r^*} = C^* = \{O \in \text{int } C \mid m(O) = m^*\}, \quad \frac{1}{r^*} = 1 + \frac{1}{m^*}, \tag{2.1.9}$$

being the *critical set* (with *critical ratio* $r^* \ge 1/2$).

In Section 2.3 we will discuss Hammer's geometric construction of this decomposition $\{C_r\}_{r^* \le r \le 1}$ of C; see [Hammer 2]. Although it follows easily from convexity (and continuity) of $1/(m + 1)$, Hammer's approach will immediately imply that C_r is a convex *body* for $r^* < r \le 1$. In particular, by Proposition 1.1.10, the boundaries ∂C_r, $r^* < r \le 1$, form a topological foliation of $C \setminus C^*$ with topological spheres.

Finally, Section 2.4 will be devoted to Klee's analysis of the critical set C^* as in [Klee 2]; in particular, the proof of Klee's estimate on its dimension.

B. Exterior Points. We now define and study the distortion ratio on the exterior $\text{ext } C = \mathcal{X} \setminus C$, $C \in \mathfrak{B}$. As expected, the behavior of the distortion on the exterior is more technical due to fact that the boundary of C splits into *visible* and *invisible* parts whereas from an interior point the entire boundary is visible. (For visibility, see [Danzer–Grünbaum–Klee, p. 111] or [Berger, Vol. I, 11.7.7].)

First, let

$$(\partial C \times \text{ext } C)_0 = \{(C, O) \in \partial C \times \text{ext } C \mid \langle C, O \rangle \cap \text{int } C \ne \emptyset\},$$

where $\langle X_1, X_2 \rangle$ stands for the (affine) *line* passing through $X_1, X_2 \in \mathcal{X}$.

The complement of $(\partial C \times \text{ext}\,C)_0$ in $\partial C \times \text{ext}\,C$ is the union of pairs $(C, \mathcal{H} \setminus C) \in \partial C \times \mathcal{X}$, where $\mathcal{H} \subset \mathcal{X}$ is a hyperplane supporting C at C. (This follows from the Hahn–Banach theorem (Section 1.2) because any line disjoint from the interior of C but passing through a boundary point $C \in \partial C$ can be extended to a hyperplane $\mathcal{H}$ that supports C at C.)

By definition, for $(C, O) \in (\partial C \times \text{ext}\,C)_0$, the line $\langle C, O \rangle$ intersects the interior of C, and therefore meets ∂C in two (distinct) points, C and another point C^o. The latter is called the *antipodal* of C with respect to O. Clearly, $(C^o, O) \in (\partial C \times \text{ext}\,C)_0$ and $(C^o)^o = C$, so that the antipodal map $C \mapsto C^o$ is an *involution* of $(\partial C \times \text{ext}\,C)_0$.

Proposition 2.1.8. *$(\partial C \times \text{ext}\,C)_0$ is open and dense in $\partial C \times \text{ext}\,C$. It has two connected components which are interchanged by the antipodal map.*

Proof. Openness is clear from the definition.

Let $(C, O) \in \partial C \times \text{ext}\,C$ with $\langle C, O \rangle$ disjoint from $\text{int}\,C$, and choose disjoint open metric balls U and $V \subset \text{ext}\,C$ about C and O, respectively. We need to show that $\langle C', O' \rangle$ intersects $\text{int}\,C$ for some $C' \in \partial C \cap U$ and $O' \in V$. We may assume that $\mathcal{X}$ is two-dimensional. (Otherwise we intersect this configuration with an affine plane that contains C and O and an interior point of C.) If the lines $\langle C, O' \rangle$, $O' \in V$, are still disjoint from $\text{int}\,C$ then C is contained in an angular region with vertex at C (and angle $< \pi$). Clearly, every line $\langle C', O \rangle$ with $C' \in \partial C \cap U$, $C' \neq C$, intersects $\text{int}\,C$ (provided that U is small enough). Density follows.

The set $(\partial C \times \text{ext}\,C)_0$ naturally splits into two disjoint subsets $(\partial C \times \text{ext}\,C)_\pm$ according as $C^o \in [C, O]$ or $C \in [C^o, O]$, $C \in (\partial C \times \text{ext}\,C)_0$. The antipodal map clearly interchanges these two subsets. Path-connectedness of $(\partial C \times \text{ext}\,C)_+$, say, can be seen as follows.

First, $(C_0, O), (C_1, O) \in (\partial C \times \text{ext}\,C)_+$, are in the same path-connected component; in fact, the projection of the parametric line segment $\lambda \mapsto (1-\lambda)C_0 + \lambda C_1$ from O to ∂C gives rise to a continuous path from (C_0, O) to (C_1, O) (Proposition 1.1.10). This path stays within $(\partial C \times \text{ext}\,C)_+$ since $[C_0, C_1]$ is disjoint from $[C_0^o, C_1^o]$, the line segment connecting the respective antipodals.

It remains to show that $(C, O_0), (C, O_1) \in (\partial C \times \text{ext}\,C)_+$ are in the same path-connected component. Moving O_0 and O_1 away from C along $\langle C, O_0 \rangle$ and along $\langle C, O_1 \rangle$ (and staying in the same path-connected component), we may assume that $[O_0, O_1] \subset \text{ext}\,C$. The final path is then given by the parametrized line segment $\lambda \mapsto (1 - \lambda)O_0 + \lambda O_1$. The proposition follows. $\qquad \blacksquare$

With these preparations, the *distortion ratio* $\Lambda = \Lambda_C : (\partial C \times \text{ext}\,C)_0 \to \mathbb{R}$ can be defined in the obvious way:

$$\Lambda(C, O) = \frac{d(C, O)}{d(C^o, O)}, \quad (C, O) \in (\partial C \times \text{ext}\,C)_0.$$

As before, $\Lambda(C^o, O) = 1/\Lambda(C, O)$, $(C, O) \in (\partial C \times \text{ext}\,C)_0$, and (in place of (2.1.2)) we have:

$$C - O = \Lambda(C, O)(C^o - O).$$

Moreover, using the method in the proof of Proposition 2.1.1, it follows easily that $\Lambda : (\partial \mathcal{C} \times \text{ext}\,\mathcal{C})_0 \to \mathbb{R}$ is continuous. Clearly, the two connected components $(\partial \mathcal{C} \times \text{ext}\,\mathcal{C})_\pm$ correspond to the domains $\{\Lambda > 1\}$ and $\{\Lambda < 1\}$, respectively.

Remark. In general, the distortion $\Lambda : (\partial \mathcal{C} \times \text{ext}\,\mathcal{C})_0 \to \mathbb{R}$ cannot be extended to $\partial \mathcal{C} \times \text{ext}\,\mathcal{C}$ continuously. An example illustrating this is when the boundary point C is in the interior of a side of a triangle and O is in the extension of the side. (See Figure 2.1.6.)

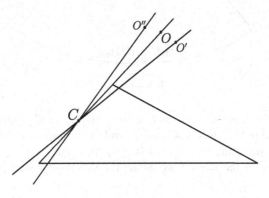

Fig. 2.1.6

For $C \in \partial \mathcal{C}$, the intersection of $(\partial \mathcal{C} \times \text{ext}\,\mathcal{C})_\pm$ by $\{C\} \times \text{ext}\,\mathcal{C}$ is denoted by $(\{C\} \times \text{ext}\,\mathcal{C})_\pm$. (See Figure 2.1.7.)

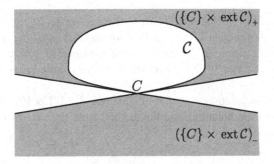

Fig. 2.1.7

The union of these two connected sets comprises the domain of $\Lambda(C, .)$. Clearly, $(\{C\} \times \text{ext}\,\mathcal{C})_-$ is convex.

Similarly, for $O \in \text{ext}\,\mathcal{C}$, the intersection of $(\partial \mathcal{C} \times \text{ext}\,\mathcal{C})_\pm$ by $\partial \mathcal{C} \times \{O\}$, is denoted by $(\partial \mathcal{C} \times \{O\})_\pm$. (See Figure 2.1.8.)

Once again, the union of these two connected sets is the domain of $\Lambda(., O)$. Both sets are *precompact* (that is, they have compact closures), and their boundaries can easily be described. Given a boundary point C of $(\partial \mathcal{C} \times \{O\})_+$, say, the intersection $[C, O] \cap \mathcal{C}$ is either a chord $[C, C']$ on the boundary of $\mathcal{C}$ with C' being a boundary

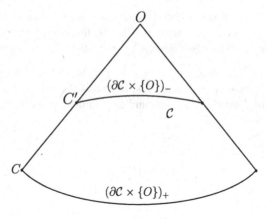

Fig. 2.1.8

point of $(\partial\mathcal{C} \times \{O\})_-$, or the singleton $C \in (\partial\mathcal{C} \times \{O\})_+ \cap (\partial\mathcal{C} \times \{O\})_-$ with $\langle C, O \rangle$ being a contact line to $\mathcal{C}$ at C. One may extend the distortion to these boundaries continuously (by declaring $\Lambda(C, O) = d(C, O)/d(C', O)$).

As in the case of the interior we define the *maximum distortion* $\mathfrak{m} = \mathfrak{m}_{\mathcal{C}}$: $\text{ext}\,\mathcal{C} \to \mathbb{R}$ by

$$\mathfrak{m}(O) = \sup_{(\partial\mathcal{C}\times\{O\})_+} \Lambda(., O) = \max_{(\partial\mathcal{C}\times\{O\})_+} \Lambda(., O), \quad O \in \text{ext}\,\mathcal{C}.$$

For the next theorem, recall that a function is *quasi-convex* if its level-sets are convex.

Theorem 2.1.9. *For $C \in \partial\mathcal{C}$, $1/|\Lambda(C, .) - 1|$ is convex on its respective domains $(\{C\} \times \text{ext}\,\mathcal{C})_\pm$. The function $1/(\mathfrak{m}-1)$ is continuous and quasi-convex on $\text{ext}\,\mathcal{C}$ and extends to $\partial\mathcal{C}$ as a continuous function by setting it equal to zero on $\partial\mathcal{C}$. (In quasi-convexity, we assume that $1/(\mathfrak{m} - 1)$ is extended to be equal to zero on $\mathcal{C}$.)*

Proof. The proof is analogous to that of Theorem 2.1.2. Let $[O_0, O_1] \subset (\{C\} \times \text{ext}\,\mathcal{C})_+$. We adopt the notations and the method from the case of interior points. (See Figure 2.1.9.)

We have

$$\rho_\lambda = \frac{d(C, O_\lambda)}{d(C, \bar{C}_\lambda^o)} \geq \frac{d(C, O_\lambda)}{d(C, C_\lambda^o)} = \frac{d(C, O_\lambda)}{d(C, O_\lambda) - d(C_\lambda^o, O_\lambda)} = 1 + \frac{1}{\Lambda(C, O_\lambda) - 1},$$

with equality for $\lambda = 0, 1$. Using Lemma 2.1.5, we obtain

$$\frac{1}{\Lambda(C, O_\lambda) - 1} \leq \frac{1 - \lambda}{\Lambda(C, O_0) - 1} + \frac{\lambda}{\Lambda(C, O_1) - 1}.$$

Convexity of $1/(\Lambda(C, .) - 1)$ on $(\{C\} \times \text{ext}\,\mathcal{C})_+$ follows.

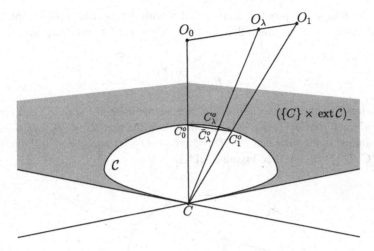

Fig. 2.1.9

The proof of convexity for $1/(1-\Lambda(C,.))$ on $(\{C\}\times \mathrm{ext}\,C)_-$ is entirely analogous. (See Figure 2.1.10.)

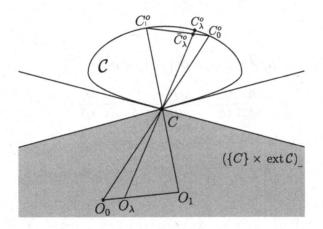

Fig. 2.1.10

Continuity and quasi-convexity of $1/(\mathfrak{m}-1)$ can be shown directly, but it will follow easily from Hammer's construction in Section 2.3.

To finish the proof of the theorem, we need the following.

Lemma 2.1.10. *On* $\mathrm{ext}\,C$, *we have*

$$\lim_{d(O,\partial C)\to 0} \mathfrak{m}(O) = \infty.$$

Proof. We mimic the proof of Lemma 2.1.6 with appropriate modifications. Fix $O \in \text{int}\,\mathcal{C}$ and $\epsilon > 0$. We claim that, for $d(O', \partial\mathcal{C}) < \epsilon$, $O' \in \text{ext}\,\mathcal{C}$, we have

$$\frac{1}{\epsilon}\frac{r_O^2}{R_O} < \mathfrak{m}(O').$$

Indeed, let B be the intersection of the line segment $[O, O']$ with $\partial\mathcal{C}$ and choose $C' \in \partial\mathcal{C}$ such that $d(C', O') < \epsilon$. We may assume that $C' \neq B$. Let $C \in \partial\mathcal{C}$ be the unique point such that $C - O = \lambda(C' - O')$ for some $\lambda < 0$. Finally, let $O'' = [C', O'] \cap \langle B, C\rangle$. (See Figure 2.1.11.)

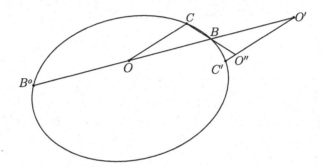

Fig. 2.1.11

The triangles $[O, B, C]$ and $[O', B, O'']$ are similar so that

$$\frac{d(B, O')}{d(B, O)} = \frac{d(O'', O')}{d(C, O)} \leq \frac{d(C', O')}{d(C, O)}.$$

Estimating, we obtain

$$d(B, O') \leq d(C', O')\frac{d(B, O)}{d(C, O)} \leq d(C', O')\frac{R_O}{r_O} < \epsilon\frac{R_O}{r_O}.$$

Hence

$$\Lambda(B, O') = \frac{d(B, O')}{d(B^o, O')} \leq \frac{d(B, O')}{d(B^o, O)} < \epsilon\frac{R_O}{r_O^2},$$

where the antipodal B^o of B is with respect to O (or O').

By construction, $B^o \in (\partial\mathcal{C} \times \{O'\})_+$ and we conclude

$$\mathfrak{m}(O') = \sup_{(\partial\mathcal{C}\times\{O'\})_+} \Lambda(., O') \geq \Lambda(B^o, O') = \frac{1}{\Lambda(B, O')} > \frac{1}{\epsilon}\frac{r_O^2}{R_O}.$$

The lemma follows.

2.2 Affine Diameters

Let $C \in \mathfrak{B}$. A *chord* of C is the intersection of a line with C. It is a (closed) line segment which we assume to be non-trivial. A chord is an *affine diameter* if C has parallel supporting hyperplanes at its endpoints. In this section we study the existence of affine diameters whose (line) extensions pass through a given point in $\mathcal{X}$.

Any boundary point is the endpoint of an affine diameter. Indeed, let $\mathcal{H}$ be any hyperplane supporting C at a given boundary point (Corollary 1.2.3). By compactness of C there exists another hyperplane $\mathcal{H}^o$ parallel to $\mathcal{H}$ and supporting C. Then any chord connecting the given point and a(ny) point in $C \cap \mathcal{H}^o$ is an affine diameter of C.

For affine diameters passing through a given interior point of C we will make a more detailed study. Recall the distortion ratio $\Lambda : \partial C \times \mathrm{int}\, C \to \mathbb{R}$ and the maximum distortion $\mathfrak{m} : \mathrm{int}\, C \to \mathbb{R}$, $\mathfrak{m}(O) = \max_{\partial C} \Lambda(., O)$, $O \in \mathrm{int}\, C$.

Proposition 2.2.1. *Let $O \in \mathrm{int}\, C$ and assume that $\Lambda(., O)$ attains a local maximum at $C_0 \in \partial C$. Setting $\mathfrak{m}_0 = \Lambda(C_0, O)$, there is an open neighborhood U of $C_0 \in \mathcal{X}$ such that we have*

$$(1 + 1/\mathfrak{m}_0)O - 1/\mathfrak{m}_0[\partial C \cap U] \subset C. \tag{2.2.1}$$

Proof. Choose a neighborhood U of C_0 in $\mathcal{X}$ such that the restriction $\Lambda(., O)|\partial C \cap U$ attains its maximum at C_0. Since $\mathfrak{m}_0 > 0$, for $C \in \partial C \cap U$, we have

$$(1 + 1/\mathfrak{m}_0)O - 1/\mathfrak{m}_0 C = O + \lambda(C^o - O), \tag{2.2.2}$$

for some $\lambda > 0$. For the inclusion in (2.2.1), we need to show that $\lambda \leq 1$. (See Figure 2.2.1.) Rearranging (2.2.2), we find $1/\mathfrak{m}_0(O - C) = \lambda(C^o - O)$ so that $\lambda = \Lambda(C, O)/\Lambda(C_0, O) \leq 1$. The proposition follows. $\quad\blacksquare$

Remark. The inclusion in (2.2.1) has a simple geometric interpretation. The similarity $S_{-1/\mathfrak{m}_0, O} : X \mapsto (1 + 1/\mathfrak{m}_0)O - 1/\mathfrak{m}_0 X$ (Section 1.1/C) maps $[\partial C \cap U]$ to the convex set on the left-hand side of the inclusion in (2.2.1). In addition, C_0 is mapped to its antipodal C_0^o. Since C_0^o is also a boundary point of C the two convex sets at the two sides of (2.2.1) meet at this common boundary point.

Corollary 2.2.2. *Given $O \in \mathrm{int}\, C$, for any $C_0 \in \partial C$ at which $\Lambda(., O)$ attains a local extremum, the chord $[C_0, C_0^o]$ is an affine diameter (passing through O). More precisely, if $\Lambda(., O)$ attains a local maximum at C_0 and $\mathcal{H}^o$ is any hyperplane supporting C at C_0^o then the hyperplane $\mathcal{H} = \mathcal{H}^o + C_0 - C_0^o$ supports C at C. In particular, through any interior point $O \in \mathrm{int}\, C$ there is an affine diameter which is split by O in the ratio $\mathfrak{m}(O)$.*

Proof. Since $\Lambda(C^o, O) = 1/\Lambda(C, O)$, $C \in \partial C$, the sets at which $\Lambda(., O)$ attains local maxima and minima are antipodals. We may therefore assume that at $C_0 \in \partial C$

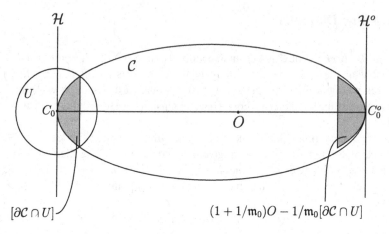

Fig. 2.2.1

we have a local maximum and Proposition 2.2.1 applies. Let $\mathcal{H}^o$ be any hyperplane that supports $\mathcal{C}$ at C_0^o. By the geometric description in the remark above it is clear that $\mathcal{H}^o$ also supports the convex set on the left-hand side of (2.2.1) at C_0^o. Applying the inverse of the similarity $S_{-1/\mathfrak{m}_0,O}$ to $\mathcal{H}^o$ we obtain another hyperplane $\mathcal{H} = (1 + \mathfrak{m}_0)O - \mathfrak{m}_0\mathcal{H}^o$ parallel to $\mathcal{H}^o$ and supporting $\mathcal{C}$ at C_0. Thus $[C_0, C_0^o]$ is an affine diameter and the corollary follows.

If (in the proof of Proposition 2.2.1) $\Lambda(., O)$ attains *global maximum* at C_0 then there is no restriction on $U \ (= \mathcal{X})$ and we obtain

$$\frac{1}{\mathfrak{m}(O)} = \frac{1}{\max_{\partial \mathcal{C}} \Lambda(., O)} = \max \{s \geq 0 \,|\, (1 + s)O - s\mathcal{C} \subset \mathcal{C}\}. \tag{2.2.3}$$

Clearly, the maximum distortion is attained at the common boundary point(s) of the convex bodies participating in the inclusion

$$C_O = (1 + 1/\mathfrak{m}(O))O - 1/\mathfrak{m}(O)\mathcal{C} \subset \mathcal{C}.$$

Remark. Corollary 2.2.2 is a local version of [Klee 2, (3.2)].

We now introduce the notion of *k-flat* point on $\partial \mathcal{C}$, $0 \leq k \leq n$. Let $C \in \partial \mathcal{C}$. We call an affine subspace $\mathcal{A} \subset \mathcal{X}$ a *supporting flat* at C if $C \in \mathcal{A}$ and $\mathcal{A}$ is contained in a hyperplane supporting $\mathcal{C}$ at C. (By the Hahn–Banach theorem (Section 1.2), an affine subspace $\mathcal{A}$ containing C is a supporting flat at C if and only if $\mathcal{A}$ is disjoint from the interior of $\mathcal{C}$.) Consider the set of supporting flats $\mathcal{A}$ at C which have the property that $\partial \mathcal{C} \cap \mathcal{A}$ is a convex body *in* $\mathcal{A}$ and C is contained in its (non-empty) relative interior. Since $\mathcal{C}$ is convex, this set has a unique maximal element denoted by $\mathcal{A}_C$. Thus, $\mathcal{C} \cap \mathcal{A}_C$ is a convex body in $\mathcal{A}_C$ with C it its relative interior, and $\mathcal{A}_C$

is maximal with respect to this property. We call C a k-*(dimensional) flat point* if $\dim \mathcal{A}_C = k$.

Clearly, C is an extremal point if and only if $k = 0$ (Section 1.2). At the other extreme, an $(n - 1)$-flat point will be called a *flat point*. (Compare this with the concept of order of a boundary point discussed after Corollary 1.2.7.)

Corollary 2.2.3. *Let C be a convex body and $C^\wedge \subset \partial C$ the set of extremal points. We have*

$$\mathfrak{m}(O) = \max_{C^\wedge} \Lambda(., O), \quad O \in \operatorname{int} C.$$

Proof. Assume that $\Lambda(., O)$ attains its maximum at a non-extremal point $C \in \partial C$. Then C is a k-flat point for some $k > 0$, $k = \dim \mathcal{A}_C$. Since $\Lambda(., O)$ attains its maximum at C, by Proposition 2.2.1, the antipodal point C^o is l-flat, $l \geq k$, and $\mathcal{A}_C$ is parallel to $\mathcal{A}_{C^o}$.

Choose a point C' on the (relative) *boundary* of the convex body $\partial C \cap \mathcal{A}_C$ in $\mathcal{A}_C$. Clearly, C' is a lower dimensional flat point than C. Since $\mathcal{A}_C$ is parallel to $\mathcal{A}_{C^o}$, $\Lambda(., O)$ is constant on $[C, C']$. We can replace C by C' without changing of the value of $\Lambda(., O)$. The corollary now follows by induction (with respect to the dimension of the flat point).

Example 2.2.4. Let $M(n, \mathbb{R})$, $n \geq 2$, be the vector space of $n \times n$-matrices with real entries. We denote by $A = (a_{ij})_{i,j=1}^n \in M(n, \mathbb{R})$ a typical element, where $a_{ij} \in \mathbb{R}$ is the entry of A in the ith row and jth column. As a vector space, $M(n, \mathbb{R})$ is isomorphic with $\mathbb{R}^{n^2}$ via the "read off map" which associates to $A = (a_{ij})_{i,j=1}^n$ the vector formed by juxtaposing the rows of A consecutively as

$$(a_{11}, \ldots, a_{1n}, a_{21}, \ldots, a_{2n}, \ldots, a_{n1}, \ldots, a_{nn}) \in \mathbb{R}^{n^2}.$$

We identify $M(n, \mathbb{R})$ with $\mathbb{R}^{n^2}$ via this isomorphism.

A matrix $A = (a_{ij})_{i,j=1}^n \in M(n, \mathbb{R})$ with *non-negative entries* is called *doubly stochastic* if the sum of the entries in each row and in each column is equal to 1, that is, for $i, j = 1, \ldots, n$, we have $\sum_{j=1}^n a_{ij} = 1$ and $\sum_{i=1}^n a_{ij} = 1$. In particular, $0 \leq a_{ij} \leq 1$ for all $i, j = 1, \ldots, n$; and hence the double stochastic matrices are contained in the unit square $[0, 1]^{n^2} \subset \mathbb{R}^{n^2}$.

Let $\mathcal{X}_n \subset M(n, \mathbb{R}) = \mathbb{R}^{n^2}$ be the *affine subspace* defined by the two sets of equations $\sum_{j=1}^n a_{ij} = 1$, $i = 1, \ldots, n$, and $\sum_{i=1}^n a_{ij} = 1$, $j = 1, \ldots, n$. Note that these constraints are dependent as each of the two sets implies $\sum_{i=1}^n \sum_{j=1}^n a_{ij} = n$; therefore one (but only one) constraint is redundant. We obtain that $\dim \mathcal{X}_n = n^2 - (2n - 1) = (n - 1)^2$.

The set $\mathcal{B}_n \subset M(n, \mathbb{R})$ of doubly stochastic matrices is the intersection of $\mathcal{X}_n$ and the convex cone consisting of all matrices with non-negative entries. Therefore $\mathcal{B}_n$ is convex. Moreover, as noted above, $\mathcal{B}_n \subset [0, 1]^{n^2}$, and therefore $\mathcal{B}_n$ is compact. Finally, $\mathcal{B}_n$ has non-empty interior in $\mathcal{X}_n$; for example, it contains the open ball of radius $1/n$ and center $E_n \in \mathcal{B}_n$, the doubly stochastic matrix all of whose entries are equal to $1/n$.

Let $\mathcal{S}_n$ denote the *symmetric group* of all permutations on $\{1, 2, \ldots, n\}$. For any permutation $\pi \in \mathcal{S}_n$, the associated *permutation matrix* $P_\pi \in M(n, \mathbb{R})$ is the matrix with all entries 0, except, for $i = 1, \ldots, n$, in the ith row the $\pi(i)$th entry is 1. Clearly, P_π is doubly stochastic; in fact, $P_\pi \in \partial \mathcal{B}_n$, $\pi \in \mathcal{S}_n$.

The symmetric group $\mathcal{S}_n$ acts naturally on $M(n, \mathbb{R})$ by linear isometries; a permutation $\pi \in \mathcal{S}_n$ sends a matrix $A \in M(n, \mathbb{R})$ to the product $P_\pi \cdot A \in M(n, \mathbb{R})$ whose rows are those of A permuted by π. The affine subspace $\mathcal{X}_n$ and its convex body $\mathcal{B}_n$ are invariant under this action. Finally, note that the action of $\mathcal{S}_n$ restricted to $\mathcal{X}_n$ has a unique fixed point; the matrix E_n.

The celebrated *Birkhoff–von Neumann theorem* asserts that $\mathcal{B}_n \subset \mathcal{X}_n$ is a *convex polytope* whose vertices are the $(n!)$ permutation matrices. (This theorem has many different proofs. Example 3 details one of the simplest proofs due to [Hurlbert] that uses the Minkowski–Krein–Milman theorem (Section 1.2). For the original accounts, see [Birkhoff, von Neumann], and also the earlier approach by [König 2, Chapter XIV, Section 3] via regular bipartite graphs.) $\mathcal{B}_n$ is called the *Birkhoff polytope*.

As an application of Corollary 2.2.3, in this example we claim that the maximum distortion

$$\mathfrak{m}_{\mathcal{B}_n}(E_n) = n - 1.$$

Since the extremal points of $\mathcal{B}_n$ are precisely the vertices, we need to calculate the distortion $\Lambda(P_\pi, E_n)$ for a(ny) permutation matrix P_π, $\pi \in \mathcal{S}_n$. (As $\mathcal{S}_n$ acts transitively (and isometrically) on the set of vertices of $\mathcal{B}_n$, all these distortions are equal.)

Let P_π^o denote the antipodal of P_π with respect to E_n. It is given by the convex combination

$$-t(P_\pi - E_n) + E_n = -tP_\pi + (1 + t)E_n$$

with the largest $t > 0$ such that this matrix has non-negative entries. In terms of the entries, this condition is $-t + (1 + t)/n \geq 0$. Thus, $t \leq 1/(n - 1)$, and equality gives

$$\Lambda(P_\pi, E_n) = \frac{|P_\pi - E_n|}{|-tP_\pi + tE_n|} = \frac{1}{t} = n - 1.$$

The claim follows.

Returning to the main line, we now consider the case of *exterior* points.

Proposition 2.2.5. *Let $O \in \text{ext}\,\mathcal{C}$ and assume that $\Lambda(., O)$ attains a local maximum at $C_0 \in \partial\mathcal{C}$ with $(C_0, O) \in (\partial\mathcal{C} \times \text{ext}\,\mathcal{C})_+$. Setting $\mathfrak{m}_0 = \Lambda(C_0, O)$, $\mathfrak{m}_0 > 1$, there is an open neighborhood U of $C_0 \in \mathcal{X}$ such that we have*

$$(1 - 1/\mathfrak{m}_0)O + 1/\mathfrak{m}_0[\partial\mathcal{C} \cap U] \subset \mathcal{X} \setminus \text{int}\,\mathcal{C}. \tag{2.2.4}$$

Proof. We choose the neighborhood U of C_0 in $\mathcal{X}$ such that $(\partial C \cap U) \times \{O\} \subset (\partial C \times \text{ext} C)_+$ and $\Lambda(., O)|\partial C \cap U$ attains its maximum at C_0. For $C \in \partial C \cap U$, we have

$$(1 - 1/m_0)O + 1/m_0 C = O + \lambda(C^o - O), \tag{2.2.5}$$

for some $\lambda > 0$. We need to show that $\lambda \leq 1$. (See Figure 2.2.2.) As before, rearranging (2.2.5), we find $1/m_0(C - O) = \lambda(C^o - O)$ so that $\lambda = \Lambda(C, O)/\Lambda(C_0, O) \leq 1$. The proposition follows.

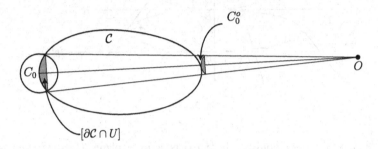

Fig. 2.2.2

Remark. Figure 2.2.2 contains the geometric interpretation of the proposition. The similarity $S_{1/m_0, O} : X \mapsto (1 - 1/m_0)O + 1/m_0 X$ maps $[\partial C \cap U]$ to the convex set on the left-hand side of the inclusion in (2.2.4). In addition, C_0 is mapped to its antipodal C_0^o. Since C_0^o is also a boundary point of C the two convex sets at the two sides of (2.2.4) meet at this common boundary point.

Corollary 2.2.6. *Given $O \in \text{ext} C$, for any $C_0 \in \partial C$ with $(C_0, O) \in (\partial C \times \text{ext} C)_0$, at which $\Lambda(., 0)$ attains a local extremum, the chord $[C, C^o]$ is an affine diameter (passing through O).*

Remark. Unlike the case of interior points, in general, global maxima of Λ may not be attained in $(\partial C \times \text{ext} C)_+$. A simple example is furnished by a triangle and O on one of the extensions of a side.

As before, for $O \in \text{ext} C$, we have

$$\frac{1}{m(O)} = \frac{1}{\sup_{(\partial C \times \{O\})_+} \Lambda(., O)} = \sup\{s \geq 0 \,|\, (1 - s)O + sC \subset \mathcal{X} \setminus C\}. \tag{2.2.6}$$

To show this, we first note that, clearly, $m(O) > 1$. The convex body

$$C_O = (1 - 1/m(O))O + 1/m(O)C \tag{2.2.7}$$

is the image of C under the similarity $S_{1/m(O),O}$ with center at O. By (2.2.4), $C_O \subset \mathcal{X} \setminus \mathrm{int}\, C$ so that the interiors of C and C_O are disjoint, and C and C_O meet at a common boundary point $B \in \partial C \cap \partial C_O$. (See Figure 2.2.3.)

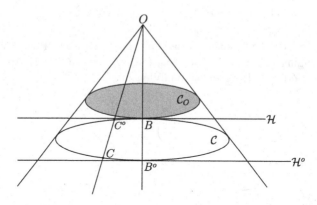

Fig. 2.2.3

Since they are both convex, there exists a hyperplane $\mathcal{H}$ (separating and) supporting C and C_O at B (Corollary 1.2.2). Applying the inverse similarity $S_{m(O),O}$, we obtain that the hyperplane $\mathcal{H}^o = (1 - m(O))O + m(O)\mathcal{H}$ is parallel to $\mathcal{H}$ and supports C at $B^o = (1-m(O))O+m(O)B \in \partial C$. (Since B may not be in $(\partial C \times \{O\})_{\pm}$, B^o may only be the antipodal of B in the limiting sense.) Thus, $[B, B^o]$ is an affine diameter of C whose line extension passes through O.

We claim that $B \in \overline{(\partial C \times \{O\})_-}$. Indeed, we have $d(B,O)/d(B^o,O) = 1/m(O) < 1$ implying that B cannot be in $(\partial C \times \{O\})_+$. In addition, being an endpoint of an affine diameter, B cannot be in the interior of any line segment on ∂C. The claim now follows from the description of the splitting of the boundary of C into visible and invisible parts (from O). (See Figure 2.1.8.)

We thus have $B^o \in \overline{(\partial C \times \{O\})_+}$ with $d(B^o,O)/d(B,O) = m(O)$. We now let $C \in (\partial C \times \{O\})_+$. Since the chord $[C, C^o]$ is between the parallel supporting hyperplanes $\mathcal{H}$ and $\mathcal{H}^o$ and the line extension of this chord passes through O, we have $\Lambda(C, O) = d(C,O)/d(C^o,O) \le d(B^o,O)/d(B,O) = m(O)$. The equality in (2.2.6) follows.

As a byproduct we also obtained the following:

Corollary 2.2.7. *Given a convex body $C \subset \mathcal{X}$, the extensions of affine diameters cover the entire space $\mathcal{X}$.*

Remark. The method above for treating the exterior points was essentially indicated by an undisclosed reviewer of [Hammer 2].

The following result gives a characterization of affine diameters in terms of their lengths.

Theorem 2.2.8. *Consider the family of all chords of C parallel to a given line. In this family the affine diameters of C are precisely the longest ones.*

Proof. Since C is between its two supporting parallel hyperplanes at the endpoints of an affine diameter, it is clear that the affine diameters are the longest in any given parallel family.

Conversely, assume now that $[C, C'] \subset C$ is the longest in a family of parallel chords. We consider the parallel translate $C' = C + C - C'$. Clearly, $C = C' + C - C' \in C \cap C'$ as $C' \in C$. We now claim that C and C' have disjoint interiors. Assume not. Let $B \in \text{int}\, C \cap \text{int}\, C'$. Since $B \in \text{int}\, C'$, we have $B = A + C - C'$, for some $A \in \text{int}\, C$. Rewriting this as $A - B = C' - C$ gives two things: $d(A, B) = d(C, C')$ and the chord $[C, C']$ and the chord passing through A an B are parallel. But B is an interior point of C so that the latter is certainly longer than $[C, C']$. This is a contradiction proving the claim.

Since C and C' have disjoint interiors but meet at $C \in \partial C \cap \partial C'$ there exists a hyperplane $\mathcal{H}$ (separating and) supporting both convex bodies at C. Now, letting $\mathcal{H}' = \mathcal{H} + C' - C$ we see that $[C, C']$ is an affine diameter of C supported at its endpoints by the parallel hyperplanes $\mathcal{H}$ and $\mathcal{H}'$.

Corollary 2.2.9. *C has an affine diameter parallel to any given line.*

Remark. The literature on affine diameters is extensive. For thorough surveys, see [Soltan] and [Martini–Swanepoel–Weiss].

The structure and mutual incidence of affine diameters within the interior of a convex body C is subtle. To begin with, improving an observation of [Dol'nikov], [Soltan, 6.1] showed that through the midpoint of an affine diameter yet another affine diameter must pass through. (See Problem 2.)

About the structure of the affine diameters, in 1963 [Grünbaum 2] posed the following:

Grünbaum Conjecture. Every convex body possesses a point through which $n + 1$ affine diameters pass through.

We will return to this question at the end of Sections 2.4 and 4.2. Related to this, [Soltan–Nguyên] proved that a convex body C either has continuum many points each belonging to (at least) three affine diameters or there is a point belonging to continuum many affine diameters. (See also [Nguyên]. For previous work on this type of result, see [Dol'nikov] and [Harazišvili].)

2.3 Hammer's Construction of the Critical Set

Let $C \in \mathfrak{B}$. Recall from Section 2.1 that the level-sets of the concave function $1/(\mathfrak{m}+1)$ (Theorem 2.1.2) define a monotonic family $\{C_r\}_{r^* \le r \le 1}$, $1/r^* = 1 + 1/\mathfrak{m}^*$, of compact convex sets with $C_1 = C$ and $C_{r^*} = C^*$, the critical set of C.

Following Hammer, using the central similarities $S_{r,C} \in \text{Dil}(\mathcal{X}) \subset \text{Aff}(\mathcal{X})$ with center C and ratio $r(\ne 0)$ introduced in Section 1.1/C, in this section we discuss a geometric description of this decomposition. As a byproduct, this will give a new insight into Yaglom–Boltyanskiǐ's approach to the Minkowski–Radon inequality (Corollary 1.4.2).

A. Interior Points. For $0 < r \leq 1$, we define

$$\mathcal{C}_r = \bigcap_{C \in \partial \mathcal{C}} S_{r,C}(\mathcal{C}) = \bigcap_{C \in \partial \mathcal{C}} ((1-r)C + r\mathcal{C}) = \bigcap_{C \in \partial \mathcal{C}} (C + r(\mathcal{C} - C)). \qquad (2.3.1)$$

With respect to the inclusion, $\{\mathcal{C}_r\}_{0 < r \leq 1}$ is an increasing family of compact convex sets with $\mathcal{C}_1 = \mathcal{C}$. Since $S_{0,C}(\mathcal{C}) = \{C\}$, $C \in \partial \mathcal{C}$, we have $\mathcal{C}_r = \emptyset$ for small $r > 0$. We define

$$r^* = \inf\{r \in (0,1] \mid \mathcal{C}_r \neq \emptyset\} \quad \text{and} \quad \mathcal{C}^* = \mathcal{C}_{r^*}. \qquad (2.3.2)$$

We call r^* the *critical ratio* and $\mathcal{C}^* = \mathcal{C}_{r^*}$ the *critical set* of $\mathcal{C}$.

To justify the overlapping terminology, we claim that (2.3.1)–(2.3.2) are equivalent to (2.1.8)–(2.1.9).

To show this, we first note that, for $0 < r \leq 1$, in (2.3.1), we have

$$\mathcal{C}_r = \{O \in \mathcal{X} \mid (1/r)O + (1 - 1/r)\mathcal{C} \subset \mathcal{C}\}. \qquad (2.3.3)$$

Indeed, $O \in \mathcal{C}_r$ in (2.3.1) if and only if $S_{r,C}^{-1}(O) \in \mathcal{C}$, $C \in \partial \mathcal{C}$, or equivalently, $(1/r)O + (1 - 1/r)\partial \mathcal{C} \subset \mathcal{C}$. Since $\mathcal{C}$ is convex $\partial \mathcal{C}$ can be replaced by $\mathcal{C}$.

Now, the parameter r in (2.3.3) is just the rescaling of the parameter s in (2.2.3). We immediately see that, for $O \in \operatorname{int}\mathcal{C}$, we have

$$O \in \mathcal{C}_r \quad \Longleftrightarrow \quad r \geq 1 - \frac{1}{\mathfrak{m}(O) + 1}. \qquad (2.3.4)$$

This is the definition of $\mathcal{C}_r$ in (2.1.8). The claim follows.

Equality holds in (2.3.4) if and only if $O \in \partial \mathcal{C}_r$. As already established in Section 2.1, the following simple geometric picture emerges: The family $\{\partial \mathcal{C}_r\}_{r^* \leq r \leq 1}$ on $\mathcal{C}_1 = \mathcal{C}$ comprises the level-sets of the *convex* function $1 - 1/(\mathfrak{m} + 1)$ (with its continuous convex extension to $\mathcal{C}$; Theorem 2.1.2). The lowest level-set (at r^*) is the (compact convex) critical set $\mathcal{C}^*$, and we have

$$r^* = 1 - \sup_{\operatorname{int}\mathcal{C}} \frac{1}{\mathfrak{m} + 1} = 1 - \frac{1}{\mathfrak{m}^* + 1} \qquad (2.3.5)$$

with the supremum attained on $\mathcal{C}^*$. For $r^* < r \leq 1$, $O \in \operatorname{int}\mathcal{C}_r$ if and only if strict inequality holds in (2.3.4); in particular, $\operatorname{int}\mathcal{C} \neq \emptyset$. We obtain that, for $r^* < r \leq 1$, $\mathcal{C}_r \in \mathfrak{B}$, and $\{\partial \mathcal{C}_r\}_{r^* < r \leq 1}$ gives rise to a topological foliation on $\mathcal{C} \setminus \mathcal{C}^*$ with the leaves being topological spheres (Proposition 1.1.10).

Finally, note that $r^* \geq 1/2$ with equality if and only if C is (centrally) symmetric with respect to the singleton C^*. Indeed, this is just a rewording of what has already been established in Section 2.1, since $m^* \geq 1$ if and only if $r^* \geq 1/2$.

At the other extreme, the proof of Corollary 1.4.2 along with the following Remark 2 shows that $C_{n/(n+1)} = \bigcap_{C \in \partial C} S_{n/(n+1),C}(C)$ is non-empty. This means that $r^* \leq n/(n+1)$, which gives again the Minkowski–Radon upper bound $m^* \leq n$. To turn the question around, in the Yaglom–Boltyanskiĭ approach to Helly's theorem (Corollary 1.4.2), the points $O \in \text{int } C$ for which the estimate $1/n \leq \Lambda(.,O) \leq n$ (on ∂C) holds fill the compact convex set $C_{n/(n+1)}$ of the Hammer decomposition.

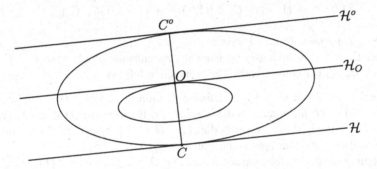

Fig. 2.3.1

A connection with affine diameters is as follows. Let $r^* < r < 1$ and $O \in \partial C_r$. Let $m(O) = \max_{\partial C} \Lambda(.,O)$ be attained at $C \in \partial C$. By Corollary 2.2.2, $[C, C^o]$ is an affine diameter (passing through O). Let $\mathcal{H}$ and $\mathcal{H}^o$ be parallel hyperplanes supporting C at C and C^o, respectively. We now claim that the hyperplane $\mathcal{H}_O$ parallel to $\mathcal{H}$ and passing through O supports C_r. (See Figure 2.3.1.)

To prove this claim, first note that $\mathcal{H}_O = S_{r,C}(\mathcal{H}^o) = C + r(\mathcal{H}^o - C)$ since the right-hand side is parallel to $\mathcal{H}^o$ and passes through O; in fact, we have $O = C + r(C^o - C)$ since $r = 1 - 1/(m + 1) = 1 - 1/(\Lambda(C,O) + 1)$. Now, $\mathcal{H}^o$ is supporting C so that $S_{r,C}(\mathcal{H}^o) = \mathcal{H}_O$ is supporting $S_{r,C}(C)$. But the latter contains C_r, and the claim follows.

B. Exterior Points. For $r \geq 1$, we define

$$C_r = \bigcup_{C \in \partial C} S_{r,C}(C) = \bigcup_{C \in \partial C} ((1-r)C + rC) = \bigcup_{C \in \partial C} (C + r(C - C)).$$

Clearly, $C \subset C_r, r \geq 1$, and $C_1 = C$.

Proposition 2.3.1. $\{C_r\}_{r \geq 1}$ *is a monotonic family of convex bodies.*

Proof. Monotonicity is clear. We need to show that C_r, $r \geq 1$, is a convex body.
First, as a consequence of the definition, we have

$$C_r = \{O \in \mathcal{X} \mid ((1/r)O + (1 - 1/r)\partial C) \cap C \neq \emptyset\}. \tag{2.3.6}$$

Indeed, for $O \in \mathcal{X}$ and $r \geq 1$, $O \in C_r$ if and only if $S_{r,C}^{-1}(O) = (1/r)O + (1-1/r)C \in C$ for some $C \in \partial C$.

Now let $O_0, O_1 \in C_r$. Then, for some $C_0, C_1 \in \partial C$, we have

$$(1/r)O_0 + (1 - 1/r)C_0 \in C \quad \text{and} \quad (1/r)O_1 + (1 - 1/r)C_1 \in C.$$

Letting $O_\lambda = (1 - \lambda)O_0 + \lambda O_1$ and $\bar{C}_\lambda = (1 - \lambda)C_0 + \lambda C_1$, $\lambda \in [0, 1]$, using these two relations we obtain $(1/r)O_\lambda + (1 - 1/r)\bar{C}_\lambda \in C$. A slight difficulty is caused by the fact that $\bar{C}_\lambda \in C$ is not necessarily on the boundary of C. To overcome this, let $C_\lambda \in \partial C$ be on the extension of the line segment $[(1/r)O_\lambda + (1 - 1/r)\bar{C}_\lambda, \bar{C}_\lambda]$ (possibly) beyond $\bar{C}_\lambda$. Then we have

$$(1/r)O_\lambda + (1 - 1/r)C_\lambda \in [(1/r)O_\lambda + (1 - 1/r)\bar{C}_\lambda, C_\lambda] \subset C,$$

and $O_\lambda \in C_r$ follows. Thus C_r is convex.

Since C_r, $r \geq 1$, is obviously compact and contains the convex body $C = C_1$, we conclude that C_r is a convex body. The proposition follows.   □

Given $O \in \mathcal{X}$ and $r \geq 1$, the defining relation in (2.3.6) clearly implies that $(1/r)O + (1 - 1/r)\partial C$ intersects the interior of C if and only if $O \in \text{int} \, C_r$. Hence $O \in \partial C_r$ if and only if the convex bodies $(1/r)O + (1 - 1/r)C$ and C meet only at their boundaries. We thus land in the situation given in (2.2.6).

Matching the parameters, we obtain that, for $O \in \text{ext} \, C$, we have $O \in \partial C_r$ if and only if

$$r = \frac{1}{1 - 1/\mathrm{m}(O)} = 1 + \frac{1}{\mathrm{m}(O) - 1},$$

where (as usual) $\mathrm{m}(O) = \sup_{(\partial C \times \{O\})_+} \Lambda(., O)$. By monotonicity, for $O \in \text{ext} \, C$, we obtain

$$O \in C_r \quad \Longleftrightarrow \quad r \geq 1 + \frac{1}{\mathrm{m}(O) - 1},$$

and equality holds if and only if $O \in \partial C_r$. Since C_r, $r \geq 1$, are convex, this means (by definition) that $1/(\mathrm{m} - 1)$ is quasi-convex, provided that it is extended to be equal to zero on C.

Going back to the boundary, we see that $\{C_r\}_{r \geq 1}$ are the level-sets of the function $1 + 1/(\mathrm{m} - 1)$. As a byproduct, we also see that m is continuous on $\text{ext} \, C$.

Indeed, since $\mathrm{m} > 1$ on $\text{ext} \, C$, it is enough to show continuity of $1 + 1/(\mathrm{m} - 1)$. Given $O \in \text{ext} \, C$, and $\epsilon > 0$ small enough, then, setting $r = 1 + 1/(\mathrm{m}(O) - 1)$, the open set $U = \text{int} \, C_{r+\epsilon} \setminus C_{r-\epsilon}$ is a neighborhood of O with $O' \in U$ and $r' = 1 + 1/(\mathrm{m}(O') - 1)$ implying $|r' - r| < \epsilon$.

Remark. The construction of the monotonic family $\{C_r\}_{r \geq r^*}$ is due to [Hammer 2]. In the last passage he follows a different approach. One of the key elements in his treatment is to prove that ∂C_r can be characterized as the set of points on affine diameters of C extended about their midpoints by the ratio $2r - 1$. The equivalence of this with our level-set approach can be seen as follows. Let $[B, B^o]$ be the affine diameter with extension containing $O \in \text{ext}\, C$ as in the discussion after formula (2.2.6). (See also Figure 2.2.3.) Letting $A = (B + B^o)/2$ be the midpoint, we calculate

$$\frac{d(A, O)}{d(A, B)} = \frac{d(B^o, O) + d(B, O)}{d(B^o, O) - d(B, O)} = \frac{\Lambda(B^o, O) + 1}{\Lambda(B^o, O) - 1} = \frac{\mathfrak{m}(O) + 1}{\mathfrak{m}(O) - 1},$$

where the last equality is because $\Lambda(., O)$ attains its maximum $\mathfrak{m}(O)$ on $(\partial C \times \{O\})_+$ at B^o. Now, the last quotient is equal to $2r - 1$ if and only if $r = 1 + 1/(\mathfrak{m}(O) - 1)$ if and only if $O \in \partial C_r$.

2.4 Klee's Inequality on the Critical Set

Let $C \in \mathfrak{B}$. The main objective of this technical section is to derive Klee's inequality

$$\mathfrak{m}_C^* + \dim C^* \leq n, \tag{2.4.1}$$

where $C^* \subset C$ is the critical set of C (Section 2.1). Note that this can be viewed as an improvement of the Minkowski–Radon upper estimate in (2.1.7) (in which the critical set is missing).

All the results in this section are due to [Klee 2]. We follow his concise and brilliant argument with simplifications (whenever possible) and detailed elaborations.

Recall the (interior) distortion ratio $\Lambda : \partial C \times \text{int}\, C \to \mathbb{R}$, and the maximal distortion $\mathfrak{m} : \text{int}\, C \to \mathbb{R}$, $\mathfrak{m}(O) = \max_{\partial C} \Lambda(., O)$, $O \in \text{int}\, C$ (Section 2.1).

We define

$$\mathcal{M}(O) = \{C \in \partial C \mid \Lambda(C, O) = \mathfrak{m}(O)\}.$$

By continuity of the functions involved (Proposition 2.1.1 and Theorem 2.1.2), $\mathcal{M}(O) \subset \partial C$ is closed and hence compact. Note also that, by Corollary 2.2.2, for every $C \in \mathcal{M}(O)$, the chord $[C, C^o]$ (passing through O) is an affine diameter.

As in Section 2.1, we let $\mathfrak{m}^* = \inf_{\text{int}\, C} \mathfrak{m}$ and $C^* = \{O \in \text{int}\, C \mid \mathfrak{m}(O) = \mathfrak{m}^*\}$ the corresponding critical set. The key to the proof of Klee's inequality is to understand the structure of (the antipodal of) $\mathcal{M}(O^*)$ with respect to a critical point $O^* \in C^*$.

We begin with a sequence of lemmas. The first lemma is illustrated in Figure 2.4.1.

Lemma 2.4.1. Let $O^* \in C^*$, and denote $\mathcal{N} = \mathcal{N}(O^*) = \mathcal{M}(O^*)^o \subset \partial C$, where the antipodal is with respect to O^*. Let $\mathfrak{G}$ be the family of closed half-spaces $\mathcal{G} \subset \mathcal{X}$ that are disjoint from $\text{int}\, C$ and $\mathcal{G} \cap \mathcal{N} \neq \emptyset$. Then $\bigcap \mathfrak{G}$ is empty.

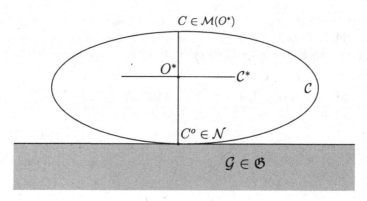

Fig. 2.4.1

Proof. We proceed by contradiction assuming $\bigcap \mathfrak{G} \neq \emptyset$. We first make a preparatory step to ensure that this relation also holds for $\mathfrak{G}$ replaced by $\tilde{\mathfrak{G}}$ obtained by changing $\mathcal{N}$ in the definition to a small neighborhood $\tilde{\mathcal{N}}$ of $\mathcal{N}$ in $\partial \mathcal{C}$.

We may assume that O^* is the origin 0. If $\mathcal{G} \in \mathfrak{G}$ then $0 \notin \mathcal{G}$ so that, for $t > 1$, we have $t\mathcal{G} \subset \mathrm{int}\,\mathcal{G}$. Thus, assuming that $\bigcap \mathfrak{G}$ is non-empty, $\bigcap \mathrm{int}\,\mathfrak{G}$ is non-empty as well, where $\mathrm{int}\,\mathfrak{G} = \{\mathrm{int}\,\mathcal{G} \mid \mathcal{G} \in \mathfrak{G}\}$. We now let $P \in \bigcap \mathrm{int}\,\mathfrak{G}$ (with $|P| \geq 1$, say).

We claim that for $\epsilon_0 > 0$ small enough, replacing $\mathcal{N}$ by its ϵ_0-neighborhood $\tilde{\mathcal{N}} = \mathcal{N}_{\epsilon_0} \cap \partial \mathcal{C}$ (where $\mathcal{N}_{\epsilon_0}$ is the open ϵ_0-neighborhood of $\mathcal{N}$ in $\mathcal{X}$; Section 1.1/A), we also have $P \in \bigcap \tilde{\mathfrak{G}}$, where $\tilde{\mathfrak{G}}$ is the family of closed half-spaces $\mathcal{G} \subset \mathcal{X}$ that are disjoint from $\mathrm{int}\,\mathcal{C}$ and $\mathcal{G} \cap \tilde{\mathcal{N}} \neq \emptyset$.

Indeed, this follows easily by contradiction using the following three steps: (1) letting $\epsilon_k = 1/k$, $k \in \mathbf{N}$; (2) assuming that P is not contained in a sequence of closed half-spaces $\mathcal{G}_{1/k}$ disjoint from $\mathrm{int}\,\mathcal{C}$ and $\mathcal{G}_{1/k} \cap \mathcal{N}_{1/k} \cap \partial \mathcal{C} \neq \emptyset$; and finally (3) using compactness and selecting convergent subsequences.

Now let $B \in \tilde{\mathcal{N}}$. We claim that $B + [0, \infty)P$ is disjoint from $\mathrm{int}\,\mathcal{C}$. By definition, there exists $\mathcal{G} \in \tilde{\mathfrak{G}}$ with $B \in \partial \mathcal{G}$. If $B + tP \in \mathrm{int}\,\mathcal{C}$ for some $t > 0$ then $(B+tP)/(1+t)$ is in the interior of $\mathcal{C}$ (since $0 \in \mathrm{int}\,\mathcal{C}$ and $\mathcal{C}$ is a convex body; Corollary 1.1.9), and, at the same time, it is also contained in $[B, P] \subset \mathcal{G}$, a contradiction. Thus, we have $(B + (\infty, 0)P) \cap \mathrm{int}\,\mathcal{C} = \emptyset$.

If $B + (-\infty, 0)P$ were also disjoint from $\mathrm{int}\,\mathcal{C}$ then, by the Hahn–Banach theorem (Section 1.2), the entire line $B + \mathbb{R}P$ could then be extended to a hyperplane supporting $\mathcal{C}$ at B. This hyperplane, in turn, would then serve as the boundary of a half-space $\mathcal{G} \in \tilde{\mathfrak{G}}$. Since $B-P, P \in \mathcal{G}$, we would then have $B = (B-P)+P \in \mathcal{G}+\mathcal{G}$. Since $0 \in \mathrm{int}\,\mathcal{C}$, we have $\mathcal{G} + \mathcal{G} \subset \mathrm{int}\,\mathcal{G}$. In particular, $B \in \mathrm{int}\,\mathcal{G}$, a contradiction.

Summarizing, we obtain that, for $B \in \tilde{\mathcal{N}}$, there exists $t_B > 0$ such that $B - t_B P \in \mathrm{int}\,\mathcal{C}$. (See Figure 2.4.2.)

To uniformize this, once again a compactness argument (in the use of the precompactness of $\tilde{\mathcal{N}}$ and possibly a smaller $\epsilon_0 > 0$) shows that there exists $\tau > 0$ such that, for $B \in \tilde{\mathcal{N}}$, we have $B - \tau P \in \mathrm{int}\,\mathcal{C}$.

After these preparatory steps we now arrive at the crux of the argument. Let $\tilde{\mathcal{M}} = \tilde{\mathcal{N}}^o$, where (as before) the antipodal is with respect to $0 = O^* \in C^*$. Since $\mathcal{N} = \mathcal{M}(0)^o \subset \tilde{\mathcal{N}}$, we have $\mathcal{M}(0) = \mathcal{N}^o \subset \tilde{\mathcal{N}}^o = \tilde{\mathcal{M}}$. In addition, as $\epsilon_0 \to 0$, we have $\tilde{\mathcal{N}} \to \mathcal{N}$ so that $\tilde{\mathcal{M}} \to \mathcal{M}(0)$.

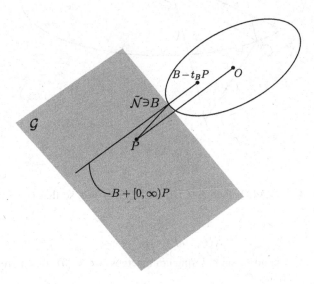

Fig. 2.4.2

Recall that $\Lambda(C, 0) = \max_{\mathcal{M}(0)} \Lambda(C, 0) = \mathfrak{m}^*$, $C \in \mathcal{M}(0)$. We now claim that for any

$$0 < \epsilon < \tau \inf_{\tilde{\mathcal{M}}} \frac{\Lambda(., 0)}{\Lambda(., 0) + 1}, \tag{2.4.2}$$

there exists $\delta > 0$ such that

$$\Lambda(C, -\epsilon P) < \mathfrak{m}^* - \delta, \quad C \in \tilde{\mathcal{M}}, \tag{2.4.3}$$

provided that $\epsilon_0 > 0$ (controlling the size of $\tilde{\mathcal{N}}$) is small enough.

Let $C \in \tilde{\mathcal{M}}$ so that $C^o \in \tilde{\mathcal{N}}$ and, by the above (with $B = C^o$), $C^o - \tau P \in \mathrm{int}\, C$. Using (2.4.2), we have

$$d(0, -\epsilon P) = \epsilon < \tau \inf_{\tilde{\mathcal{M}}} \frac{\Lambda(., 0)}{\Lambda(., 0) + 1} \le d(C^o, C^o - \tau P)\frac{d(0, C)}{d(C, C^o)}$$

Using similar triangles, we conclude that $-\epsilon P$ is in the interior of the triangle $[C, C^o, C^o - \tau P]$ so that $-\epsilon P \in \mathrm{int}\, C$ (see Figure 2.4.3).

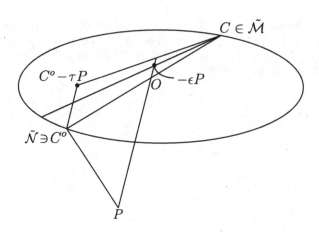

Fig. 2.4.3

Moreover, if $C \in \mathcal{M}(0)$ then $\Lambda(C, 0) = \mathfrak{m}(0) = \mathfrak{m}^*$, so that $C^o - \tau P \in \operatorname{int} C$ implies

$$\Lambda(C, -\epsilon P) < \mathfrak{m}^* - \delta_C$$

for some $\delta_C > 0$ depending on ϵ. Using compactness of $\mathcal{M}(0)$, we can make $\delta > 0$ uniform in $C \in \mathcal{M}(0)$:

$$\Lambda(C, -\epsilon P) < \mathfrak{m}^* - \delta, \quad C \in \mathcal{M}(0),$$

with $\delta > 0$ still depending on ϵ. Since $\tilde{\mathcal{M}} \to \mathcal{M}(0)$ as $\epsilon_0 \to 0$, choosing $\epsilon_0 > 0$ small enough, we arrive at (2.4.3).

On the other hand, by definition, $\Lambda(., 0)$ attains its maximum $\mathfrak{m}^*$ on $\mathcal{M}(0)$. Hence, for every $\epsilon' > 0$ there exists $\delta' > 0$ such that for $C \in \partial C \setminus \mathcal{M}(0)_{\epsilon'}$ (with $\mathcal{M}(0)_{\epsilon'}$ the (relatively open) ϵ'-neighborhood of $\mathcal{M}(0)$ in ∂C) and $d(O, 0) < \delta'$, we have $\Lambda(C, O) < \mathfrak{m}^* - \delta'$.

We now match ϵ' and δ' with ϵ and δ as follows. First, for $\epsilon' > 0$ small enough, we have $\mathcal{M}(0)_{\epsilon'} \cap \partial C \subset \tilde{\mathcal{M}}$. For this $\epsilon' > 0$ we choose $\delta' > 0$ as above. Finally, we choose $\epsilon < \delta'/d(P, 0)$. With these choices, we have

$$\Lambda(C, -\epsilon P) < \mathfrak{m}^* - \delta', \quad C \in \partial C \setminus \mathcal{M}(0)_{\epsilon'}. \tag{2.4.4}$$

Since $\partial C \setminus \tilde{\mathcal{M}} \subset \partial C \setminus \mathcal{M}(0)_{\epsilon'}$, comparing (2.4.3) with (2.4.4), we obtain

$$\max_{\partial C} \Lambda(., -\epsilon P) = \mathfrak{m}(-\epsilon P) < \mathfrak{m}^*.$$

This, however, contradicts to $\mathfrak{m}^* = \inf_{\operatorname{int} C} \mathfrak{m}$. The lemma follows.

Remark. Note that an immediate consequence of Lemma 2.4.1 is that $\mathcal{M}(O^*)$, $O^* \in C^*$, consists of *at least three* points. (Since $\bigcap \mathfrak{G} \neq \emptyset$, the family $\mathfrak{G}$ must consist of at least three closed half-spaces.) This is the best possible for $n = 2$ (equilateral triangle), but, for $n \geq 3$, will be improved later (Corollary 2.4.14).

The next lemma asserts that the critical set C^* can be mapped to ∂C by a suitable central similarity from a boundary point of C.

Lemma 2.4.2. *Let $O^* \in \operatorname{int} C^*$, the relative interior of C^* with respect to the affine span $\langle C^* \rangle$, and $C \in \mathcal{M}(O^*)$. Then*

$$S_{1+1/\mathfrak{m}^*,C}(C^*) = C + \left(1 + \frac{1}{\mathfrak{m}^*}\right)(C^* - C) \subset \partial C, \tag{2.4.5}$$

where the set on the left-hand side of the inclusion is the set of antipodals of C with respect to points in C^*. Moreover, $C \in \mathcal{M}(O'^*)$ for any $O'^* \in \operatorname{int} C^*$.

Proof. Since $C \in \partial C$, by Hammer's definition of the critical set C^* in (2.3.1)–(2.3.2), we have

$$C^* = C_{r^*} \subset S_{r^*,C}(C).$$

Inverting, and using $1/r^* = 1 + 1/\mathfrak{m}^*$, we obtain

$$S_{1/r^*,C}(C^*) = S_{1+1/\mathfrak{m}^*,C}(C^*) = C + \left(1 + \frac{1}{\mathfrak{m}^*}\right)(C^* - C) \subset C. \tag{2.4.6}$$

Comparing (2.4.5) and (2.4.6), by convexity of the sets involved (and Proposition 1.1.7), it remains to show that a point in the *relative interior* of $S_{1+1/\mathfrak{m}^*,C}(C^*)$ in (2.4.6) is in the boundary ∂C. We choose this point to be $S_{1+1/\mathfrak{m}^*,C}(O^*)$.

Since $\Lambda(C, O^*) = \mathfrak{m}(O^*) = \mathfrak{m}^*$, by the definition of antipodal in (2.1.2) (with respect to O^*), we have

$$C - O^* = -\Lambda(C, O^*)(C^o - O^*) = -\mathfrak{m}^*(C^o - O^*).$$

Rewriting this, we obtain

$$C + \left(1 + \frac{1}{\mathfrak{m}^*}\right)(O^* - C) = C^o \in \partial C.$$

(2.4.5) follows.

Finally, the last statement follows from (2.4.5) since $\Lambda(C, .) = \mathfrak{m}^*$ is constant on C^*. (See Figure 2.4.4.)

Lemma 2.4.2 implies that $\mathcal{M}(O^*)$ is *independent* of $O^* \in \operatorname{int} C^*$. Henceforth we will denote this set by $\mathcal{M}^*$. On the other hand, $\mathcal{N} = \mathcal{N}(O^*)$ does depend on $O^* \in \operatorname{int} C^*$ (even though it may be suppressed in the notation) since it is the antipodal of $\mathcal{M}^*$ with respect to O^*.

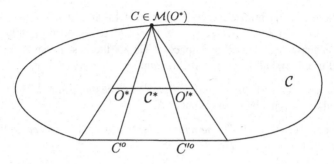

Fig. 2.4.4

The following simple example, due to [Hammer–Sobczyk], is a good illustration of Lemma 2.4.2.

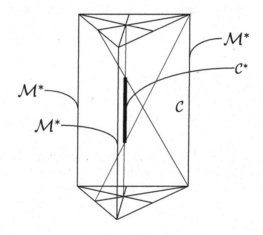

Fig. 2.4.5

Example 2.4.3. Let $\Delta = [C_1, C_2, C_3] \subset \mathbb{R}^2$ be an equilateral triangle and $\mathcal{C} = \Delta \times [-h/2, h/2] \subset \mathbb{R}^2 \times \mathbb{R} = \mathbb{R}^3$ the vertical cylinder of height $h > 0$ on Δ. We can write $\mathcal{C} = [\Delta^-, \Delta^+]$, where $\Delta^\pm = [C_1^\pm, C_2^\pm, C_3^\pm]$, $C_i^\pm = (C_i, \pm h/2)$, $i = 1, 2, 3$.

The critical set Δ^* of the *planar* convex body Δ is comprised only by the centroid $(C_1 + C_2 + C_3)/3$ with $\mathfrak{m}_\Delta^* = 2$. We may assume that this point is the origin. The horizontal planar slices $\Delta \times \{t\}$, $|t| \leq h/2$, of $\mathcal{C}$ are isometric copies of Δ.

It follows directly from Hammer's decomposition (2.3.1) that $\mathfrak{m}_\mathcal{C}^* = \mathfrak{m}_\Delta^* = 2$ and $\mathcal{C}^* = [O^-, O^+]$, where $O^\pm = (0, 0, \pm h/6)$, the middle third of the vertical axis.

Finally, $\mathcal{M}^*$ is comprised by the three vertical edges. Note that in Klee's estimate (2.4.1) equality holds. (See Figure 2.4.5.)

From now on, for a convex set $\mathcal{C}$, will write $\operatorname{int} \mathcal{C}^*$ for the *relative* interior of $\mathcal{C}^*$ in its affine span $\langle \mathcal{C}^* \rangle$.

Corollary 2.4.4. *Let $O^* \in \operatorname{int} C^*$, $\mathcal{N} = \mathcal{N}(O^*) = \mathcal{M}(O^*)^o = (\mathcal{M}^*)^o \subset \partial C$ and $\mathfrak{G}$ as in Lemma 2.4.1. If $\mathcal{G} \in \mathfrak{G}$ has a common point with $\mathcal{N}$ at $C^o \in \mathcal{N}$, $C \in \mathcal{M}^*$, then $\mathcal{G}$ contains the affine subspace $C + (1 + 1/\mathfrak{m}^*)(\langle C^* \rangle - C)$ (parallel to $\langle C^* \rangle$) on its boundary. In particular, $\operatorname{int} \mathcal{G}$ contains an affine subspace parallel to $\langle C^* \rangle$.*

Proof. The point to note here is that C^o is in the (relative) interior of the convex set on the left-hand side of the inclusion in (2.4.5). (See Figure 2.4.6.)

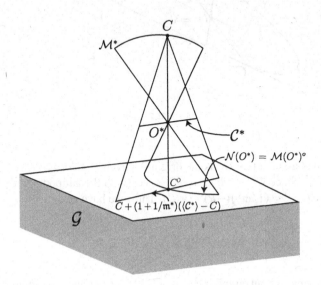

Fig. 2.4.6

Therefore, the hyperplane $\partial \mathcal{G}$ supporting C at C^o must contain the entire convex set $C + (1 + 1/\mathfrak{m}^*)(C^* - C)$. The corollary follows.

Corollary 2.4.5. *Let $O^* \in \operatorname{int} C^*$ and $C \in \mathcal{M}^*$. Then, for the affine diameter $[C, C^o]$, there exist parallel hyperplanes $\mathcal{H}$ and $\mathcal{H}^o$ supporting C at C and C^o which are also parallel to $\langle C^* \rangle$.*

Proof. By construction, the affine span of the left-hand side in (2.4.5) is disjoint from $\operatorname{int} C$. Hence, by the Hahn–Banach theorem (Section 1.2), it extends to a hyperplane $\mathcal{H}^o$ supporting C at C^o. The existence of $\mathcal{H}$ now follows from Corollary 2.2.2. □

Lemma 2.4.6. *Let $O^* \in \operatorname{int} C^*$ and $\mathcal{E} \subset \mathcal{X}$ an affine subspace containing O^* and complementary to $\langle C^* \rangle$. Let $\pi : \mathcal{X} \to \mathcal{E}$ be the projection onto $\mathcal{E}$ with kernel (parallel to) $\langle C^* \rangle$. Then we have*

$$\mathfrak{m}_{\pi(C)}(O^*) = \mathfrak{m}_C^* \quad \text{and} \quad \mathcal{M}_{\pi(C)}(O^*) = \pi(\mathcal{M}_C(O^*)) = \pi(\mathcal{M}_C^*), \qquad (2.4.7)$$

where the dependence on the respective convex sets is indicated by subscripts. (See Figure 2.4.7.)

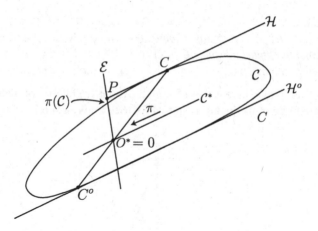

Fig. 2.4.7

Proof. We may assume that O^* is the origin 0. Given $P \in \partial\pi(\mathcal{C})$, let $C \in \partial\mathcal{C}$ such that $\pi(C) = P$. Since $C^o = (1/\mathfrak{m}^*)(-C) \in \partial\mathcal{C}$ (with the antipodal with respect to 0), we have $\pi(C^o) = (1/\mathfrak{m}^*)(-P) \in \pi(\mathcal{C})$. Thus, we have

$$\Lambda_{\pi(\mathcal{C})}(P, 0) \leq \Lambda_{\mathcal{C}}(C, 0) \leq \mathfrak{m}_{\mathcal{C}}^*$$

with equality (throughout) implying $C \in \mathcal{M}_{\mathcal{C}}(0)$. Taking the maximum in $P \in \partial\mathcal{C}$, we obtain $\mathfrak{m}_{\pi(\mathcal{C})}(0) \leq \mathfrak{m}_{\mathcal{C}}^*$.

On the other hand, for $C \in \mathcal{M}_{\mathcal{C}}(0)$, the chord $[C, C^o]$ is an affine diameter, and by Corollary 2.4.5, there exist parallel hyperplanes $\mathcal{H}$ and $\mathcal{H}^o$ supporting $\mathcal{C}$ at C and C^o which are also parallel to $\langle C^* \rangle$. Since the kernel of π is $\langle C^* \rangle$, the intersections $\mathcal{H} \cap \mathcal{E}$ and $\mathcal{H}^o \cap \mathcal{E}$ are parallel hyperplanes supporting $\pi(\mathcal{C})$ at $\pi(C)$ and $\pi(C^o)$.

Thus, we have

$$\Lambda_{\pi(\mathcal{C})}(\pi(C), 0) = \mathfrak{m}_{\mathcal{C}}^* \quad \text{and} \quad \mathfrak{m}_{\pi(\mathcal{C})}(0) \geq \mathfrak{m}_{\mathcal{C}}^*.$$

The lemma follows. ·

Lemma 2.4.7. *Let* $O \in \mathrm{int}\, \mathcal{C}$ *and* $\mathcal{A} \subset \partial\mathcal{C}$ *such that* $O \in [\mathcal{A}]$. *Assume that, for some* $m \in \mathbb{N}$, *we have* $\Lambda(C, O) > m$ *for all* $C \in \mathcal{A}$. *Then there exists* $\mathcal{A}_0 \subset \mathcal{A}$ *such that* $\dim[\mathcal{A}_0] > m$ *and* $O \in \mathrm{int}\,[\mathcal{A}_0]$. *Moreover* $\mathcal{A}_0$ *can be chosen as the set of vertices of a k-simplex,* $k > m$.

Proof. As before, for computational simplicity, we may assume that O is the origin 0. Let $k \geq 1$ be the *smallest* integer such that $0 \in [C_0, \ldots, C_k]$, where $\{C_0, \ldots, C_k\} \subset \mathcal{A}$.

The existence of k ($\leq \dim \mathcal{C}$) is a direct consequence of Carathéodory's theorem (Section 1.3). Minimality of k implies that $\{C_0, \ldots, C_k\}$ are the vertices of a k-simplex and $0 \in \text{int}\,[C_0, \ldots, C_k]$. It remains to show that $k > m$.

We have $\sum_{i=0}^{k} \lambda_i C_i = 0$, $\sum_{i=0}^{k} \lambda_i = 1$, $\{\lambda_0, \ldots, \lambda_k\} \subset (0, 1)$. Let $B_i = -C_i^o = (1/\Lambda(C_i, 0))C_i$, $i = 0, \ldots, k$, where the antipodal is with respect to 0. We have $\sum_{i=0}^{k} \lambda_i \Lambda(C_i, 0)B_i = 0$, so that

$$B_0 = -\sum_{i=1}^{k} \mu_i B_i, \tag{2.4.8}$$

where $\mu_i = \lambda_i \Lambda(C_i, 0)/(\lambda_0 \Lambda(C_0, 0)) > 0$, $i = 1 \ldots, k$. Comparing $C_i = -\Lambda(C_i, 0)C_i^o \in \partial\mathcal{C}$ and $mB_i = -mC_i^o$, the assumption $m < \Lambda(C_i, 0)$ implies that $mB_i \in \text{int}\,\mathcal{C}$, $i = 0, \ldots, k$. We therefore rewrite (2.4.8) as

$$-B_j = \frac{1}{m\mu_j}mB_0 + \sum_{1 \leq i \leq k;\, i \neq j} \frac{\mu_i}{m\mu_j}mB_i, \quad j = 1 \ldots, k.$$

Noting that $-B_j = C_j^o \in \partial\mathcal{C}$, this implies that

$$1 < \frac{1}{m\mu_j} + \sum_{1 \leq i \leq k;\, i \neq j} \frac{\mu_i}{m\mu_j}, \quad j = 1 \ldots, k,$$

or equivalently

$$m\mu_j < 1 + \sum_{1 \leq i \leq k;\, i \neq j} \mu_i, \quad j = 1 \ldots, k.$$

Summing up with respect to $j = 1, \ldots, k$, we obtain

$$m\mu < k + (k-1)\mu, \quad \mu = \sum_{j=1}^{k} \mu_j,$$

or equivalently

$$\frac{\mu}{\mu+1} < \frac{k}{m+1}. \tag{2.4.9}$$

On the other hand, rewriting (2.4.8) again, we have

$$-B_0 = \sum_{i=1}^{k} \frac{\mu_i}{m}mB_i,$$

and the same argument applies yielding $m < \mu$. This gives $m/(m+1) < \mu/(\mu+1)$. Using (2.4.9), we finally obtain $m < k$. The lemma follows.

Remark. A very simple interpretation of Lemma 2.4.7 will be given in Section 4.1 in terms of a new mean Minkowski measure.

The first step in proving Klee's inequality is the following:

Lemma 2.4.8. *Assume that $\mathcal{M}^*$ is not contained in any open half-space (whose boundary is) parallel to $\langle C^* \rangle$ but not containing C^*. Then Klee's inequality in (2.4.1) holds.*

Proof. This follows from Lemmas 2.4.6–2.4.7. For clarity we indicate the respective convex sets by subscripts. We may assume that $0 \in \operatorname{int} C^*$ so that $\mathcal{M}_C(0) = \mathcal{M}_C^*$. Let $\mathcal{E} \subset \mathcal{X}$ and $\pi : \mathcal{X} \to \mathcal{E}$ be as in Lemma 2.4.6. Then, for any $P \in \mathcal{M}_{\pi(C)}(0) = \pi(\mathcal{M}_C(0)) = \pi(\mathcal{M}_C^*)$, we have $\mathfrak{m}_{\pi(C)}(0) = \Lambda_{\pi(C)}(P, 0) = \mathfrak{m}_C^*$.

The assumption on $\mathcal{M}_C^*$ means that $0 \in [\pi(\mathcal{M}_C^*)]$. Applying Lemma 2.4.7 to $\mathcal{A} = \pi(\mathcal{M}_C^*) = \mathcal{M}_{\pi(C)}(0) \subset \partial \pi(C)$ we obtain

$$\mathfrak{m}_C^* > m \quad \Rightarrow \quad \dim \pi(C) > m, \quad m \in \mathbf{N}.$$

Since the dimension is an integer, it follows that $\dim \pi(C) \geq \mathfrak{m}_C^*$. Now, $\dim \pi(C) = \dim C - \dim C^*$, and (2.4.1) follows. $\qquad\square$

In view of future applications, we formulate the next lemma for a compact convex set $\mathcal{K} \subset \mathcal{X}$, and work in the *linear* span of $\mathcal{K}$.

Lemma 2.4.9. *Let $\mathcal{K} \subset \mathcal{X}$ be a compact convex set, $0 < t < 1/\mathfrak{m}_{\mathcal{K}}^*$, $C = [\mathcal{K} \cup (-t\mathcal{K})]$, $O^* \in \mathcal{K}^* \setminus \{0\}$, and $s > t(1 + \mathfrak{m}_{\mathcal{K}}^*)/(1 - t\mathfrak{m}_{\mathcal{K}}^*)$. Then $-sO^*$ is contained in every closed half-space that intersects $-t\mathcal{K}$ but disjoint from $\operatorname{int} C$. (See Figure 2.4.8.)*

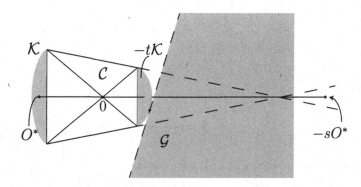

Fig. 2.4.8

Proof. For brevity, we set $\mathfrak{m}^* = \mathfrak{m}_{\mathcal{K}}^*$. By Hammer's decomposition, we have

$$\left(1 + \frac{1}{\mathfrak{m}^*}\right) O^* - \frac{1}{\mathfrak{m}^*} \mathcal{K} \subset \mathcal{K}. \tag{2.4.10}$$

Let $\mathcal{G} \subset \mathcal{X}$ be a closed half-space that contains $-tC$ for some $C \in \mathcal{K}$, and disjoint from int C. (Note that if $\mathcal{G}$ does not exist then we are automatically done.) Let ϕ : $\mathcal{X} \to \mathbb{R}$ be a non-zero *linear* functional such that

$$\mathcal{G} = \{X \in \mathcal{X} \mid \phi(X) \le \phi(-tC)\}.$$

Note that $\phi(-tC) \le \phi(0) = 0$ since $0 \in C$.

Applying (2.4.10) to $C \in \mathcal{K}$ and noting that $\mathcal{K} \subset C$, we obtain

$$\left(1 + \frac{1}{\mathfrak{m}^*}\right)\phi(O^*) - \frac{1}{\mathfrak{m}^*}\phi(C) \ge \phi(-tC).$$

Rearranging, we find

$$0 \ge \phi(-tC) \ge \frac{t(1 + \mathfrak{m}^*)}{1 - t\mathfrak{m}^*}\phi(-O^*).$$

If $\phi(O^*) = 0$ then $\phi(-tC) = 0$ and hence $\phi(-sO^*) = 0$ implies $-sO^* \in \mathcal{G}$. Otherwise, $\phi(O^*) > 0$ and, by the above, we have

$$\phi(-tC) \ge \frac{t(1 + \mathfrak{m}^*)}{(1 - t\mathfrak{m}^*)}\phi(-O^*) \ge s\phi(-O^*) = \phi(-sO^*),$$

and $-sO^* \in \mathcal{G}$ follows again.

Lemma 2.4.10. *Let $C \in \mathfrak{B}$ and assume that*

$$n - 1 \le \mathfrak{m}^*_C + \dim C^*. \tag{2.4.11}$$

Then $\mathcal{M}^*$ is not contained in any open half-space parallel to $\langle C^*\rangle$ but not containing C^*.

Proof. By continuity of the data involved, we may assume that strict inequality holds in (2.4.11), that is, we have $\mathfrak{m}^*_C > n - k - 1$, where $k = \dim C^*$. As usual, we may assume $0 \in \operatorname{int} C^*$. We let $\mathcal{N} = \mathcal{M}(0)^o = (\mathcal{M}^*)^o$ and $\mathfrak{G}$ as in Lemma 2.4.1. Setting $\mathfrak{I} = \operatorname{int} \mathfrak{G}$, by (the proof of) Lemma 2.4.1, we have $\bigcap \mathfrak{I} = \emptyset$.

Assume now that the statement of the lemma is false, that is, there exists an open half-space containing $\mathcal{M}^*$ such that it is parallel to $\langle C^*\rangle$ but not containing C^*. This means that there also exists an open half-space $\mathcal{I}$ such that $0 \notin \mathcal{I}$, $-\mathcal{M}^* \subset \mathcal{I}$, and $\partial \mathcal{I}$ is parallel to $\langle C^*\rangle$. (See Figure 2.4.9.)

The family $\mathfrak{I}$ is interior-complete, and, by Corollary 2.4.4, each element of $\mathfrak{I}$ contains a translate of $\langle C^*\rangle$. We now claim that any $n - k$ members of $\mathfrak{I}$ have common intersection with $\mathcal{I}$ arbitrarily far away from $\partial \mathcal{I}$. According to Klee's Helly-type theorem (Section 1.6; actually, Corollary 1.6.4), this means $\bigcap \mathfrak{I} \ne \emptyset$, a contradiction.

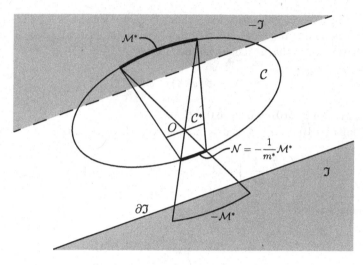

Fig. 2.4.9

To show the claim, let $\mathcal{I}_1, \ldots, \mathcal{I}_{n-k} \in \mathfrak{I}$. Let $C_1, \ldots, C_{n-k} \in \mathcal{M}^*$ such that $C_i^o \in \partial \mathcal{I}_i$, $i = 1, \ldots, n-k$. The convex hull $\mathcal{K} = [C_1, \ldots, C_{n-k}]$ is at most $(n-k-1)$-dimensional and hence, by the universal upper bound in (2.1.7), we have $\mathrm{m}_\mathcal{K}^* \leq n-k-1$. We are now in the position to apply Lemma 2.4.9 to $t = 1/\mathrm{m}_\mathcal{C}^* < 1/(n-k-1) \leq 1/\mathrm{m}_\mathcal{K}^*$. ($0 \notin \mathcal{K}$ since $\mathcal{K} \subset [\mathcal{M}^*] \subset -\mathcal{I}$ and $0 \notin -\mathcal{I}$. Hence any $O^* \in \mathcal{K}^*$ is different from the origin 0.) We obtain that $\mathcal{I}_1, \ldots, \mathcal{I}_{n-k}$ have a common point (sO^* with $s > t(1+\mathrm{m}_\mathcal{K}^*)/(1-t\mathrm{m}_\mathcal{K}^*)$) arbitrarily far away from $\partial \mathcal{I}$. The lemma follows.

We now realize that Klee's inequality in (2.4.1) follows. Indeed, Lemmas 2.4.8 and 2.4.10 give (2.4.1) provided that $n-1 \leq \mathrm{m}_\mathcal{C}^* + \dim \mathcal{C}^*$. On the other hand, if $n-1 > \mathrm{m}_\mathcal{C}^* + \dim \mathcal{C}^*$ then (2.4.1) is automatic.

Corollary 2.4.11. *Let $\mathcal{C} \in \mathfrak{B}$ and assume that (2.4.11) holds. Then, for the relative interiors, we have*

$$\mathrm{int}\, \mathcal{C}^* \subset \mathrm{int}\,[\mathcal{C}^* \cup \mathcal{M}^*]. \qquad (2.4.12)$$

Proof. Let $O^* \in \mathrm{int}\, \mathcal{C}^*$. Assume, on the contrary, that $O^* \in \partial[\mathcal{C}^* \cup \mathcal{M}^*]$ in $\langle \mathcal{C}^* \cup \mathcal{M}^* \rangle$. Since $[\mathcal{C}^* \cup \mathcal{M}^*]$ is a convex body in $\langle \mathcal{C}^* \cup \mathcal{M}^* \rangle$, there is a hyperplane $\mathcal{H}_0 \subset \langle \mathcal{C}^* \cup \mathcal{M}^* \rangle$ supporting $[\mathcal{C}^* \cup \mathcal{M}^*]$ at O^*. Since $O^* \in \mathrm{int}\, \mathcal{C}^* \subset [\mathcal{C}^* \cup \mathcal{M}^*]$, this hyperplane must contain $\mathrm{int}\, \mathcal{C}^*$ and therefore $\langle \mathcal{C}^* \rangle$. Let $\mathcal{H} \subset \mathcal{X}$ be any hyperplane extension of $\mathcal{H}_0$ and $\mathcal{G}$ the closed half-space containing $[\mathcal{C}^* \cup \mathcal{M}^*]$ with $\partial \mathcal{G} = \mathcal{H}$. Clearly, $\mathrm{int}\, \mathcal{G}$ is an open half-space parallel to $\langle \mathcal{C}^* \rangle$ but not containing $\mathcal{C}^*$ and $\mathrm{int}\, \mathcal{G}$ contains $\mathcal{M}^*$ (by convexity of $\mathcal{C}$). This contradicts to Lemma 2.4.10. □

Corollary 2.4.12. *Let $\mathcal{C} \in \mathfrak{B}$. Then $\dim \mathcal{C}^* \leq n-2$.*

Proof. Since $m_\mathcal{C}^* \geq 1$ the dimension estimate on $\mathcal{C}^*$ follows from Klee's inequality in (2.4.1) unless $m_\mathcal{C}^* = 1$. In the latter case, however, $\mathcal{C}$ is symmetric with respect to the singleton $\mathcal{C}^*$ and the inequality also holds.

Let $\mathcal{C} \in \mathfrak{B}$ and consider the compact convex set $\mathcal{K} = [\mathcal{M}^*]$. We claim that

$$m_\mathcal{C}^* \leq m_\mathcal{K}^*. \tag{2.4.13}$$

Indeed, if $m_\mathcal{C}^* > m_\mathcal{K}^*$ were true then applying Lemma 2.4.9 to $t = 1/m_\mathcal{C}^* < 1/m_\mathcal{K}^*$ and with a choice of $O^* \in \mathcal{K}^* \setminus \{0\}$, we would get contradiction with Lemma 2.4.1. (If O^* did not exist then $\mathcal{K}^* = \{0\}$ and we would get $m_\mathcal{K}^* = m_\mathcal{K}(0) = \max_{\partial\mathcal{K}} \Lambda(.,0) \geq \max_{\partial\mathcal{C}} \Lambda_\mathcal{C}(.,0) = m_\mathcal{C}^*$ since $\mathcal{M}^* \subset \partial\mathcal{K}$; a contradiction again.) The inequality in (2.4.13) follows.

We note two consequences:

Corollary 2.4.13. *If $m_\mathcal{C}^* > n - 1$ then $\mathcal{C}^*$ is a singleton, $\langle \mathcal{M}^* \rangle = \mathcal{X}$ and $\mathcal{C}^* \subset$ int $[\mathcal{M}^*]$.*

Proof. The first statement is a direct consequence of Klee's inequality (2.4.1). Using the notations above, let dim $\mathcal{K} = m$, where $\mathcal{K} = [\mathcal{M}^*]$. By (2.1.7), we have $m_\mathcal{K}^* \leq m$. Using the assumption and (2.4.13), we have $n - 1 < m_\mathcal{C}^* \leq m_\mathcal{K}^* \leq m$. Thus, we have $m = n$, and $\langle \mathcal{M}^* \rangle = \mathcal{X}$ follows. The last statement now is a consequence of (2.4.12).

For the next lemma recall that the *ceiling* $\lceil x \rceil$ of $x \in \mathbb{R}$ is the smallest integer $\geq x$.

Corollary 2.4.14. $\mathcal{M}^*$ *contains at least* $\lceil m_\mathcal{C}^* + 1 \rceil$ *points.*

Proof. As in the previous proof, let dim $\mathcal{K} = m$, where $\mathcal{K} = [\mathcal{M}^*]$; in particular, $\mathcal{M}^*$ consists of at least $m + 1$ points. Once again, by (2.1.7), we have $m_\mathcal{K}^* \leq m$. Using (2.4.13), we have $m + 1 \geq m_\mathcal{K}^* + 1 \geq m_\mathcal{C}^* + 1$ and the corollary follows.

Remark 1. By Corollary 2.2.2, given $O^* \in \text{int}\,\mathcal{C}^*$, for every $C \in \mathcal{M}^*$, $[C, C^o]$ is an affine diameter containing O^*. Therefore Corollary 2.4.14 implies that the number of affine diameters passing through O^* is at least $\lceil m^* + 1 \rceil$. In particular, if $m^* > n - 1$ then there are $n + 1$ affine diameters passing through O^*, $\mathcal{C}^* = \{O^*\}$. Note that this is a partial answer to the Grünbaum Conjecture stated in Section 2.2. We will return to this in Section 4.2.

Remark 2. Klee's proof of Corollary 2.4.14 is somewhat more involved as follows. If $n - 1 < m_\mathcal{C}^*(\leq n)$ then (as above) $\mathcal{C}^*$ is a singleton and $\mathcal{C}^* \subset$ int $[\mathcal{M}^*]$. Hence, $\mathcal{M}^*$ contains at least $n + 1 \geq \lceil m_\mathcal{C}^* + 1 \rceil$ points, and the corollary follows in this case. It remains to consider the case $m_\mathcal{C}^* + 1 \leq n$. Assume, on the contrary, that $\mathcal{M}^*$ consists of $m + 1$ points with $m + 1 < \lceil m_\mathcal{C}^* + 1 \rceil \leq n$. By the definition of $\mathcal{M}^*$ it follows that the convex hull $\mathcal{K} = [\mathcal{M}^*]$ is a j-simplex with $j \leq m$. The rest follows easily by explicit computation.

2.5   Convex Bodies of Constant Width

Convex bodies of constant width form a special class of convex bodies whose geometric invariants can often be explicitly calculated. In this final section we give an extended treatise on them with the purpose of providing transparent and beautiful examples to the general theory developed so far.

A convex body $C \in \mathfrak{B}$ is said to have *constant width* if any pair $\mathcal{H}$ and $\mathcal{H}'$ of parallel supporting hyperplanes of C are at the same distance apart. In this case the minimal width $d = d_C$ is called the *width* of C.

By (1.5.2), C has constant width if and only if $d_C = D_C$, the diameter of C.

A simple comparison of the support functions of a convex body C and its Minkowski symmetral $\tilde{C} = (C - C)/2$ shows that C is of constant width d if and only if $\tilde{C} = \bar{B}_{d/2}$, the closed ball of radius $d/2$. (See Problem 7. For properties of the Minkowski symmetral, see Section 3.2.)

Although unexpected at the first sight, and as Euler was already aware of, there is an abundance of convex sets with constant width way beyond the obvious examples of closed metric balls. For many characterizations and examples of convex bodies of constant width, see [Chakerian–Groemer, 2].

In the study of convex bodies of constant width the notion of (diametric) completeness plays a key role. A convex body $C \in \mathfrak{B}$ is called *(diametrically) complete* if $C \subset C', C' \in \mathfrak{B}$, and $D_C = D_{C'}$ imply $C = C'$.

Now, *a convex body is complete if and only if it is of constant width.* (This result is usually referred to as Meissner's theorem; see [Bonnesen-Fenchel, 64] and [Chakerian–Groemer, 4].)

The "if" part is straightforward (see Problem 9).

The "only if" part can be seen though yet another characteristic property. A simple argument shows that C is complete if and only if it satisfies the *spherical intersection property*

$$C = \bigcap_{X \in C} \bar{B}_d(X), \quad d = D_C.$$

(Indeed, one needs to observe that $Y \in \mathcal{X}$ belongs to the intersection of the closed metric balls above if and only if $\sup_{X \in C} d(X, Y) \leq D_C$ if and only if $D_{[Y,C]} = D_C$.) In particular, along with any two of its points C' and C'' of a *complete* convex body C, it also contains *any* proper circular arc of radius $\geq D_C$ connecting C' and C''. (Here proper means that the arc is smaller than a semi-circle.)

The "only if" part now follows by contradiction: If a complete convex body C has $d_C < D_C$ then consider the endpoints $C, C' \in C$ of a double normal realizing d_C, and yet another point $C'' \in C$ with D_C distance from C. (The existence of C'' follows from completeness of C.) Finally, consider a circular arc of radius D_C connecting C' and C'' (within the affine planar span of the entire configuration), and realize that this arc crosses the given hyperplane supporting C at C'. This is a contradiction.

As a byproduct we see that a convex body C is of constant width d if and only if C satisfies the spherical intersection property. (For a direct proof of this, see [Maehara, Corollary 1].) (Note that, in Minkowski space, the concepts of constant width and completeness both make sense, but they are not equivalent; see [Chakerian–Groemer, 6].)

Every convex body C has a *(diametric) completion*, a complete convex body $C^\sharp \supset C$ such that $D_C = D_{C^\sharp}$. (This result is usually referred to as Pál's Theorem ($n = 2$). See again [Bonnesen-Fenchel, 64] and [Chakerian–Groemer, 4].)

Given $C \in \mathfrak{B}$ of diameter $D_C = d$, one is tempted to consider the *wide spherical hull*

$$\eta(C) = \{X \in \mathcal{X} \mid D_{[X,C]} = D_C\} = \bigcap_{X \in C} \bar{B}_d(X)$$

of C, but, as the example of a regular tetrahedron shows, this, in general, is not complete. (See Example 2.5.2 below.)

A quick fix of this problem is the following construction in [Eggleston 1, Theorem 54, p. 126]. Let

$$\rho(C) = \sup_{X \in \eta(C)} d(X, C),$$

and

$$\tilde{\eta}(C) = \{X \in \mathcal{X} \mid X \in \eta(C) \text{ and } d(X, C) = \rho(C)\}.$$

(Clearly, $C \subset \eta(C)$, and C is complete if and only if $\eta(C) = C$ if and only if $\rho(C) = 0$.) Assume now that C is not complete, and define the sequences $\{X_k\}_{k \geq 1} \subset \mathcal{X}$ and $\{C_k\}_{k \geq 1} \subset \mathfrak{B}$ inductively as follows: $X_1 \in \tilde{\eta}(C)$, and $C_1 = [X_1, C]$. If X_k and C_k are defined then let $X_{k+1} \in \tilde{\eta}(C_k)$ and $C_{k+1} = [X_{k+1}, C_k]$. It is easy to see that (1) $\lim_{k \to \infty} C_k = C_\infty$ exists; (2) $C^\sharp = \bar{C}_\infty$ is convex; and (3) $D_{C^\sharp} = D_C$.

We claim that $C^\sharp$ is complete. Assuming the contrary, let $Y \notin C^\sharp$ such that $D_{[Y,C^\sharp]} = D_C$. Let $\delta = d(Y, C^\sharp) > 0$. Given $l > k \geq 1$, we have $d(X_k, X_l) > \rho(C_{l-1})$. On the other hand, we have $d(Y, C_l) \geq \delta$ so that $\rho(C_{l-1}) \geq \delta$. We obtain $d(X_k, X_l) > \delta, k \geq l \geq 1$, which is impossible since the sequence $\{X_k\}_{k \geq 1}$ is bounded.

A more refined approach of [Scott] shows that a convex body C has a completion $C^\sharp$ *contained in the circumball of* C. (An independent proof of this by [Vrećica] is outlined in Problem 14.)

An explicit and direct construction of a completion due to [Maehara] is as follows. For $C \in \mathfrak{B}$ of diameter $d = D_C$, we apply η above to the wide spherical hull $\eta(C)$ again to obtain the *tight spherical hull*

$$\eta^2(C) = \bigcap_{X \in \eta(C)} \bar{B}_d(X)$$

of C. Clearly, $C \subset \eta^2(C) \subset \eta(C)$; in fact, in the first inclusion, $\eta^2(C)$ is the intersection of all closed balls of radius d that contain C.

The arithmetic mean

$$\mu(\mathcal{C}) = \frac{\eta(\mathcal{C}) + \eta^2(\mathcal{C})}{2},$$

the *Maehara body* of $\mathcal{C}$, is a convex body of constant width d, therefore a completion of $\mathcal{C}$.

Finally, one may be tempted to iterate Maehara's procedure, but, with obvious notation, we have $\eta^3(\mathcal{C}) = \eta(\mathcal{C})$. (For details, see [Maehara], and also [Moreno–Schneider 2].)

The completion of a convex body is not unique, however. (See, for example, the two Meissner tetrahedra in Example 2.5.2 below.) Moreover, [Groemer 4] showed that a *non-regular* tetrahedron has infinitely many completions. (For more information, see the extensive survey of [Chakerian–Groemer] as well as the additions by E. Heil and H. Martini in [Gruber–Wills, pp. 363–368].)

On a positive note [Moreno–Schneider 1, Moreno–Schneider 2] studied Lipschitz continuous (with respect to the Hausdorff metric) single-valued selections of the multi-valued *completion map*, the map that associates to $\mathcal{C}$ all its completions $\mathcal{C}^\sharp$.

We now begin with low dimensional examples:

Example 2.5.1. Given $d > 0$, a *Reuleaux polygon of width d* is a convex body $\mathcal{C} \subset \mathbb{R}^2$ whose boundary $\partial\mathcal{C}$ is composed of finitely many arcs of circles of radius d, and the centers of these circles are also the vertices of $\mathcal{C}$ (as well as the endpoints of the circular arcs).

Clearly a Reuleaux polygon is a convex set of constant width d. The number p of circular arcs contributing to the boundary of a p-Reuleaux polygon must be odd. (See Problem 8.) The extremal radial segments connecting the endpoints of the circular arcs with the corresponding (antipodal) centers form a star-shaped set with equal side lengths d.

If the boundary arcs are all congruent then the Reuleaux polygon is said to be *regular*.

An explicit construction of a regular $(2m + 1)$-Reuleaux polygon $\mathcal{Q}_{2m+1}$, $m \geq 1$, is as follows. The vertices of $\mathcal{Q}_{2m+1}$ are the same as the vertices of the regular $(2m + 1)$-gon $\mathcal{P}_{2m+1} = \{2m + 1\}$ inscribed in the unit circle $\mathcal{S}$:

$$V_k = \left(\cos\left(\frac{2k\pi}{2m + 1} \right), \sin\left(\frac{2k\pi}{2m + 1} \right) \right) \in \mathcal{S}, \ k = 0, \ldots, 2m.$$

For $k = 0, \ldots, 2m$, the circle with center at V_k and radius $d = 2\sin(m\pi/(2m + 1))$ passes through V_{k+m} and V_{k+m+1}, and the shorter circular arc connecting these two vertices forms the kth part of the boundary of $\mathcal{Q}_{2m+1}$ antipodal of V_k. Note that the star-shaped set of extremal radii of $\mathcal{Q}_{2m+1}$ form an inscribed regular star-polygon $\{\frac{2m+1}{m}\}$. Clearly, $\mathcal{Q}_{2m+1} = \mathcal{P}^\sharp_{2m+1}$.

According to the *Blaschke–Lebesgue theorem*, the Reuleaux polygons have the least area among all convex bodies with a given constant width d. (For a simple proof of this, see [Eggleston 1, 7].) For corresponding stability results, see H. Groemer's article in [Gruber–Wills].

Example 2.5.2. Let $\Delta = [V_0, V_1, V_2, V_3] \subset \mathbb{R}^3$ be a regular tetrahedron of diameter (edge length) d with vertices V_i, $i = 0, \ldots, 3$. The *Reuleaux tetrahedron* is the intersection

$$\mathcal{C} = \bigcap_{i=0}^{3} \bar{B}_d(V_i)$$

of the four closed balls with radius d and centers V_i, $i = 0, \ldots, 3$. An elementary geometric reasoning shows that $\mathcal{C}$ does *not* have constant width. (See Problem 10.) From $\mathcal{C}$ [Meissner] constructed a convex body $\mathcal{M}$ of constant width d (named after him) by "shaving off" three edges of $\mathcal{C}$ as follows. Let $\mathcal{E}$ denote the union of the three edges of $\mathcal{C}$ that meet in V_0. We then define the *Meissner tetrahedron* as

$$\mathcal{M} = \bigcap_{C \in \mathcal{E}} \bar{B}_d(C) \cap \mathcal{C}.$$

The three edges of $\mathcal{C}$ disjoint from V_0 (and composing a triangle) are replaced here by curved patches which are formed as surfaces of revolution of circular arcs. (See Problem 11.)

Dually, replacing $\mathcal{E}$ by the three edges that are disjoint from V_0, one can similarly shave off the three edges that meet at the common vertex V_0 to obtain another (non-congruent) *Meissner tetrahedron* of constant width d.

Note that both Meissner tetrahedra are completions of the regular simplex Δ, showing, in particular, non-unicity of the completion for $n \geq 3$.

Finally, according to the (still open) Bonnesen–Fenchel conjecture, the two Meissner tetrahedra have minimum volume among all convex bodies of constant width d.

Remark. [Lanchand-Robert–Oudet] gave the following "dimension raising" construction for convex bodies of constant width.

Let $\mathcal{H} \subset \mathcal{X}$ be a hyperplane and $\mathcal{C}_0 \subset \mathcal{H}$ a convex body of constant width d in $\mathcal{H}$. The construction results in a convex body $\mathcal{C} \subset \mathcal{X}$ of constant width d in $\mathcal{X}$ such that $\mathcal{C}_0 = \mathcal{C} \cap \mathcal{H}$.

First, let $\mathcal{G}'$ and $\mathcal{G}''$ be the two half-spaces with common boundary $\partial\mathcal{G}' = \partial\mathcal{G}'' = \mathcal{H}$. The construction depends on the choice of a set $\tilde{\mathcal{C}}$ satisfying

$$\mathcal{C}_0 \subset \tilde{\mathcal{C}} \subset \mathcal{G}' \cap \bigcap_{X \in \mathcal{C}_0} \bar{B}_d(X),$$

where $\bar{B}_d(X)$ is the closed ball (of radius d and center X) *in $\mathcal{X}$*. Using this, we define

$$\mathcal{C}'' = \mathcal{G}'' \cap \bigcap_{X \in \tilde{\mathcal{C}}} \bar{B}_d(X) \quad \text{and then} \quad \mathcal{C}' = \mathcal{G}' \cap \bigcap_{X \in \mathcal{C}''} \bar{B}_d(X).$$

Finally, we set $\mathcal{C} = \mathcal{C}' \cup \mathcal{C}''$.

Note that if $C_0 = \tilde{C} \subset \mathbb{R}$ is an interval (of width d) then $C \subset \mathbb{R}^2$ is a Reuleaux triangle. If $C_0 = \tilde{C} \subset \mathbb{R}^2$ is a Reuleaux triangle then $C \subset \mathbb{R}^3$ is a Meissner tetrahedron. (For details, see [Lanchand-Robert–Oudet].)

We now begin to explore the geometry of convex bodies of constant width.

Theorem 2.5.3. *Let $C \in \mathfrak{B}$ be of constant width d. Then the critical set C^* consists of a single point O^*.*

Proof. First note that C is *strictly convex*, that is, every boundary point is extremal.

Indeed, assume, on the contrary, that $[C_0, C_1] \subset \partial C$, $C_0 \neq C_1$, is a non-trivial line segment on the boundary. Let $\mathcal{H}$ be a hyperplane that supports C at an interior point of this line segment. Clearly, $[C_0, C_1] \subset \mathcal{H}$. Let $\mathcal{H}' \neq \mathcal{H}$ be the second parallel hyperplane supporting C, say, at a point C. Then one of the chords $[C, C_0]$ or $[C, C_1]$ must have length $> d$, contradicting to $D_C = d_C$ in (1.5.2). Thus, C is strictly convex.

Moreover, by Lemma 2.4.2, an affine (in fact, homothetic) copy of the (convex) critical set C^* is contained in the boundary of C. Thus, C^* must consist of a single point.

Theorem 2.5.4. *Let $C \in \mathfrak{B}$ be of constant width d. Then the circumcenter (of the unique circumball) coincides with the critical point O^*. Moreover, the inball is unique, and its incenter also coincides with O^*. The circumradius R_C and inradius r_C can be expressed in terms of the Minkowski measure as*

$$R_C = \frac{\mathrm{m}_C^*}{\mathrm{m}_C^* + 1} d \quad and \quad r_C = \frac{1}{\mathrm{m}_C^* + 1} d. \tag{2.5.1}$$

In particular, we have $R_C + r_C = d$ and

$$\mathrm{m}_C^* = \frac{R_C}{r_C}. \tag{2.5.2}$$

Proof. Let $C \in \mathfrak{B}$. Given $O \in \mathrm{int}\, C$, we define

$$R_C(O) = \inf\{s > 0 \,|\, C \subset \bar{B}_s(O)\},$$
$$r_C(O) = \sup\{s > 0 \,|\, \bar{B}_s(O) \subset C\}.$$

Note that, by definition, we have $R_C = \inf_{O \in \mathrm{int}\, C} R_C(O)$ and $r_C = \sup_{O \in \mathrm{int}\, C} r_C(O)$. (The infima and suprema are clearly attained.)

We first claim that

$$R_C(O) = \frac{\mathrm{m}(O)}{\mathrm{m}(O) + 1} d \quad and \quad r_C(O) = \frac{1}{\mathrm{m}(O) + 1} d, \quad O \in \mathrm{int}\, C. \tag{2.5.3}$$

Given $O \in \mathrm{int}\, C$ and $C \in \mathcal{M}(O)$ (Section 2.4), the chord $[C, C^o]$ (passing through O) is an affine diameter (Corollary 2.2.2) so that there exist parallel hyperplanes $\mathcal{H}$ and

$\mathcal{H}^o$ supporting $\mathcal{C}$ at C and C^o, respectively. Since $d(C, O)/d(C^o, O) = \Lambda(C, O) = m(O)$ and $d(C, O) + d(C^o, O) = d(C, C^o) = d$, we obtain

$$d(C, O) \doteq \frac{m(O)}{m(O) + 1}d. \tag{2.5.4}$$

If $[C', C'^o]$ is any chord passing through O then we have

$$\frac{d(C', O)}{d(C', C'^o) - d(C', O)} = \frac{d(C', O)}{d(C'^o, O)} \leq m(O).$$

Hence

$$d(C', O) \leq \frac{m(O)}{m(O) + 1}d(C', C'^o) \leq \frac{m(O)}{m(O) + 1}d.$$

Since $C' \in \partial\mathcal{C}$ was arbitrary, this means that the closed ball with center O and radius $m(O)/(m(O) + 1)d$ contains $\mathcal{C}$. By (2.5.4), no smaller closed ball (with center O) contains $\mathcal{C}$, so that the first equality in (2.5.3) follows.

Let $C \in \partial\mathcal{C}$ such that $r_\mathcal{C}(O) = d(C, O)$ and $\mathcal{H}$ a hyperplane supporting $\mathcal{C}$ at C. Since $\mathcal{H}$ also supports the closed ball with center O and radius $r_\mathcal{C}(O)$, we see that $[C, C^o]$ is a *normal* cord since it is perpendicular to $\mathcal{H}$ at C. Since $\mathcal{C}$ is of constant width, every normal is a double normal (and a metric diameter). (See Problem 9.) Therefore, the supporting hyperplane $\mathcal{H}^o$ parallel to $\mathcal{H}$ supports $\mathcal{C}$ at C^o. As a byproduct, we also see that $d(C, O) + d(C^o, O) = d$.

We now claim that

$$\frac{d(C^o, O)}{d(C, O)} = \Lambda(C^o, O) = m(O). \tag{2.5.5}$$

(Note the reverse roles of C and C^o.)

As before, if $[C', C'^o]$ is any chord passing through O then $d(C', O) \geq r_\mathcal{C}(O) = d(C, O)$ and $d(C', O) + d(C'^o, O) = d(C', C'^o) \leq d = d(C, O) + d(C^o, O)$, and we obtain $d(C'^o, O) \leq d(C^o, O)$. Taking ratios, we get $d(C^o, O)/d(C, O) \geq d(C'^o, O)/d(C', O)$. Since $C' \in \partial\mathcal{C}$ was arbitrary, (2.5.5) follows.

Using the previous argument, we obtain

$$d(C, O) = \frac{1}{m(O) + 1}d.$$

The second equality in (2.5.3) also follows.

To finish the proof of the theorem, we note that uniqueness of the critical point O^* implies that, for *any* $O \in \text{int}\,\mathcal{C} \setminus \{O^*\}$, we have $m^* = m(O^*) < m(O)$. Thus, by (2.5.3), we also have

$$R_\mathcal{C}(O^*) < R_\mathcal{C}(O) \quad \text{and} \quad r_\mathcal{C}(O^*) > r_\mathcal{C}(O), \quad O \in \text{int}\,\mathcal{C} \setminus \{O^*\}.$$

Thus, $R_\mathcal{C}(O^*) = R_\mathcal{C}$ and $r_\mathcal{C}(O^*) = r_\mathcal{C}$, and (2.5.3) (for $O = O^*$) implies (2.5.1). The rest of the theorem follows.

Remark 1. As a byproduct of the proof, for $O \in \text{int}\,C$, we have $R_C(O) + r_C(O) = d$ and $\mathfrak{m}(O) = R_C(O)/r_C(O)$. Note that parts of Theorem 2.5.4 are in [Bonnesen-Fenchel, 63] and [Eggleston 1, Theorem 53 and its Corollary, p. 125].

Remark 2. As shown by [Sallee], in Minkowski space it is the spherical intersection property which implies that the circumball and inball are concentric.

Example 2.5.1 (continued). For a regular $(2m+1)$-sided polygon $\mathcal{P}_{2m+1} = \{2m+1\}$, the unique critical point is the origin. (This can be seen directly, or by noting that the symmetry group of $\mathcal{P}_{2m+1}$, the dihedral group $D_{2(2m+1)}$, must leave the critical set invariant.) By Corollary 2.2.3 (and simple geometry), we have

$$\mathfrak{m}^*_{\mathcal{P}_{2m+1}} = \frac{1}{\cos\left(\frac{\pi}{2m+1}\right)}.$$

(The maximum distortion $\Lambda(.,0)$ is attained at the vertex V_0, say, and the antipodal of V_0 is $V_0^o = (V_m + V_{m+1})/2$.)

For a regular $(2m + 1)$-Reuleaux polygon, the unique critical point is also the origin (for the same reason as above), and (2.5.2) gives

$$\mathfrak{m}^*_{\mathcal{Q}_{2m+1}} = \frac{1}{d-1}, \quad d = 2\sin\left(\frac{m\pi}{2m+1}\right).$$

According to a result of [Guo–Jin], this Minkowski measure (of a regular $(2m + 1)$-Reuleaux polygon) is *minimal* among the Minkowski measures of *all* $(2m + 1)$-Reuleaux polygons.

Theorem 2.5.5 ([Jin–Guo 1]). *If $C \in \mathfrak{B}$ is a convex body of constant width then we have*

$$1 \le \mathfrak{m}^*_C \le \frac{n + \sqrt{2n(n+1)}}{n+2}. \tag{2.5.6}$$

The lower bound is attained if and only if C is a closed ball. The upper bound is attained if C is a completion of a regular simplex.

Proof. By (2.1.7), the lower bound holds for any $C \in \mathfrak{B}$. If $\mathfrak{m}^*_C = 1$ for a convex body $C \in \mathfrak{B}$ of constant width then, by (2.5.2), we have $R_C = r_C$, so that C is a closed metric ball. Since the converse is obvious, it remains only to treat the upper bound.

Let $C \in \mathfrak{B}$ be of constant width d. Using Jung's upper estimate (1.5.3) with $D_C = d$ along with (2.5.2), we calculate

$$\mathfrak{m}^*_C = \frac{R_C}{r_C} = \frac{R_C}{d - R_C} \le \frac{\sqrt{\frac{n}{2(n+1)}}\,d}{d - \sqrt{\frac{n}{2(n+1)}}\,d}$$

$$= \frac{\sqrt{n}}{\sqrt{2(n+1)} - \sqrt{n}} = \frac{n + \sqrt{2n(n+1)}}{n+2}.$$

The upper bound in (2.5.6) follows.

Finally, let Δ be a regular simplex of diameter (edge length) d and $\Delta^\sharp$ a completion of Δ. Since $\Delta \subset \Delta^\sharp$, we have

$$\sqrt{\frac{n}{2(n+1)}}d = R_\Delta \le R_{\Delta^\sharp} \le \sqrt{\frac{n}{2(n+1)}}d,$$

where, in the last inequality, we used Jung's upper estimate (1.5.3) for the completion $\Delta^\sharp$ with $D_{\Delta^\sharp} = D_\Delta = d$. Thus, equality holds everywhere, and the upper bound in (2.5.6) is attained for $\Delta^\sharp$.

Remark 1. The last part of the proof above shows that we have $R_\Delta = R_{\Delta^\sharp}$ for a regular simplex Δ and any of its completions $\Delta^\sharp$. Therefore the same holds for the respective circumballs; that is, the circumball of a *regular* simplex is also the circumball of any of its completions. (For a different proof of this involving the "Reuleaux simplex" of Δ, see [Jin–Guo 4]. Finally, note that regularity is crucial here. As noted previously, a convex body may have many completions, and the circumballs of the completions may be different.)

Remark 2. The material presented here is given in [Jin–Guo 1]. In a subsequent paper [Jin–Guo 3] observed that the upper bound in (2.5.6) is attained *precisely* on completions of regular simplices. Indeed, this follows directly from a theorem of [Melzak] asserting that a convex body $C \in \mathfrak{B}$ of constant width d for which equality holds in Jung's upper estimate (1.5.3) must contain a regular simplex of diameter d. (This has already been noted in Remark 2 after Theorem 1.5.1. The special case $n = 2$ is also treated in [Jin–Guo 2].)

For a classical account of convex sets of constant width, see [Bonnesen-Fenchel, pp. 135–147]. For a modern treatment, see [Chakerian–Groemer], and the subsequent additions by E. Heil and H. Martini in [Gruber–Wills, pp. 363–368].

Example 2.5.2 (continued). By construction, the Meissner tetrahedron $\mathcal{M}$ is a completion of the regular tetrahedron Δ. By Theorem 2.5.5, we have

$$\mathrm{m}_\mathcal{M}^* = \frac{3 + 2\sqrt{6}}{5}.$$

(This can also be obtained by elementary geometrical reasoning using the threefold symmetry of $\mathcal{M}$ about V_0.)

Note also that the unique critical point of $\mathcal{M}$ is the centroid of Δ, and, by (2.5.1), we also have

$$R_\mathcal{M} = \frac{\sqrt{6}}{4}d \quad \text{and} \quad r_\mathcal{M} = \left(1 - \frac{\sqrt{6}}{4}\right)d.$$

Exercises and Further Problems

1.* Let $C \in \mathfrak{B}$ and $O \in \text{int}\,C$. If C is symmetric with respect to O then, clearly, every chord in C through O is an affine diameter of C. Prove the converse: If every chord in C through O is an affine diameter of C then C is symmetric with respect to O.

2.* Let $C \in \mathfrak{B}$ and $[C, C'] \subset C$ an affine diameter. Show that C has another affine diameter through the midpoint $O = (C + C')/2$.

3.* Proceed along the following argument to obtain a proof of the Birkhoff–von Neumann theorem due to [Hurlbert]. (1) Use Theorem 1.2.6 (along with transitivity of the action of the symmetric group $\mathcal{S}_n$ on the set of permutation matrices), to realize that it is enough to show that any extremal point of $\mathcal{B}_n$ is a permutation matrix. Let $A \in \mathcal{B}_n$ be a doubly stochastic matrix and assume that A is *not a permutation matrix*. Prove that A is not an extremal point of $\mathcal{B}_n$ as follows. (2) Letting $A = (a_{ij})_{i,j=1}^n$, there exists an entry $a_{i_1 j_1} \in (0, 1)$ for some $i_1, j_1 = 1, \ldots, n$. Since the sum of the entries in each row is equal to 1, there exists another entry $a_{i_1 j_2} \in (0, 1)$, $j_2 = 1, \ldots, n$, $j_1 \neq j_2$. Since the sum of the entries in each column is equal to 1, there exists another entry $a_{i_2 j_2} \in (0, 1)$, $i_2 = 1, \ldots, n$, $i_1 \neq i_2$. Continuing this way we obtain a sequence of entries $a_{i_1 j_1}, a_{i_1 j_2}, a_{i_2 j_2}, a_{i_2 j_3}, a_{i_3 j_3}, \ldots$ whose *consecutive* members are different. Since there are finitely many entries, this sequence repeats itself. A sequence of entries with the first and last entry being the same is called a *cycle*. A cycle is called *minimal* if it contains the least amount of elements. Clearly, in a minimal cycle all members are distinct except the first and the last. Show that a minimal cycle must contain an *even number of distinct members*. (3) Let $a_{i_1 j_1}, a_{i_1 j_2}, a_{i_2 j_2}, \ldots, a_{i_m j_m}$ be a minimal cycle ($i_1 = i_m$ and $j_1 = j_m$), and denote by d_0 the distance of this cycle as a subset in $(0, 1)$ from the boundary $\{0, 1\}$. Let $0 < \epsilon < d_0$, and define $A^+(\epsilon)$ as A with the entries participating in the cycle replaced according to the rule $a_{i_k j_k} \mapsto a_{i_k j_k} + \epsilon$ and $a_{i_k j_{k+1}} \mapsto a_{i_k j_{k+1}} - \epsilon$, $k = 1, \ldots, m-1$. Similarly, define $A^-(\epsilon)$ as A with the entries participating in the cycle replaced according to the rule $a_{i_k j_k} \mapsto a_{i_k j_k} - \epsilon$ and $a_{i_k j_{k+1}} \mapsto a_{i_k j_{k+1}} + \epsilon$, $k = 1, \ldots, m-1$. Show that $A^\pm(\epsilon) \in \mathcal{B}_n$ and $A = (A^+(\epsilon) + A^-(\epsilon))/2$. Conclude that A is not an extremal point of $\mathcal{B}_n$.

4. Let $C \in \mathfrak{B}$ be a *planar* convex body. This problem addresses the question of non-unicity of line extensions of affine diameters passing through a given point in the exterior of C (Corollary 2.2.7). For simplicity, we call these lines *extended affine diameters*. All the results here are due to [Hammer 2]. (a) Given a point $C \in \mathcal{X} \setminus C$, if two extended affine diameters pass through C then ∂C has a pair of parallel line segments which contain the respective endpoints of the two affine diameters. (b) If C has two parallel extended affine diameters then ∂C has a pair of parallel line segments which contain the respective endpoints of the two affine diameters. (c) Fix a metric ball $B_r(O)$ that contains C and for each $C \in \mathcal{S}_r(O)$ consider the family of extended affine

diameters that pass through C. Use (a) to show that this family either consists of a unique extended affine diameter or it is between two extremal extended affine diameters. In each case we select a unique extended affine diameter and call it *essential* as follows. In the first case we choose the unique extended affine diameter, and in the second, we choose one of the extremal extended affine diameters consistently with respect to a fixed orientation. Let $\mathfrak{D}$ denote the family of essential affine diameters obtained this way.

Show that $\mathfrak{D}$ has the following properties: (1) through every point $C \in \mathcal{X} \setminus C$ there passes the line extension of a unique essential diameter; (2) given a unit vector $N \in \mathcal{S}$, there is a unique element in $\mathfrak{D}$ parallel to $\mathbb{R} \cdot N$; (3) every pair of essential diameters intersect in a unique point in C; and (4) through every point in C there passes at least one essential diameter.

5. In the setting of Problem 4, show that the essential diameters of C_r, $r^* < r \leq 1$, in Hammer's decomposition are extensions of the essential diameters of C by the ratio $2r - 1$ about their midpoints.

6.* Let $\mathcal{X} = \mathcal{X}_0 \times \mathbb{R}$ and $C = [C_0, V] \subset \mathcal{X}$ a convex cone, a convex body in $\mathcal{X}$ with base $C_0 \subset \mathcal{X}_0$ and vertex $V \in \mathcal{X} \setminus \mathcal{X}_0$. Show that $\mathfrak{m}_C^* = \mathfrak{m}_{C_0}^* + 1$.

7. For $C \in \mathfrak{B}$, define the *width function* $w_C : \mathcal{X} \to \mathbb{R}$ by $w_C(X) = h_C(X) + h_C(-X)$, $X \in \mathcal{X}$, where $h_C : \mathcal{X} \to \mathbb{R}$ is the support function of C (Section 1.2). Show that C is of constant width d if and only if the width function is constant on $\mathcal{S}$: $w_C|_\mathcal{S} = d$. Use the properties of the support function to prove that C is of constant width d if and only if the Minkowski symmetral $\tilde{C} = \bar{B}_{d/2}$. Finally, given a convex body C of constant width d, show that, for $r > 0$, the closed r-neighborhood $\bar{C}_r = C + \bar{B}_r$ (Section 1.1) is of constant width $d + 2r$.

8.* Show that the number of vertices of a Reuleaux polygon must be odd.

9. The construction of a double normal with minimal width in Section 1.5 shows that if C is of constant width then every *normal* (a chord of C with the orthogonal hyperplane at one of its endpoints supporting C) is a double normal, and every double normal is a metric diameter. In particular, any point of C is contained in a double normal, and C has a double normal parallel to any given line.

Prove the converse statements as follows. Given $C \in \mathfrak{B}$, the following properties are equivalent: (1) C is of constant width; (2) every double normal of C is a metric diameter; (3) any point in C is contained in a double normal; and (4) C has a double normal parallel to any given line.

10. Show that the distance between the midpoints of two antipodal edges of the Reuleaux tetrahedron is $(\sqrt{6} - 1)/\sqrt{2} \cdot d > d$.

11. Show that the Meissner tetrahedron $\mathcal{M}$ constructed in Example 2.5.2 has constant width d. (Note the threefold symmetry about the vertex V_0.)

12.* Let $C \in \mathfrak{B}$. Show that if C is of constant width d then C has the spherical intersection property: $C = \bigcap_{X \in C} \bar{B}_d(X)$.

13.* Let $\Delta = [V_0, \dots, V_n] \subset \mathcal{X}$ be a regular simplex with circumball $\bar{B}_R(O)$, where $R = R_\Delta = \sqrt{n/(2(n+1))}\, d$ and $d = D_\Delta$ is the diameter (edge length) of Δ. Construct the "Reuleaux simplex" $C = \bigcap_{i=0}^n \bar{B}_d(V_i)$. Show that $C \subset \bar{B}_R(O)$.

14.* Let $C \in \mathfrak{B}$ be of diameter d and with circumball $\bar{B}_R(O)$ and circumradius $R = R_C(< d)$. Use the following steps to derive a theorem of [Vrećica]: there exists a completion $C^\sharp$ of C such that $C \subset C^\sharp \subset \bar{B}_R(O)$. (1) Consider the family

$$\mathfrak{F} = \{C' \in \mathfrak{B} \mid C \subset C' \subset \bar{B}_R(O),\ D_{C'} = d\},$$

partially ordered by the inclusion relation. (2) Show that every chain in $\mathfrak{F}$ has an upper bound so that the Zorn lemma applies. (3) Let $C^\sharp \in \mathfrak{F}$ be a maximal element. Show that $C^\sharp$ is complete.

15. Let $C \in \mathfrak{B}$ and $C^\sharp$ the extension constructed in Problem 14. Use $r_{C^\sharp} + R_{C^\sharp} = D_{C^\sharp}$ (Theorem 2.5.4) to show that in general, we have $r_C + R_C \leq D_C$.

Chapter 3
Measures of Symmetry and Stability

3.1 The Minkowski Measure and Duality

In this section we give another interpretation of the Minkowski measure of a convex body (Section 2.1) in terms of supporting hyperplanes. We show that the associated "support ratio" corresponds to our earlier distortion ratio through the concept of duality (or polarity) between convex bodies. We introduce and discuss duality by means of "musical correspondences" which prove to be transparent and convenient technical tools to derive various simple facts such as the so-called Bipolar Theorem.

Let $\mathcal{X}$ be a Euclidean space, and $\mathcal{C} \in \mathfrak{B}$, a convex body. Recall from Section 2.1 the distortion ratio $\Lambda_{\mathcal{C}} : \partial\mathcal{C} \times \mathrm{int}\,\mathcal{C} \to \mathbb{R}$ defined by

$$\Lambda_{\mathcal{C}}(C, O) = \frac{d(C, O)}{d(C^o, O)}, \quad C \in \partial\mathcal{C}, \ O \in \mathrm{int}\,\mathcal{C},$$

where $C^o \in \partial\mathcal{C}$ is the antipodal of C with respect to O. (In this section it will be important to keep track of the dependence of various objects on the convex body by subscript.)

For the *involution* of $\partial\mathcal{C}$ given by $C \mapsto C^o$ (with $(C^o)^o = C$), we have $\Lambda_{\mathcal{C}}(C^o, O) = 1/\Lambda_{\mathcal{C}}(C, O)$, $C \in \partial\mathcal{C}$.

The maximum distortion is the function $\mathfrak{m}_{\mathcal{C}} : \mathrm{int}\,\mathcal{C} \to \mathbb{R}$ given by

$$\mathfrak{m}_{\mathcal{C}}(O) = \max_{C \in \partial\mathcal{C}} \Lambda_{\mathcal{C}}(C, O) \geq 1, \quad O \in \mathrm{int}\,\mathcal{C}.$$

Another notion of ratio can be introduced using supporting hyperplanes as follows. Let $\delta\mathcal{C}$ denote the set of all hyperplanes supporting $\mathcal{C}$. Associating to each $\mathcal{H} \in \delta\mathcal{C}$ the unit normal vector that points outward $\mathcal{C}$ establishes a one-to-one correspondence between $\delta\mathcal{C}$ and the unit sphere $\mathcal{S}$ in $\mathcal{X}$. This correspondence endows $\delta\mathcal{C}$ a natural (compact) topology homeomorphic with the unit sphere $\mathcal{S}$.

© Springer International Publishing Switzerland 2015
G. Toth, *Measures of Symmetry for Convex Sets and Stability*,
Universitext, DOI 10.1007/978-3-319-23733-6_3

For $\mathcal{H} \in \delta C$, there is a unique $\mathcal{H}^o \in \delta C$ such that $\mathcal{H}$ and $\mathcal{H}^o$ are parallel; we call $\mathcal{H}^o$ the *antipodal* of $\mathcal{H}$ (Section 2.1).

We define the *support ratio* $\mathcal{R}_C : \delta C \times \operatorname{int} C \to \mathbb{R}$ by

$$\mathcal{R}_C(\mathcal{H}, O) = \frac{d(\mathcal{H}, O)}{d(\mathcal{H}^o, O)}, \quad \mathcal{H} \in \delta C, \ O \in \operatorname{int} C.$$

(See Figure 3.1.1.)

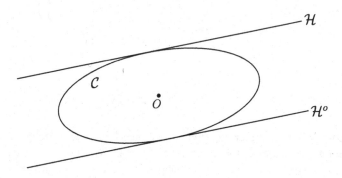

Fig. 3.1.1

For the *involution* of δC given by $\mathcal{H} \mapsto \mathcal{H}^o$ (with $(\mathcal{H}^o)^o = \mathcal{H}$), we have $\mathcal{R}_C(\mathcal{H}^o, O) = 1/\mathcal{R}_C(\mathcal{H}, O), \mathcal{H} \in \delta C$.

In analogy with the maximum distortion (ratio), we can introduce the maximum support ratio. The next proposition shows that these two concepts are the same.

Proposition 3.1.1. *We have*

$$m_C(O) = \sup_{\mathcal{H} \in \delta C} \mathcal{R}_C(\mathcal{H}, O), \ O \in \operatorname{int} C, \tag{3.1.1}$$

where the supremum is attained.

Proof. Let $O \in \operatorname{int} C$. Choose a point $C_0 \in \partial C$ at which the distortion $\Lambda(., O)$ attains its maximum.

By Corollary 2.2.2, given any hyperplane $\mathcal{H}_0$ supporting C at the antipodal C_0^o, the parallel hyperplane $\mathcal{H}_0' = \mathcal{H}_0 + C_0 - C_0^o$ supports C at C_0, and $[C_0, C_0^o]$ is an affine diameter. (In particular, $\mathcal{H}_0' = \mathcal{H}_0^o$ is the antipodal of $\mathcal{H}_0$.)

We thus have

$$m_C(O) = \max_{C \in \partial C} \Lambda_C(C, O) = \frac{d(C_0, O)}{d(C_0^o, O)} = \mathcal{R}_C(\mathcal{H}_0', O) \leq \sup_{\mathcal{H} \in \delta C} \mathcal{R}_C(\mathcal{H}, O).$$

(See Figure 3.1.2.)

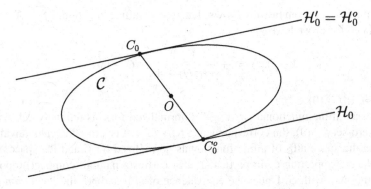

Fig. 3.1.2

For the opposite inequality in (3.1.1), let $\mathcal{H}, \mathcal{H}^o \in \delta\mathcal{C}$ be a parallel pair of hyperplanes supporting $\mathcal{C}$. Let $C \in \partial\mathcal{C} \cap \mathcal{H}$ and $C' \in \mathcal{H}^o$ the corresponding point with $O \in [C, C']$. Since $\mathcal{H}$ and $\mathcal{H}'$ are both supporting, $\mathcal{C}$ lies between these parallel hyperplanes. It follows that the antipodal $C^o \in \partial\mathcal{C}$ of C is contained in the line segment $[O, C']$. (See Figure 3.1.3.)

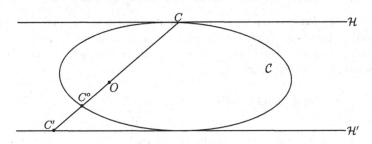

Fig. 3.1.3

We have

$$\mathcal{R}_{\mathcal{C}}(\mathcal{H}, O) = \frac{d(C, O)}{d(C', O)} \leq \frac{d(C, O)}{d(C^o, O)} = \Lambda_{\mathcal{C}}(C, O) \leq \mathfrak{m}_{\mathcal{C}}(O).$$

Taking the supremum in $\mathcal{H} \in \delta\mathcal{C}$, the opposite inequality in (3.1.1) follows.

A technically more convenient reformulation of the support ratio is by normalized affine functionals introduced in Section 1.2. Recall that an affine functional $f : \mathcal{X} \to \mathbb{R}$ is normalized for $\mathcal{C} \in \mathfrak{B}$ if $f(\mathcal{C}) = [0, 1]$. In this case, $f^{-1}(0)$ and $f^{-1}(1)$ are parallel hyperplanes both supporting $\mathcal{C}$.

A simple ratio comparison shows that, for an affine functional $f : \mathcal{X} \to \mathbb{R}$ normalized for C, we have

$$1 - f(O) = \frac{1}{\mathcal{R}_C(\mathcal{H}, O) + 1}, \quad O \in \text{int}\, C,$$

where $\mathcal{H} = f^{-1}(0) \in \delta C$.

Since an affine functional $f : \mathcal{X} \to \mathbb{R}$ normalized for C is uniquely determined by its zero-set $f^{-1}(0)$, the correspondence $f \mapsto f^{-1}(0)$ is a topological equivalence between the space aff_C of affine functionals normalized for C and the space δC of hyperplanes supporting C. In particular, aff_C carries a natural (compact) topology homeomorphic with S. Under the equivalence of aff_C and δC, the *involution* $f \mapsto 1 - f$ in aff_C corresponds to the involution $\mathcal{H} \mapsto \mathcal{H}^o$ in δC.

Taking infima in the equation above, Proposition 3.1.1 gives

$$\inf_{f \in \text{aff}_C} f(O) = \inf_{f \in \text{aff}_C} (1 - f(O)) = \frac{1}{\sup_{\mathcal{H} \in \delta C} \mathcal{R}_C(\mathcal{H}, O) + 1} = \frac{1}{\mathsf{m}_C(O) + 1}, \quad O \in \text{int}\, C. \tag{3.1.2}$$

Remark. Since the infimum of any set of affine functionals is concave, (3.1.2) implies that the function $1/(\mathsf{m}_C + 1)$ is concave on $\text{int}\, C$. This was established in Theorem 2.1.2 by different means. Note that Corollary 2.1.3 also follows.

The second interpretation of the maximum distortion $\mathsf{m}_C : \text{int}\, C \to \mathbb{R}$ (as a maximum support ratio) can be naturally interpreted in terms of the *dual* of C.

To define the dual, we first let $C_0 \in \mathfrak{B}$ be a convex body with $0 \in \text{int}\, C_0$. The *dual* of C_0 with respect to 0 is defined by

$$C_0^0 = \{X \in \mathcal{X} \mid h_C(X) \le 1\} = \{X \in \mathcal{X} \mid \sup_{C \in C} \langle C, X \rangle \le 1\}, \tag{3.1.3}$$

where $h_C : \mathcal{X} \to \mathbb{R}$ is the support function of C (Section 1.2).

Clearly, $C_0^0 \in \mathfrak{B}$ and $0 \in \text{int}\, C_0^0$.

The general case is reduced to this by using the group of translations $\mathcal{T}(\mathcal{X}) \subset \text{Aff}(\mathcal{X})$ in the affine group of $\mathcal{X}$. A typical translation is denoted by $T_V : \mathcal{X} \to \mathcal{X}$, $V \in \mathcal{X}$, where $T_V(X) = X + V, X \in \mathcal{X}$.

We define the *dual* of $C \in \mathfrak{B}$ with respect to an interior point $O \in \text{int}\, C$ by

$$C^O = T_O(C_0^0) = (C - O)^0 + O, \quad C_0 = T_{-O}(C) = C - O.$$

Clearly, $C^O \in \mathfrak{B}$ and $O \in \text{int}\, C^O$.

Let $C, C' \in \mathfrak{B}$ with $O \in \text{int}\, C$. The dual satisfies the following properties:

(1) $(\lambda C)^O = (1/\lambda) C^O, \lambda > 0$;
(2) $C \subset C'$ implies $C'^O \subset C^O$;
(3) $(C^O)^O = C$;
(4) $(C \cap C')^O = [C^O \cup C'^O], O \in \text{int}\,(C \cap C')$, and $[C \cup C']^O = C^O \cap C'^O, O \in \text{int}\,(C^O \cap C'^O)$.

The first two properties are obvious. The third property can be derived directly (as in [Eggleston 1, 1.9]), or it will follow as the byproduct of the musical equivalencies to be discussed shortly. The fourth property follows from (2) (applied to $\mathcal{C} \cap \mathcal{C}' \subset \mathcal{C}, \mathcal{C}' \subset \mathcal{C} \cup \mathcal{C}'$) and (3).

Remark. The dual is often called *polar*, or even *polar dual*. For this reason, the third property is sometimes called the "Bipolar Theorem."

Example 3.1.2. A general ellipsoid with center at the origin 0 can be defined as

$$\mathcal{E} = \{X \in \mathcal{X} \mid \langle QX, X \rangle \leq 1\},$$

where $Q : \mathcal{X} \to \mathcal{X}$ is a *symmetric positive definite* linear endomorphism of $\mathcal{X}$.

Since the defining inequality $\langle QX, X \rangle \leq 1$ is equivalent to $|Q^{1/2}X|^2 \leq 1$, we see that the dual ellipsoid (with respect to the center, the origin) is given by

$$\mathcal{E}^0 = \{X \in \mathcal{X} \mid \langle Q^{-1}X, X \rangle \leq 1\}.$$

Let $\mathcal{C} \in \mathfrak{B}$ and $O \in \operatorname{int} \mathcal{C}$. As a technical tool to study duality, we now introduce the *musical equivalencies*

$$\flat = \flat_{\mathcal{C},O} : \partial\mathcal{C} \to \operatorname{aff}_{\mathcal{C}^O} \quad \text{and} \quad \sharp = \sharp_{\mathcal{C},O} : \operatorname{aff}_{\mathcal{C}} \to \partial\mathcal{C}^O,$$

which are inverses of each other in the following sense:

$$\sharp_{\mathcal{C}^O,O} \circ \flat_{\mathcal{C},O} = \operatorname{id}_{\partial\mathcal{C}} \quad \text{and} \quad \flat_{\mathcal{C}^O,O} \circ \sharp_{\mathcal{C},O} = \operatorname{id}_{\operatorname{aff}_{\mathcal{C}}}. \tag{3.1.4}$$

Remark. The first equality in (3.1.4) implies $\partial\mathcal{C} = \partial(\mathcal{C}^O)^O, \mathcal{C} \in \mathfrak{B}$, so that the third property of the dual above follows.

In addition, we will show that the musical equivalencies are compatible with the involutions of the respective spaces:

$$(\mathcal{C}^O)^\flat = 1 - \mathcal{C}^\flat \quad \text{and} \quad (f^\sharp)^O = (1 - f)^\sharp, \quad \mathcal{C} \in \partial\mathcal{C}, f \in \operatorname{aff}_{\mathcal{C}}, \tag{3.1.5}$$

and that they satisfy the following equations

$$\mathcal{C}^\flat(O) = \frac{1}{\Lambda_{\mathcal{C}}(\mathcal{C}, O) + 1} \quad \text{and} \quad f(O) = \frac{1}{\Lambda_{\mathcal{C}^O}(f^\sharp, O) + 1}, \quad \mathcal{C} \in \partial\mathcal{C}, f \in \operatorname{aff}_{\mathcal{C}}, O \in \operatorname{int}\mathcal{C}. \tag{3.1.6}$$

(Whenever convenient, the subscripts will be suppressed.)

To define the musical equivalencies with these properties, as in the case of the definition of the dual, we reduce the general case to the case when the base point is the origin.

First, the translation group $\mathcal{T}(\mathcal{X})$ of $\mathcal{X}$ acts on the space $\operatorname{aff}(\mathcal{X})$ of affine functionals via $T_V^o(f) = f \circ T_{-V}, V \in \mathcal{X}, f \in \operatorname{aff}(\mathcal{X})$. Using our previous notations for the dual, for $O \in \operatorname{int} C$, the linear map T_O^o restricts to $T_O^o : \operatorname{aff}_{C_0} \to \operatorname{aff}_C$ between the spaces of affine functionals normalized for C_0 and C. Since $C_0^0 = T_{-O}(C^O)$, we also have the restriction $T_O^o : \operatorname{aff}_{C_0^0} \to \operatorname{aff}_{C^o}$ between the spaces of affine functionals for the duals C_0^0 and C^O.

We now let

$$\flat_{C,o} = T_O^o \circ \flat_{C_0,0} \circ T_{-O} \quad \text{and} \quad \sharp_{C,o} = T_O \circ \sharp_{C_0,0} \circ T_{-O}^o.$$

This reduces the general case to C_0 with $0 \in \operatorname{int} C_0$. For brevity, we now suppress the subscript 0, and set $C = C_0$ with $0 \in \operatorname{int} C$.

For $C \in \partial C$, we let $C^\flat : \mathcal{X} \to \mathbb{R}$ be the affine functional

$$C^\flat(X) = \frac{1}{\Lambda_C(C,0) + 1}(1 - \langle C, X \rangle), \quad X \in \mathcal{X}. \tag{3.1.7}$$

Evaluating this at the origin 0, the first equality in (3.1.6) follows.

The antipodal of $C \in \partial C$ with respect to the origin 0 is $C^o = -C/\Lambda_C(C,0)$. Replacing C by C^o in the definition (3.1.7), a simple computation gives the first formula in (3.1.5).

Finally, in this sequence, the definition of the dual in (3.1.3) shows that $C^\flat$ is normalized for the dual C^0. We conclude that the musical map $\flat = \flat_{C,0} : \partial C \to \operatorname{aff}_{C^0}$ is well-defined and satisfies its respective properties.

Let $f \in \operatorname{aff}_C$. Since f is a non-constant affine functional, we have

$$f(X) = \langle A, X \rangle + a, \ X \in \mathcal{X}, \quad 0 \neq A \in \mathcal{X}, \ a \in \mathbb{R}.$$

Since f is normalized (and $0 \in \operatorname{int} C$), we have $0 < a < 1$. We define

$$f^\sharp = -\frac{A}{a}. \tag{3.1.8}$$

Once again, since f is normalized, we have $f(C) = [0, 1]$, so that

$$0 \leq \langle A, C \rangle + a \leq 1, \quad \text{for all } C \in C$$

(with both equalities attained).

The lower bound here and the definition of the dual in (3.1.3) now shows that $-A/a \in \partial C^0$. Thus, the musical map $\sharp = \sharp_{C,0} : \operatorname{aff}_C \to \partial C^0$ is well-defined.

The upper bound, along with $1 - f$ in place of f in (3.1.8), now gives

$$(1 - f)^\sharp = \frac{A}{1 - a} = (f^\sharp)^o.$$

The second equality in (1.3.5) follows.

Since $-A/a$ and $A/(1 - a)$ are antipodals in $\mathcal{C}^0$, we have $\Lambda(f^\sharp, 0) = 1/a - 1$. Thus, $f(0) = a$, and the second equality in (1.3.6) also follows.

Finally, it remains to show (3.1.4). Combining (3.1.7) and (3.1.8), we obviously have $(\mathcal{C}^\flat)^\sharp = C, C \in \partial\mathcal{C}$. The first equality in (3.1.4) follows. For the second, as before, we let $f(X) = \langle A, X \rangle + a, X \in \mathcal{X}$. Using the second equality in (3.1.6) along with (3.1.7) and (3.1.8), we have $(f^\sharp)^\flat(X) = a(1 + \langle A/a, X \rangle) = f(X)$. The second equality in (3.1.4) also follows. This finishes the construction of the musical equivalencies.

Taking the infima in the respective sets in the two equalities in (3.1.6), for $O \in \operatorname{int}\mathcal{C}$, we obtain

$$\inf_{C \in \partial\mathcal{C}} C^\flat(O) = \frac{1}{\mathfrak{m}_\mathcal{C}(O) + 1} \quad \text{and} \quad \inf_{f \in \operatorname{aff}\mathcal{C}} f(O) = \frac{1}{\mathfrak{m}_{\mathcal{C}^0}(O) + 1}, \quad O \in \operatorname{int}\mathcal{C}.$$

This shows that the maximum distortion and the maximum support ratio are *dual constructions*. In addition, the corresponding two measures are equal by (3.1.2), so that, as a byproduct, we obtain

$$\mathfrak{m}_\mathcal{C}(O) = \mathfrak{m}_{\mathcal{C}^0}(O), \quad O \in \operatorname{int}\mathcal{C}. \tag{3.1.9}$$

3.2 Stability of the Minkowski Measure and the Banach–Mazur Distance

The Minkowski measure studied in the previous section serves as a prominent example and primary motivation to introduce the concept of stability for geometric inequalities. We choose here a gradual development by first deriving Groemer's stability estimate for the natural *lower bound* for the Minkowski measure in terms of the Hausdorff distance.

Seeking a notion of distance better fit to stability naturally leads to the affine invariant Banach–Mazur metric. A large central part of this section is devoted to the discussion of this concept along with the Banach–Mazur compactum and John's original approach to prove boundedness of this space.

Next, to complete the circle, we return to our original purpose, and derive a simple stability estimate for the lower bound of the Minkowski measure in terms of the Banach–Mazur metric. We do this by relating the (global) maxima of the distortion ratio discussed in Section 2.2 to the Banach–Mazur distance of a convex body from its Minkowski symmetral.

Finally, to complete yet another circle, we derive Schneider's stability estimate for the Minkowski–Radon *upper bound* of the Minkowski measure once again in terms of the Banach–Mazur metric.

Recall from the previous section (Proposition 3.1.1) that, for the *Minkowski measure*, we have

$$\mathfrak{m}_C^* = \inf_{O\in \mathrm{int}\,C} \mathfrak{m}_C(O) = \inf_{O\in \mathrm{int}\,C} \max_{C\in\partial C} \Lambda_C(C,O) = \inf_{O\in\mathrm{int}C} \max_{\mathcal{H}\in\delta C} \mathcal{R}_C(\mathcal{H},O), \ \ C\in\mathfrak{B}.$$

$$(3.2.1)$$

Based on the last equality, $\mathfrak{m}_C^*$ can conveniently be expressed by the support function h_C of C as follows. Recall that $h_C(N)$ is the signed distance of the hyperplane supporting C (with outward normal vector $N \in S$) *from the origin*. It is positive if and only if N points into the respective open half-space *disjoint* from the origin.

Now, given $O \in \mathrm{int}\,C$, *measuring distances from O* in the definition of h_C (or moving O to the origin by a suitable translation of C), we thus have

$$\mathcal{R}_C(\mathcal{H},O) = \frac{h_C(N)}{h_C(-N)} \geq 1, \quad O \in \mathrm{int}\,C,$$

with *unique* choice of the normal vector $N \in S$ to $\mathcal{H}$ (except when the ratio is 1). Using (3.2.1), we thus have

$$\mathfrak{m}_C(O) = \max_{N\in S} \frac{h_C(N)}{h_C(-N)}, \tag{3.2.2}$$

where h_C is once again with respect to $O \in \mathrm{int}\,C$. Finally, if the infimum in (3.2.2) is attained at a point O^* (that is, O^* is in the critical set C^* of C) then this formula gives the Minkowski measure

$$\mathfrak{m}_C^* = \max_{N\in S} \frac{h_C(N)}{h_C(-N)}, \tag{3.2.3}$$

where h_C is with respect to O^*.

Remark. In (3.2.3) the dependence on the base point $O \in \mathrm{int}\,C$ can be explicitly incorporated in the support function $h_{C,O} : \mathcal{X} \to \mathbb{R}$ defined by

$$h_{C,O}(X) = \sup_{C\in C}\langle C - O, X\rangle = h_C(X) - \langle O, X\rangle, \ X \in \mathcal{X}.$$

With this (3.2.3) rewrites as

$$\mathfrak{m}_C^* = \inf_{O\in\mathrm{int}\,C} \max_{N\in S} \frac{h_{C,O}(N)}{h_{C,O}(-N)}.$$

Recall from Section 2.1 that the Minkowski measure $\mathfrak{m}^*$ attains its minimum value $\mathfrak{m}^* = 1$ in $\mathfrak{B}$ at symmetric bodies. (See the inequality in (2.1.7) and the subsequent discussion.) A natural question arises: If a convex body $C \in \mathfrak{B}$ has Minkowski measure $\mathfrak{m}_C^*$ close to this minimum value 1, how closely does C approximate a *symmetric* convex body?

The answer is a *stability estimate*. This question can easily be generalized to a much wider setting of geometric inequalities which have a well-characterized class of geometric objects for which equality holds.

We will see many examples of stability estimates in the rest of this book.

Finally, note that, to make the question more precise, one needs to specify the metric on $\mathfrak{B}$ with respect to which the close approximation is understood.

As a simple application of (3.2.3), we have the following stability estimate for the lower bound of the Minkowski measure:

Theorem 3.2.1. *Let* $0 \le \epsilon \le n - 1$. *If* $C \in \mathfrak{B}$ *satisfies*

$$\mathfrak{m}_C^* \le 1 + \epsilon$$

then there exists a symmetric convex body $\tilde{C} \in \mathcal{B}$ *such that*

$$d_H(C, \tilde{C}) \le \frac{D_C}{2} \frac{n}{n+1} \epsilon, \tag{3.2.4}$$

where D_C *is the diameter of* C.

Proof. For simplicity, we may assume that the origin is a critical point. Then, by (3.2.3) and our assumption, we have

$$\frac{h_C(N)}{h_C(-N)} \le 1 + \epsilon, \quad N \in \mathcal{S} \tag{3.2.5}$$

(where h_C is with respect to the origin).

The symmetric body in (3.2.4) is the *Minkowski symmetral* of C defined by

$$\tilde{C} = (C - C)/2 = \{(X - X')/2 \,|\, X, X' \in C\}.$$

(We briefly encountered this in Section 2.5; see, in particular, Problem 7 at the end of Chapter 2.) It follows directly from the definition that $\tilde{C}$ is compact and convex. Moreover, the Minkowski symmetral of an inball $\bar{B}_r(O) \subset C$ (with inradius $r = r_C$) is $\bar{B}_r \subset C$ (with center at the origin). We conclude that $\tilde{C} \in \mathfrak{B}$ is a convex body.

Using the algebraic properties of the support function (Section 1.2), we have

$$\left| h_C(N) - h_{\tilde{C}}(N) \right| = \left| h_C(N) - \frac{1}{2}(h_C(N) + h_C(-N)) \right|$$

$$= \frac{1}{2} |h_C(-N)| \left| \frac{h_C(N)}{h_C(-N)} - 1 \right|.$$

Taking the supremum in $N \in \mathcal{S}$, by (1.2.2), we have

$$d_H(C, \tilde{C}) = \sup_{N \in \mathcal{S}} \left| h_C(N) - h_{\tilde{C}}(N) \right| \le \frac{1}{2} \sup_{N \in \mathcal{S}} h_C(N) \cdot \epsilon, \tag{3.2.6}$$

where we used (3.2.5) (for $\pm N \in \mathcal{S}$).

To estimate the supremum of h_C on the unit sphere S, we first make use of (1.5.2) along with the subsequent Remark 1:

$$\sup_{N \in S} (h_C(N) + h_C(-N)) = D_C.$$

Using (3.2.5) in the supremum once again, we obtain

$$h_C(N) \leq \frac{1+\epsilon}{2+\epsilon}(h_C(N) + h_C(-N)) \leq \frac{1+\epsilon}{2+\epsilon}D_C \leq \frac{n}{n+1}D_C, \quad N \in S, \quad (3.2.7)$$

where the last inequality is due to our restriction $0 \leq \epsilon \leq n - 1$. Substituting this into (3.2.6), we arrive at (3.2.4).

Remark. By the Minkowski–Radon upper bound in (2.1.7), we have $\mathfrak{m}_C^* \leq n$. This corresponds to $\epsilon = n - 1$. Therefore Theorem 3.2.1 can be concisely paraphrased by the single inequality

$$d_H(C, \tilde{C}) \leq \frac{D_C}{2}\frac{n}{n+1}(\mathfrak{m}_C^* - 1), \quad C \in \mathfrak{B}.$$

Theorem 3.2.1 (and its proof) is essentially due to [Groemer 1] with the *centroid* of C as the base point. (See also the discussion in the next section.) Note also that by narrowing down the range of ϵ, the first upper bound in (3.2.7) gives much sharper estimate on the Hausdorff distance; for example, for $0 \leq \epsilon \leq 1$, we obtain $d_H(C, \tilde{C}) \leq (D_C/3)\epsilon$.

The use of the Hausdorff metric in a stability estimate as above is exceptional. A suitable concept of distance should be defined on the quotient of $\mathfrak{B}_{\mathcal{X}}$ by the group Aff$(\mathcal{X})$ of affine transformations of $\mathcal{X}$ (with respect to the natural action of Aff$(\mathcal{X})$ on $\mathfrak{B}_{\mathcal{X}}$). The (initial) difficulty in using the Hausdorff metric stems from the fact that the topology on the natural domain $\mathfrak{C}_{\mathcal{X}}$ of compact subsets of $\mathcal{X}$ (Section 1.1) with the Hausdorff metric induces a *non-metrizable* topology on the quotient $\mathfrak{C}_{\mathcal{X}}$/Aff$(\mathcal{X})$. (See Problem 4.) As shown by [Webster], however, the Hausdorff metric can be used to define a *metric* on this quotient, and in this new topology $\mathfrak{C}_{\mathcal{X}}$/Aff$(\mathcal{X})$ is compact. The reason that this Webster metric topology is different from the natural (non-metrizable) quotient topology is the varying dimensions of the convex hulls of the elements in $\mathfrak{C}_{\mathcal{X}}$. Restricting to $\mathfrak{B}_{\mathcal{X}} \subset \mathfrak{C}_{\mathcal{X}}$, these difficulties disappear and on the quotient $\mathfrak{B}_{\mathcal{X}}$/Aff$(\mathcal{X})$ the two topologies coincide. Moreover, as shown by [Macbeath] this topology on the quotient is *compact*.

Even though it is constructed from the Hausdorff distance, the complexity of the Webster metric on the quotient $\mathfrak{B}_{\mathcal{X}}$/Aff$(\mathcal{X})$ makes it difficult to use. Instead, we now introduce another, very transparent concept of distance.

The (extended) *Banach–Mazur distance* function $d_{BM} : \mathfrak{B} \times \mathfrak{B} \to \mathbb{R}$ is defined, for $C, C' \in \mathfrak{B}$ as

$$d_{BM}(C, C') = \inf\{\lambda \geq 1 \,|\, C' \subset \phi(C) \subset \lambda C' + X \text{ with } \phi \in \text{Aff}(\mathcal{X}) \text{ and } X \in \mathcal{X}\}.$$
$$(3.2.8)$$

The infimum is clearly attained. d_{BM} satisfies the following properties:

(1) $d_{BM}(C, C') = 1$ for $C, C' \in \mathfrak{B}$ if and only if $C' = \phi(C)$ for some $\phi \in \text{Aff}(\mathcal{X})$;
(2) $d_{BM}(C, C') = d_{BM}(C', C)$ for any $C, C' \in \mathfrak{B}$;
(3) $d_{BM}(C, C'') \leq d_{BM}(C, C') \cdot d_{BM}(C', C'')$ for any $C, C', C'' \in \mathfrak{B}$;
(4) $d_{BM}(\phi(C), \psi(C')) = d_{BM}(C, C')$ for any $C, C' \in \mathfrak{B}$ and $\phi, \psi \in \text{Aff}(\mathcal{X})$.

Property (1) is a direct consequence of the definition since, for $\phi \in \text{Aff}(\mathcal{X})$, $C, C' \in \mathfrak{B}$, the inclusions $C' \subset \phi(C) \subset C' + X$ imply $X = 0$, so that $C' = \phi(C)$.

Properties (2)–(4) all depend on the fact that, in addition to the group of dilatations $\text{Dil}(\mathcal{X})$ being normal in $\text{Aff}(\mathcal{X})$ (Section 1.1/C), conjugation of a similarity $S_{\lambda, X}$ by any affine transformation does not change the ratio λ. In fact, for $\phi \in \text{Aff}(\mathcal{X})$, we have

$$\phi^{-1} \circ S_{\lambda, C} \circ \phi = S_{\lambda, A^{-1}(C-V)},$$

where $\phi(X) = A \cdot X + V, X \in \mathcal{X}, A \in GL(\mathcal{X}), V \in \mathcal{X}$.

To show (2), for $\lambda \geq 1$, assuming

$$C' \subset \phi(C) \subset \lambda C' + X, \quad \phi \in \text{Aff}(\mathcal{X}), X \in \mathcal{X},$$

we have

$$\phi(C) \subset \lambda C' + X \subset \lambda \phi(C) + X.$$

Applying the inverse of ϕ, we obtain

$$C \subset \psi(C') \subset \lambda C + Y,$$

for some $Y \in \mathcal{X}$, where $\psi \in \text{Aff}(\mathcal{X})$ is the affine transformation defined by $\psi(X') = \phi^{-1}(\lambda X' + X), X' \in \mathcal{X}$. Hence $d_{BM}(C', C) \leq d_{BM}(C, C')$. Switching the roles of C and C', we obtain (2). The proof of (3) is similar. For (4), first note that $d_{BM}(\phi(C), C') = d_{BM}(C, C')$ obviously holds for any $\phi \in \text{Aff}(\mathcal{X})$. Now, using (2), we have

$$d_{BM}(\phi(C), \psi(C')) = d_{BM}(C, \psi(C')) = d_{BM}(\psi(C'), C) = d_{BM}(C', C) = d_{BM}(C, C').$$

Thus, (4) follows.

It follows that $\ln(d_{BM})$ is a distance function on the quotient $\mathfrak{B}_{\mathcal{X}} / \text{Aff}(\mathcal{X})$. The simple "algebraic" character of the Banach–Mazur distance in (3.2.8) makes it very convenient to use in explicit computations. Whenever suitable, we will also use the more geometric definition

$$d_{BM}(C, C') = \inf\{\lambda \geq 1 \mid \phi(C) \subset C' \subset S_{\lambda, X}(\phi(C)) \text{ with } \phi \in \text{Aff}(\mathcal{X}) \text{ and } X \in \text{int } \phi(C)\}, \tag{3.2.9}$$

where $S_{\lambda, X}$ is the similarity with center $X \in \mathcal{X}$ and ratio $\lambda \neq 0$ as defined in (1.1.4).

Remark. The extended Banach–Mazur distance function is sometimes called the *Minkowski distance* or *affine distance*. Originally, d_{BM} was defined only for symmetric convex bodies. Following several authors, we retained the original name.

There are several striking results in connection with the Banach–Mazur metric properties of $\mathfrak{B}_{\mathcal{X}} / \operatorname{Aff}(\mathcal{X})$.

Given $\mathcal{C} \in \mathfrak{B}$, let $\mathcal{E} = \mathcal{E}_{\mathcal{C}}$ be the unique ellipsoid of *maximal* volume contained in $\mathcal{C}$. (For existence and uniqueness, see Problem 5.) In 1948 in a pioneering work [John] proved that

$$\mathcal{E} \subset \mathcal{C} \subset S_{n,O}(\mathcal{E}) = n(\mathcal{E} - O) + O, \tag{3.2.10}$$

where $O = O_{\mathcal{C}}$ is the center (centroid) of $\mathcal{E}$. In addition, for $\mathcal{C}$ symmetric, he also showed that the scaling factor n can be improved to $\sqrt{n}$.

Dually (see the following remark), $\mathcal{C}$ is contained in an ellipsoid $\mathcal{E}' = \mathcal{E}'_{\mathcal{C}}$ of *minimal* volume such that

$$(1/n)(\mathcal{E}' - O') + O' = S_{1/n,O'}(\mathcal{E}') \subset \mathcal{C} \subset \mathcal{E}',$$

where $O' = O'_{\mathcal{C}}$ is the center of $\mathcal{E}'$. As before, for $\mathcal{C}$ symmetric, the scaling factor $1/n$ can be improved to $1/\sqrt{n}$.

Although the terminology varies slightly, $\mathcal{E} = \mathcal{E}_{\mathcal{C}}$ is usually called the *John ellipsoid* of $\mathcal{C}$, and $\mathcal{E}'$ the *Löwner ellipsoid* of $\mathcal{C}$. The latter is to recognize Löwner's unpublished contributions, such as the unicity of the circumscribed minimum volume ellipsoid. (See [Busemann 1, Busemann 2].)

Remark. Let $\mathcal{C} \in \mathfrak{B}$. Let O be the center of the John ellipsoid $\mathcal{E}_{\mathcal{C}}$ of $\mathcal{C}$. Taking duals with respect to O, we see that $\mathcal{E}_{\mathcal{C}} \subset \mathcal{C}$ implies $\mathcal{C}^O \subset (\mathcal{E}_{\mathcal{C}})^O$. By Example 3.1.2, $(\mathcal{E}_{\mathcal{C}})^O$ is an ellipsoid. It is, in fact, the Löwner ellipsoid of the dual $\mathcal{C}^O$, that is, we have $\mathcal{E}'_{\mathcal{C}^O} = (\mathcal{E}_{\mathcal{C}})^O$. (See [Tomczak-Jaegerman].) It follows that the John and Löwner ellipsoid constructions are dual. This property of the centers of the John and Löwner ellipsoids is generalized by [Meyer–Schütt–Werner 1] to the *dual affine invariant points*.

We now make a detour here to discuss the original approach of [John] with some variations. His estimate (3.2.10) is an *application* of his general result on *non-linear optimization with inequality constraints*.

We will treat the Löwner ellipsoid variant when $\mathcal{C} = [V_1, \ldots, V_N] \subset \mathbb{R}^n, N > n$, is a convex polytope. (The general case $\mathcal{C} \in \mathfrak{B}$ follows by approximation.)

First, the Löwner ellipsoid is the solution of the following optimization problem:

$$\min_{Q,O} : \ -\ln(\det Q),$$

subject to: $(V_i - O)^\top Q(V_i - O) \le 1, \ i = 1, \ldots, N, \quad \text{and} \quad Q > 0,$

with variables $Q(> 0)$, a symmetric (positive definite) $n \times n$-matrix, and $O \in \mathbb{R}^n$. If Q', O' solves this problem then $\mathcal{E}' = \{X \in \mathbb{R}^n \mid (X - O')^\top Q'(X - O') \leq 1\}$ is the Löwner ellipsoid of $\mathcal{C}$ with center O'.

The hard part of John's proof is to show that for this optimization problem the Karush–Kuhn–Tucker (KKT) conditions are not only necessary but also sufficient for the solution.

They are as follows:

$$-Q'^{-1} + \sum_{i=1}^{N} \lambda_i(V_i - O')(V_i - O')^\top = 0, \quad Q' > 0,$$

$$\sum_{i=1}^{N} \lambda_i Q'(V_i - O') = 0, \quad \lambda_i \geq 0, \quad i = 1, \ldots, N,$$

$$(V_i - O')^\top Q'(V_i - O') \leq 1, \quad i = 1, \ldots, N,$$

$$\lambda_i(V_i - O')^\top Q'(V_i - O') = \lambda_i, \quad i = 1, \ldots, N.$$

(The use of the logarithmic scaling factor in $-\ln(\det Q)$ is justified as the derivative (with respect to the matrix coefficients) gives $-Q^{-1}$.)

Using these KKT conditions, the lower bound $(1/n)(\mathcal{E}' - O') + O' \subset \mathcal{C}$ (as well as its $(1/\sqrt{n})$-refinement for symmetric $\mathcal{C}$) can now be derived as follows.

Introducing the affine transformation $\phi \in \text{Aff}(\mathbb{R}^n)$ by

$$\phi(X) = (Q')^{1/2}(X - O'), \quad X \in \mathbb{R}^n,$$

we realize that ϕ transforms the ellipsoid $\mathcal{E}'$ to the closed unit ball $\bar{\mathcal{B}}$. Performing ϕ on all elements of our configuration and *retaining the original notation*, we arrive at the following scenario: The minimal volume ellipsoid containing the convex body $\mathcal{C} = [V_1, \ldots, V_N]$ is the closed unit ball $\bar{\mathcal{B}}$, and we have

$$\sum_{i=1}^{N} \lambda_i V_i \cdot V_i^\top = I \quad \text{and} \quad \sum_{i=1}^{N} \lambda_i V_i = 0,$$

where $\{V_1, \ldots, V_N\} \subset \bar{\mathcal{B}}$, and $\lambda_i \geq 0$, $i = 1, \ldots, N$. In addition, if $\lambda_i > 0$ then $|V_i| = 1$, and therefore $V_i \in \mathcal{S} \cap \partial\mathcal{C}$, $\mathcal{S} = \partial\bar{\mathcal{B}}$.

The first equality gives $\sum_{i=1}^{N} \lambda_i = n$ (by taking the traces in both sides), and $X = \sum_{i=1}^{N} \lambda_i \langle X, V_i \rangle V_i$, $X \in \mathbb{R}^n$. The second equality gives $0 \in \text{int}[V_1, \ldots, V_N]$.

For John's estimates we need to determine the largest radius $r > 0$ such that $\bar{\mathcal{B}}_r \subset [V_1, \ldots, V_N]$. Let $X_0 \in \mathcal{S}_r \cap \partial[V_1, \ldots, V_N]$, $\mathcal{S}_r = \partial\bar{\mathcal{B}}_r$. The hyperplane $\mathcal{H}$ that supports $[V_1, \ldots, V_N]$ at X_0 also supports $\bar{\mathcal{B}}_r$ so that X_0 is orthogonal to $\mathcal{H}$. Letting $X = X_0/|X_0| \in \mathcal{S}$, we clearly have $r = \max_{0 \leq i \leq N} \langle X, V_i \rangle$.

We now follow the method of [Belloni–Freund] (with simplifications) which, for *general* $X \in \mathcal{S}$, aims to estimate $\max_{0 \leq i \leq N} \langle X, V_i \rangle$.

For $i = 1, \ldots, N$, letting $w_i = \langle X, V_i \rangle$, the two equations above give

$$\sum_{i=1}^{N} \lambda_i w_i^2 = \sum_{i=1}^{N} \lambda_i \langle X, V_i \rangle^2 = |X|^2 = 1 \quad \text{and} \quad \sum_{i=1}^{N} \lambda_i w_i = \sum_{i=1}^{n} \lambda_i \langle X, V_i \rangle = 0.$$

Setting $w_{\max} = \max_{1 \leq i \leq N} w_i$ and $w_{\min} = \min_{1 \leq i \leq N} w_i$, and using the equations above, we now calculate

$$0 \leq \sum_{i=1}^{N} \lambda_i (w_{\max} - w_i)(w_i - w_{\min})$$

$$= \sum_{i=1}^{N} \lambda_i \cdot w_{\max}(-w_{\min}) - \sum_{i=1}^{N} \lambda_i w_i^2$$

$$= n \cdot w_{\max}(-w_{\min}) - 1.$$

Thus

$$w_{\max}(-w_{\min}) \geq \frac{1}{n}.$$

On the other hand, by the definition of the maximal distortion $\mathfrak{m}_{[V_1,\ldots,V_N]} = \mathfrak{m}_{\mathcal{C}}$, we have

$$-w_{\min} = -\min_{1 \leq i \leq N} \langle X, V_i \rangle \leq \mathfrak{m}_{\mathcal{C}}(0) \max_{1 \leq i \leq N} \langle X, V_i \rangle = \mathfrak{m}_{\mathcal{C}}(0) \cdot w_{\max}.$$

Combining these, we obtain

$$w_{\max} \geq \frac{1}{\sqrt{n\mathfrak{m}_{\mathcal{C}}(0)}}.$$

Taking now $X = X_0/|X_0|$ as a point of common support of $\bar{\mathcal{B}}_r$ and $\mathcal{C}$ as above, we arrive at

$$r \geq \frac{1}{\sqrt{n\mathfrak{m}_{\mathcal{C}}(0)}}.$$

Reverting to the original notation, we obtain the estimate

$$(1/\sqrt{n\mathfrak{m}_{\mathcal{C}}(O')})(\mathcal{E}' - O') + O' \subset \mathcal{C} \subset \mathcal{E}',$$

For $\mathcal{C}$ symmetric, O' coincides with the center of symmetry of $\mathcal{C}$ (since otherwise $\mathcal{E}'$ could be reflected to this center, and we could construct a circumscribed ellipsoid of smaller volume). Thus, we have $\mathfrak{m}(O') = 1$, and the estimate above reduces to John's $(1/\sqrt{n})$-estimate in the symmetric case.

For the non-symmetric case, first note that, by the above, we have

$$-(C - O') \subset -(\mathcal{E}' - O') = \mathcal{E}' - O' \subset \sqrt{n\,\mathfrak{m}(O')}(C - O').$$

On the other hand, $-(C - O') \subset \mathfrak{m}(O')(C - O')$ with the *best* constant $\mathfrak{m}(O')$. (See also Lemma 3.2.3 below.) Combining these, we obtain that $\mathfrak{m}(O') \leq \sqrt{n\,\mathfrak{m}(O')}$. This gives $\mathfrak{m}(O') \leq n$, and once again, the Banach–Mazur estimate above reduces to John's $(1/n)$-estimate in the non-symmetric case.

Finally, note that to obtain the $(1/n)$-estimate directly is simpler; see [Ball 1]. In the next section we will derive John's estimates pursuing a different approach.

We now return to the Banach–Mazur distance. In (3.2.9), we have $d_{BM}(\mathcal{E}, C) \leq n$ or $\leq \sqrt{n}$, for symmetric C. Since any two ellipsoids are affine equivalent (and also to the closed unit ball $\bar{B}$), the properties in (3) and (4) of the Banach–Mazur distance above imply that $d_{BM}(C, C') \leq n^2$ for *any* $C, C' \in \mathfrak{B}$. In addition, $d_{BM}(C, C') \leq n$ provided that $C, C' \in \mathfrak{B}$ are both symmetric.

Finally, note that an estimate for the mixed case was given by [Lassak] in 1983 who proved that $d_{BM}(C, C') \leq 2n - 1$, if *one of the convex bodies* $C, C' \in \mathfrak{B}$ is symmetric.

As noted above, we will derive these estimates in the next section.

John's estimates $d_{BM}(\bar{B}, C) \leq n$, $C \in \mathfrak{B}$, and $\leq \sqrt{n}$, for symmetric C, are sharp. Indeed, if $C = \Delta$ is a simplex then $d_{BM}(\bar{B}, \Delta) = n$, and if $C = \mathcal{Q}$, is a cube then $d_{BM}(\bar{B}, \mathcal{Q}) = \sqrt{n}$. (See Problem 6.) Conversely, if $d_{BM}(\bar{B}, C) = n$ then C is a simplex. This has been proved by [Leichtweiss 2] and rediscovered by [Palmon].

As noted by [Hug–Schneider], no stability result seems to be known for the upper bound $d_{BM}(\bar{B}, C) \leq n$, $C \in \mathfrak{B}$, but they prove a weaker stability for the *volume quotient*:

$$\mathfrak{volq}\,(C) = \left(\frac{\mathrm{vol}\,(\mathcal{E}'_C)}{\mathrm{vol}\,(\mathcal{E}_C)} \right)^{1/n}, \quad C \in \mathfrak{B},$$

where $\mathcal{E} = \mathcal{E}_C$ and $\mathcal{E}' = \mathcal{E}'_C$ are the John and Löwner ellipsoids of C, respectively.

By (3.2.8), we have

$$\mathfrak{volq}\,(C) \leq d_{BM}(\bar{B}, C) \leq n, \quad C \in \mathfrak{B},$$

and equalities hold throughout if and only if C is a simplex. (See also [Leichtweiss 2].)

For the volume quotient $\mathfrak{volq}$ [Hug–Schneider] proved the following stability estimate: There exist constants $c_0(n) = O(n^{13/2}) > 0$ and $\epsilon_0(n) > 0$ such that, for $\epsilon \in [0, \epsilon_0(n)]$, we have

$$\mathfrak{volq}\,(C) \geq (1 - \epsilon)n \quad \Rightarrow \quad d_{BM}(C, \Delta) \leq 1 + c_0(n)\epsilon^{1/4}, \quad C \in \mathfrak{B},$$

where Δ is a simplex.

For other volume quotients, see [Ball 2, Barthe 1, Barthe 2, Schmuckenschlänger].

Recently there has been extensive work in finding the best possible upper bound in John's n^2-estimate $d_{BM}(C, C') \leq n^2$, $C, C' \in \mathfrak{B}$. Most noteworthy is the estimate of [Rudelson] replacing n^2 with $\alpha \cdot n^{4/3} \log^\beta n$, where α and β are universal constants.

For lower bounds, one should also note the result of [Gluskin]: For any $n \geq 2$, there exist symmetric convex bodies $C, C' \in \mathfrak{B}$ such that $d_{BM}(C, C') \geq c \cdot n$, where $c > 0$ is a universal constant.

Equipped with the Banach–Mazur distance, $\mathfrak{B}_{\mathcal{X}} / \mathrm{Aff}(\mathcal{X})$ becomes a *compact* metric space. A quick proof of this is as follows.

Since, for metrizable spaces, compactness is equivalent to sequential compactness, it is enough to show that any sequence $\{[|C_k|]\}_{k \geq 1} \subset \mathfrak{B}_{\mathcal{X}} / \mathrm{Aff}(\mathcal{X})$ subconverges. (Here $[|C|] \in \mathfrak{B}_{\mathcal{X}} / \mathrm{Aff}(\mathcal{X})$ stands for the equivalence class (orbit) of $C \in \mathfrak{B}_{\mathcal{X}}$ by the affine group $\mathrm{Aff}(\mathcal{X})$, that is, $[|C|]$ is the set of all convex bodies that are affine equivalent to C.) Since any two ellipsoids are affine equivalent, we may assume that the John ellipsoid of each representative C_k, $k \geq 1$, is the closed unit ball $\bar{B}$. By John's estimate, we have $\bar{B} \subset C_k \subset n\bar{B}$, $k \geq 1$, so that the Hausdorff distance $d_H(C_k, \bar{B}) \leq n - 1$, $k \geq 1$. In particular, the sequence $\{C_k\}_{k \geq 1} \subset \mathfrak{B}_{\mathcal{X}}$ is Hausdorff bounded. By Blaschke's selection theorem (Section 1.1/B), this sequence subconverges *in* $\mathfrak{K}_{\mathcal{X}} (\supset \mathfrak{B}_{\mathcal{X}})$. Selecting a convergent subsequence and adjusting the notation, we may assume that the Hausdorff limit $\lim_{k \to \infty} C_k = C \in \mathfrak{K}_{\mathcal{X}}$.

Let $\epsilon_k = d_H(C_k, C)$, $k \geq 1$, so that $\lim_{k \to \infty} \epsilon_k = 0$. Since $\bar{B} \subset C_k \subset C + \epsilon_k \bar{B}$, letting $k \to \infty$, we conclude that $\bar{B} \subset C$. In particular, we have $C \in \mathfrak{B}_{\mathcal{X}}$. (In fact, $\bar{B}$ is the John ellipsoid of C but we do not need this fact.) Now we have

$$C_k \subset C + \epsilon_k \bar{B} \subset (1 + \epsilon_k) C$$
$$C \subset C_k + \epsilon_k \bar{B} \subset (1 + \epsilon_k) C_k.$$

Combining these, we obtain

$$C \subset (1 + \epsilon_k) C_k \subset (1 + \epsilon_k)^2 C.$$

By the definition of the Banach–Mazur distance, this means that

$$d_{BM}([|C_k|], [|C|]) \leq (1 + \epsilon_k)^2, \quad k \geq 1.$$

Thus the sequence $\{[|C_k|]\}_{k \geq 1} \subset \mathfrak{B}_{\mathcal{X}} / \mathrm{Aff}(\mathcal{X})$ converges to $[|C|]$ in the Banach–Mazur distance, and the stated (sequential) compactness follows.

Remark. [Schneider 4] gave a direct proof for the compactness of $\mathfrak{B}_{\mathcal{X}} / \mathrm{Aff}(\mathcal{X})$ without using John's ellipsoid as follows.

Given $C \in \mathfrak{B}$, let $\Delta = [V_0, \ldots, V_n] \subset C$ be a simplex of *maximal volume* inscribed in C. Given a vertex V_i, $i = 0, \ldots, n$, with opposite face F_i, let $\mathcal{H}_i$ be a hyperplane containing V_i and parallel to F_i. Due to the maximality of the volume of Δ, $\mathcal{H}_i$ *supports* C. (Indeed, if it did not then, choosing a point $V \in \partial C$ in the interior

3.2 The Minkowski Measure and Banach–Mazur Distance

of the other side of $\mathcal{H}_i$ (which does not contain C), we would get vol $[F_i, V] >$ vol Δ.)
The hyperplanes $\mathcal{H}_i$, $i = 0, \ldots, n$, enclose the simplex $-n(\Delta - O) + O$, where O is
the centroid of Δ. By construction, we have $\Delta \subset C \subset -n(\Delta - O) + O$.

Let Δ_0 be a *regular* simplex with centroid at the origin 0, and $\Delta_0' = -n\Delta_0$.
The previous chain of inclusions can then be written as $\Delta_0 \subset \phi(C) \subset \Delta_0'$ with a
suitable affine transformation $\phi \in \mathrm{Aff}(\mathcal{X})$. We obtain that $\mathfrak{B}$ is the $\mathrm{Aff}(\mathcal{X})$ orbit of
the compact set $\{C_0 \in \mathfrak{B} \mid \Delta_0 \subset C_0 \subset \Delta_0'\}$. Compactness of $\mathfrak{B}_{\mathcal{X}}/\mathrm{Aff}(\mathcal{X})$ follows.

We now return to the main line. How well the Banach–Mazur distance is adapted
to stability (as in Theorem 3.2.1) is illustrated by the following:

Proposition 3.2.2. *For $C \in \mathfrak{B}$, we have*

$$d_{BM}(C, \tilde{C}) \le \mathrm{m}_C^*, \tag{3.2.11}$$

where $\tilde{C} = (C - C)/2 \in \mathfrak{B}$ is the Minkowski symmetral of C.

Clearly, (3.2.11) reduces stability of the lower bound for the Minkowski measure
to tautology.

Before the proof of Proposition 3.2.2, we need a simple expression for m^* as
follows:

Lemma 3.2.3. *Let $C \in \mathfrak{B}$. We have*

$$\mathrm{m}_C^* = \inf\{\lambda > 0 \mid C + X \subset -\lambda C \text{ for some } X \in \mathcal{X}\}. \tag{3.2.12}$$

Proof. Using the substitutions $\lambda = 1/s$ and $sX = -(s + 1)O$ in the defining
inclusion in (3.2.12), the *reciprocal* of the infimum can be written as

$$\sup\{s > 0 \mid (s + 1)O - sC \subset C \text{ for some } O \in \mathcal{X}\}. \tag{3.2.13}$$

By Corollary 1.1.9, if $(s + 1)O - sC \subset C$ for some $s > 0$ then $O \in \mathrm{int}\,C$. Therefore
the range of O in (3.2.13) can be restricted to $\mathrm{int}\,C \subset \mathcal{X}$. Using (2.2.3), we obtain

$$\sup\{s > 0 \mid (s+1)O - sC \subset C \text{ for some } O \in \mathrm{int}\,C\} = \sup_{O \in \mathrm{int}\,C} \frac{1}{\mathrm{m}_C(O)} = \frac{1}{\mathrm{m}_C^*}.$$

The lemma follows.

Remark. Although we do not need it at present, for $C \in \mathfrak{B}$, a more careful tracking
of the base point $O \in \mathrm{int}\,C$ in the proof above also gives the following refinement of
(3.2.12):

$$\mathrm{m}_C(O) = \sup\{\lambda > 0 \mid -(C - O) \subset \lambda(C - O)\}.$$

(Alternatively, this can also be proved directly using the definition of maximal
distortion.)

Proof of Proposition 3.2.2. As in (3.2.12), let $\lambda > 0$ such that $C + X \subset -\lambda C$. Adding to both sides of this inclusion first λC then $-C$ and using $-\tilde{C} = \tilde{C} = (C - C)/2$, a short computation gives

$$C \subset \frac{2\lambda}{\lambda + 1}\tilde{C} - \frac{1}{\lambda + 1}X \subset \lambda C + \frac{\lambda - 1}{\lambda + 1}X.$$

Now (3.2.11) follows from the definition of the Banach–Mazur distance in (3.2.8).

We close this long section by proving stability of the upper bound of the Minkowski measure in the Minkowski–Radon inequality (2.1.7) with respect to the Banach–Mazur distance:

Theorem 3.2.4 ([Schneider 1]). *Let $0 \leq \epsilon < 1/n$. If $C \in \mathfrak{B}$ satisfies*

$$\mathfrak{m}_C^* > n - \epsilon$$

then, for an n-simplex $\Delta(\subset C)$, we have

$$d_{BM}(C, \Delta) < 1 + \frac{(n + 1)\epsilon}{1 - n\epsilon}, \tag{3.2.14}$$

Remark. This stability result in its present complete form is due to [Schneider 1]. We follow his treatment with minor modifications. Note that [Böröczky 1, Böröczky 2] and [Guo 1] proved similar stability results albeit with weaker estimates.

Note also that if $\mathfrak{m}_C^* = n$ then Theorem 3.2.4 (with $\epsilon \to 0$) along with the first property of the Banach–Mazur distance d_{BM} imply that C is a simplex.

Proof of Theorem 3.2.4. Given $C \in \mathfrak{B}$ as in the theorem, we will find a suitable n-simplex $\Delta \subset C$ close to C with respect to d_{BM}. This is yet another application of Helly's Theorem which, in turn, borrows the idea of the proof of Corollary 1.4.2 by [Yaglom–Boltyanskiĭ].

For $0 \leq \nu \leq n$ and $A \in C$, we define

$$C_{A,\nu} = S_{\nu/(\nu+1),A}(C) = \frac{1}{\nu + 1}A + \frac{\nu}{\nu + 1}C.$$

We first claim that, for fixed $0 \leq \nu \leq n$, we have

$$X \in \bigcap_{A \in C} C_{A,\nu} \quad \text{if and only if} \quad -(C - X) \subset \nu(C - X). \tag{3.2.15}$$

This is straightforward since, for $A \in C$, we have $X \in C_{A,\nu}$ if and only if $-(A - X) \in \nu(C - X)$.

Turning to the construction of the n-simplex Δ, we let $0 \leq \epsilon < 1/n$ and set

$$\nu = n - \epsilon.$$

By assumption, we have $m_C^* > v$. By (3.2.12), for $\lambda \leq v$, we have $C + X \not\subset -\lambda C$ for any $X \in \mathcal{X}$. For $\lambda = v$ (replacing X by $(1+v)X$) we obtain that $-(C-X) \not\subset v(C-X)$ for any $X \in \mathcal{X}$. Now (3.2.15) implies that the intersection $\bigcap_{A \in C} C_{A,v}$ is empty.

Helly's Theorem applies yielding $\{C_0, \ldots, C_n\} \subset C$ such that the finite intersection $\bigcap_{i=0}^n C_{C_i,v}$ is also empty.

This is an open condition in the $(n + 1)$-fold product C^{n+1}, so that the set $\{C_0, \ldots, C_n\}$ can be chosen to be affinely independent. We now set $\Delta = [C_0, \ldots, C_n] \subset C$, an n-simplex with vertices $C_0, \ldots, C_n$.

Applying a translation to the entire configuration if necessary, we may assume that the centroid of Δ is at the origin:

$$\sum_{i=0}^n C_i = 0. \tag{3.2.16}$$

In the next step, we need a reference point $Y \in \bigcap_{i=1}^n C_{C_i,v}$ which, by construction, is *not* in $C_{C_0,v}$. The simplest choice is

$$Y = \frac{v - n}{v + 1} C_0. \tag{3.2.17}$$

Indeed, using (3.2.16), for a *fixed* $i = 1, \ldots, n$, Y can be represented by the nested convex combinations

$$Y = \frac{v - n + 1}{v} \left(\frac{v}{v + 1} C_0 + \frac{1}{v + 1} C_i \right) + \sum_{j=1, j \neq i}^n \frac{1}{v} \left(\frac{v}{v + 1} C_j + \frac{1}{v + 1} C_i \right).$$

This shows that

$$Y \in \frac{1}{v + 1} C_i + \frac{v}{v + 1} \Delta = \Delta_{C_i,v} \subset C_{C_i,v}, \quad i = 1, \ldots, n,$$

as stated. By Helly's theorem, we have $Y \notin C_{C_0,v}$. Playing this back to C in the use of (3.2.16) and (3.2.17), we obtain the scaled point

$$Y_0 = -\kappa C_0 = \kappa \sum_{i=1}^n C_i \notin C, \tag{3.2.18}$$

where

$$\kappa = \frac{n - v + 1}{v} \geq \frac{1}{n}.$$

With these preparations we now turn to the proof of the theorem. Setting

$$\mu = 1 + \frac{(n + 1)\epsilon}{1 - n\epsilon} > 1$$

as in the theorem, for the upper estimate in the Banach–Mazur distance in (3.2.14), it is enough to show that

$$\mathcal{C} \subset \operatorname{int} \mu\Delta. \tag{3.2.19}$$

Assume not. Then $\mathcal{C}$ meets a face of $\mu\Delta$ opposite to the vertex μC_0, say, in a point $C \in \mathcal{C} \cap \mu\Delta$, and we have

$$C = \sum_{i=1}^{n} \alpha_i \mu C_i, \quad \sum_{i=1}^{n} \alpha_i = 1, \quad \{\alpha_1, \dots, \alpha_n\} \subset [0, 1]. \tag{3.2.20}$$

Since $\mu > 1$, the set $\{C, C_1, \dots, C_n\}$ is affinely independent, so that the point $Y_0 \notin \mathcal{C}$ constructed above has the unique convex combination

$$Y_0 = \beta_0 C + \sum_{i=1}^{n} \beta_n C_n, \quad \sum_{i=0}^{n} \beta_i = 1. \tag{3.2.21}$$

Combining this with (3.2.20), we have

$$Y_0 = \sum_{i=1}^{n} (\beta_0 \alpha_i \mu + \beta_i) C_i.$$

Comparing this with (3.2.18), we obtain

$$\beta_0 \alpha_i \mu + \beta_i = \kappa, \quad i = 1, \dots, n. \tag{3.2.22}$$

We now solve this system (along with the second equation in (3.2.21)) for $\beta_0, \dots, \beta_n$. Summing (3.2.22) over $i = 1, \dots, n$, we have

$$\beta_0 \mu + \sum_{i=1}^{n} \beta_i = 1 + \beta_0(\mu - 1) = n\kappa,$$

so that

$$\beta_0 = \frac{n\kappa - 1}{\mu - 1} \geq 0.$$

Substituting this back to (3.2.22), for $i = 1, \dots, n$, we obtain

$$\beta_i = \kappa - (n\kappa - 1)\alpha_i \frac{\mu}{\mu - 1} = (n\kappa - 1)(1 - \alpha_i) \frac{\mu}{\mu - 1} \geq 0,$$

since $\kappa(\mu - 1) - (n\kappa - 1)\mu = 0$.

Since all the coefficients β_i, $i = 0, \ldots, n$, are non-negative, the expansion in (3.2.21) is a convex combination. We obtain that $Y_0 \in [C, C_1, \ldots, C_n] \subset C$. This contradicts to (3.2.18). The theorem follows.

3.3 The Theorems of John and Lassak

In the previous section we sketched the original proof of John's theorem. The purpose of the present section is to give a complete proof of the estimates of John and Lassak following the unified approach of [Guo–Kaijser 2] with simplifications and corrections. This ingenious method is based on constructing one-parameter families of volume increasing affine transformations.

The main result of this section is the following:

Theorem 3.3.1. *Let* $C, C' \in \mathcal{B}$. *We have:*

(1) $d_{BM}(C, C') \leq (n - 1) \min(\mathfrak{m}_C^*, \mathfrak{m}_{C'}^*) + n$;
(2) $d_{BM}(C, C') \leq n$, *if* C *and* C' *are both symmetric;*
(3) $d_{BM}(C, C') \leq n$, *if* C *is an ellipsoid;*
(4) $d_{BM}(C, C') \leq \sqrt{n}$, *if* C *is an ellipsoid and* C' *is symmetric.*

Remark 1. The inequality in (1) is due to Qi Guo. In the original proof in [Guo–Kaijser 2] there is a minor gap which is emended below based on the author's communication with him. If, in (1), one of the convex bodies $C, C' \in \mathcal{B}$ is symmetric then $\min(\mathfrak{m}_C^*, \mathfrak{m}_{C'}^*) = 1$, and (1) reduces to Lassak's estimate $d_{BM}(C, C') \leq 2n - 1$ noted in the previous section.

Note also that, even though (4) clearly implies (2), for the proof it is more natural to follow the sequence given above; in fact, (4) will be a quick byproduct of the proof of (3).

Finally, by affine invariance of the Banach–Mazur distance, in (3) and (4) C can be replaced by the closed unit ball $\bar{B}$. The notation here is purely of technical convenience to streamline the sequence (1)–(4).

Remark 2. In yet another paper [Guo–Kaijser 3] also gave a general upper bound on the Banach–Mazur distance in terms of the Minkowski measures:

$$d_{BM}(C, C') \leq n \frac{\mathfrak{m}_C^* + \mathfrak{m}_{C'}^*}{2}, \quad C, C' \in \mathfrak{B}.$$

In view of (2.1.7) and the subsequent discussion, this inequality also recovers John's estimate.

Remark 3. Recall that the [Belloni–Freund] method, discussed in the previous section, gives

$$C \subset \mathcal{E}' \subset \sqrt{n \, \mathfrak{m}(O')}(C - O') + O',$$

where $\mathcal{E}' = \mathcal{E}'_\mathcal{C}$ is the Löwner ellipsoid of $\mathcal{C}$ and $O' = O'_\mathcal{C}$ is its center. In particular, this gives the interesting Banach–Mazur estimate

$$d_{BM}(\mathcal{C}, \bar{B}) \le \sqrt{n \, \mathfrak{m}_\mathcal{C}(O'_\mathcal{C})}.$$

Since the Banach–Mazur distance is multiplicative, we obtain

$$d_{BM}(\mathcal{C}, \mathcal{C}') \le n\sqrt{\mathfrak{m}_\mathcal{C}(O'_\mathcal{C})\mathfrak{m}_{\mathcal{C}'}(O'_{\mathcal{C}'})}, \quad \mathcal{C}, \mathcal{C}' \in \mathfrak{B},$$

where $O'_\mathcal{C}$ and $O'_{\mathcal{C}'}$ are the centers of the Löwner ellipsoids of $\mathcal{C}$ and $\mathcal{C}'$, respectively.

[Belloni–Freund] conjectured that, in the Banach–Mazur estimates above, the maximum distortions can be replaced by the respective infima, the Minkowski measures:

$$d_{BM}(\mathcal{C}, \bar{B}) \le \sqrt{n \, \mathfrak{m}^*_\mathcal{C}}, \quad \mathcal{C} \in \mathfrak{B},$$

and (consequently)

$$d_{BM}(\mathcal{C}, \mathcal{C}') \le n\sqrt{\mathfrak{m}^*_\mathcal{C}\mathfrak{m}^*_{\mathcal{C}'}}, \quad \mathcal{C}, \mathcal{C}' \in \mathfrak{B}.$$

Note that, using the arithmetic-geometric means inequality, the latter would clearly imply the Guo-Kaijser estimate in Remark 2.

Remark 4. Finally, note that [Giannopoulos–Perissinaki–Tsolomitis] generalized John's theorem to a pair of convex bodies $\mathcal{C} \subset \mathcal{C}'$, where $\mathcal{C}$ has maximal volume in $\mathcal{C}'$. Aside from some smoothness conditions that their proof required, they recovered the theorems of John and Lassak as special cases. In addition, they gave a direct proof of the upper estimate $(\mathrm{vol}\,(\mathcal{C}')/\mathrm{vol}\,(\mathcal{C}))^{1/n} \le n$ for the respective volume ratio. (For the original proof, see [Ball 2].)

We now return to the main line and prove Theorem 3.3.1.

The general plan is as follows. Given $\mathcal{C}, \mathcal{C}' \in \mathfrak{B}$, we place an affine copy $\phi(\mathcal{C})$, $\phi \in \mathrm{Aff}(\mathcal{X})$, of $\mathcal{C}$ of *maximal volume inside* $\mathcal{C}'$, then write (as in (3.2.9))

$$\phi(\mathcal{C}) \subset \mathcal{C}' \subset S_{\lambda,O}(\phi(\mathcal{C})), \quad \lambda \ge 1, \quad O \in \phi(\mathcal{C})^*,$$

where the asterisk indicates the critical set. We then construct a continuous one-parameter family of affine transformations $\{\psi_r\}_{r \in (r_0,1]} \subset \mathrm{Aff}(\mathcal{X})$, $r_0 \in (0,1)$, $\psi_1 = I$, and derive a condition for ψ_r, $r \in (r_0, 1)$, to be *volume increasing*. This will then give various estimates for λ.

Note that this method will automatically imply the *existence* of John's ellipsoid (satisfying (3.2.10)) in both the general and symmetric cases. (For a direct proof of the existence and uniqueness of John's ellipsoid, see Problem 5.)

We work backwards and begin with the second step:

Lemma 3.3.2. *Let $C \in \mathfrak{B}$ and $C' = C + Z \in \mathfrak{B}$, $Z \in \mathcal{X}$, a translated copy of C. Let $\mathcal{H}$ be a hyperplane with both C and C' on one side of $\mathcal{H}$ such that the translation vector Z is not parallel to $\mathcal{H}$. Choose $C_0 \in C$ such that $d(C_0, \mathcal{H}) = max_{C \in C} d(C, \mathcal{H})$, and assume that Z points away from $\mathcal{H}$ in the sense that $d(C_0, \mathcal{H}) < d(C_0 + Z, \mathcal{H})$. Let $\phi \in \mathrm{Aff}(\mathcal{X})$ be the affine transformation that fixes $\mathcal{H}$ (pointwise) and maps C_0 to $C_0 + Z$. Then we have*

$$\phi(C) \subset [C, C'].$$

Proof. By assumption, the line with direction vector Z through C_0 intersects the hyperplane $\mathcal{H}$ at a point X_0. Setting this point as the origin of $\mathcal{X}$ (by a suitable translation), ϕ becomes linear. Comparing distances we find that $C \in [0, C_0] \Rightarrow \phi(C) \in [C, C + Z] \subset [C, C']$. The lemma now follows by writing an arbitrary $C \in C$ as a sum $C = tC_0 + H$, $t \in [0, 1]$, $H \in \mathcal{H}$, and applying ϕ. (See Figure 3.3.1.)

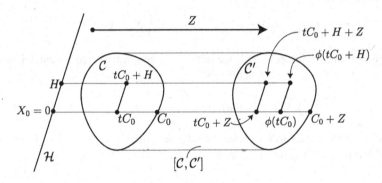

Fig. 3.3.1

According to our plan, we now proceed with the construction of the one-parameter family of affine transformations in the following setting. We let $C \in \mathfrak{B}$ be a given convex body, and a pair $\mathcal{H}, \mathcal{H}'$ of parallel hyperplanes enclosing C. We choose X, resp. X', from the (closed) half-space of $\mathcal{H}$, resp. $\mathcal{H}'$, which does not contain C. We first construct a continuous one-parameter family of affine transformations $\{\psi_r\}_{r \in (0,1]}$, $\psi_1 = I$, such that $\psi_r(C) \subset [X, C, X']$, $r \in (0, 1]$.

For $r \in (0, 1]$, let Y_r, resp. Y_r', be the intersection point of the hyperplane $S_{r,X}(\mathcal{H}')$, resp. $S_{r,X'}(\mathcal{H}')$, with the line segment $[X, X']$. (Note that $Y_1 = Y_1' = \mathcal{H}' \cap [X, X']$.) We consider the affine transformation ϕ_r, which fixes the hyperplane $\mathcal{H}_r = S_{r,X}(\mathcal{H})$ (pointwise) and maps Y_r to Y_r'. Clearly, $\phi_1 = I$. Finally, let C_0 be the portion of $[X, C, X']$ that lies in the closed slab bounded by the parallel hyperplanes $\mathcal{H}$ and $\mathcal{H}'$. (See Figure 3.3.2.)

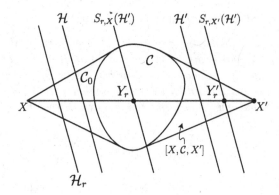

Fig. 3.3.2

We are now in the position to apply Lemma 3.3.2 to $S_{r,X}(\mathcal{C}_0)$, *its translated copy* $S_{r,X'}(\mathcal{C}_0)$, the hyperplane $\mathcal{H}_r$, and ϕ_r. We obtain

$$\phi_r(S_{r,X}(\mathcal{C}_0)) \subset [S_{r,X}(\mathcal{C}_0) \cup S_{r,X'}(\mathcal{C}_0)] \subset [X, \mathcal{C}, X'].$$

Thus, setting $\psi_r = \phi_r \circ S_{r,X}$, we obtain the continuous one-parameter family $\{\psi_r\}_{r \in (0,1]} \subset \mathrm{Aff}(\mathcal{X})$, $\psi_1 = I$, satisfying

$$\psi_r(\mathcal{C}) \subset \psi_r(\mathcal{C}_0) \subset [X, \mathcal{C}, X'], \quad r \in (0, 1]. \tag{3.3.1}$$

Given $r \in (0, 1)$, we now ask under what condition will ψ_r be volume increasing. By the definition of ψ_r, the volume changing ratio is

$$\Delta(r) = \frac{d(Y'_r, \mathcal{H}_r)}{d(Y_r, \mathcal{H}_r)} r^n, \quad \Delta(1) = 1.$$

We calculate the distance ratio as follows

$$\frac{d(Y'_r, \mathcal{H}_r)}{d(Y_r, \mathcal{H}_r)} = \frac{d(S_{r,X'}(\mathcal{H}'), S_{r,X}(\mathcal{H}))}{d(S_{r,X}(\mathcal{H}'), S_{r,X}(\mathcal{H}))}$$

$$= \frac{d(X, \mathcal{H}) + d(\mathcal{H}, \mathcal{H}') + d(X', \mathcal{H}') - d(X, S_{r,X}(\mathcal{H})) - d(X', S_{r,X'}(\mathcal{H}'))}{rd(\mathcal{H}, \mathcal{H}')}$$

$$= \frac{1 - r}{r} \frac{d(X, \mathcal{H}) + d(\mathcal{H}, \mathcal{H}') + d(X', \mathcal{H}')}{d(\mathcal{H}, \mathcal{H}')} + 1.$$

Putting everything together, we see that the volume changing ratio is a degree n polynomial in r:

$$\Delta(r) = r^{n-1}(1 - r)\rho + r^n, \tag{3.3.2}$$

where

$$\rho = \rho(X, X', \mathcal{H}, \mathcal{H}') = \frac{d(X, \mathcal{H}) + d(\mathcal{H}, \mathcal{H}') + d(X', \mathcal{H}')}{d(\mathcal{H}, \mathcal{H}')}. \tag{3.3.3}$$

Differentiating (3.3.2) (at $r = 1$), we obtain

$$\Delta'(1) = n - \rho.$$

Summarizing, we see that if $\rho > n$ then $\Delta'(1) < 0$; in particular, there exists $r_0 \in (0, 1)$ such that, for $r \in (r_0, 1)$, we have $\Delta(r) > 1$, that is ψ_r is *volume increasing*.

Proof of Theorem 3.3.1. Let $C, C' \in \mathfrak{B}$. Choose an affine transformation $\phi \in \text{Aff}(\mathcal{X})$ such that $\phi(C) \subset C'$, and $\phi(C)$ has maximal volume. Since the Banach–Mazur distance is affine invariant, we may replace C by $\phi(C)$ and retain the original notation. With this, $C \subset C'$ has maximum volume, and, by (3.2.9), $d_{BM}(C, C')$ is bounded (from above) by those $\lambda \geq 1$ for which $C' \subset S_{\lambda, X}(C)$, $X \in \text{int } C$.

Let O be a critical point of C'. Since C has maximal volume in C', we have $O \in C$. Given a boundary point $X \in \partial C'$, the line segment $[O, X]$ intersects the boundary of C at a unique point $C \in \partial C$. Let $\mathcal{H}$ be a hyperplane supporting C at C, and $\mathcal{H}'$ the other supporting hyperplane for C parallel to $\mathcal{H}$. Finally, let $X' \in \mathcal{H}'$ be on the line extension of $[O, X]$, that is $O \in [X, X']$.

We are tempted to apply the previous construction for $X, X', \mathcal{H}, \mathcal{H}'$, and C, but X' may be outside of C'. (See Figure 3.3.3.)

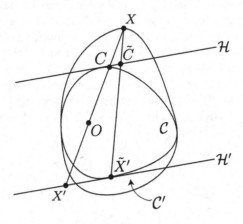

Fig. 3.3.3

We remedy this with replacing X' by $\tilde{X}' \in C \cap \mathcal{H}'$ and C by $\tilde{C}$, the intersection point of the line segment $[X, \tilde{X}']$ with the hyperplane $\mathcal{H}$. Clearly, we have

$$\frac{d(X, X')}{d(C, X')} = \frac{d(X, \tilde{X}')}{d(\tilde{C}, \tilde{X}')}.$$

We are now in the position to apply the previous construction to $X, \tilde{X}', \mathcal{H}, \mathcal{H}'$, and $\mathcal{C}$. We obtain a continuous one-parameter family $\{\psi_r\}_{r\in(0,1]} \subset \mathrm{Aff}(\mathcal{X})$ such that $\psi_r(\mathcal{C}) \subset [X, \mathcal{C}, \tilde{X}'] = [X, \mathcal{C}] \subset \mathcal{C}', r \in (0, 1]$.

Now, since $\mathcal{C}$ is of maximal volume in $\mathcal{C}'$, ψ_r cannot be volume increasing for any $r \in (0, 1]$. By the discussion above, we then must have $\rho \leq n$, where ρ is given in (3.3.3). We now calculate

$$n \geq \rho = \frac{d(X, \tilde{X}')}{d(\tilde{C}, \tilde{X}')} = \frac{d(X, X')}{d(C, X')} = \frac{d(X, O) + d(X', O)}{d(C, O) + d(X', O)} = \frac{\lambda + \mathcal{R}(\mathcal{H}, O)}{1 + \mathcal{R}(\mathcal{H}, O)},$$

where $\lambda = d(X, O)/d(C, O)$ and $\mathcal{R}(\mathcal{H}, O) = \mathcal{R}_\mathcal{C}(\mathcal{H}, O) = d(X', O)/d(C, O)$ (Section 3.1). This gives

$$\lambda \leq n(1 + \mathcal{R}(\mathcal{H}, O)) - \mathcal{R}(\mathcal{H}, O) = (n - 1)\mathcal{R}(\mathcal{H}, O) + n \leq (n - 1)\mathrm{m}_\mathcal{C}^* + n.$$

Since $X \in \partial \mathcal{C}'$ was arbitrary, we obtain

$$d_{BM}(\mathcal{C}, \mathcal{C}') \leq \sup_{X \in \partial \mathcal{C}'} \frac{d(X, O)}{d(C, O)} \leq (n - 1)\mathrm{m}_\mathcal{C}^* + n,$$

where the first inequality is because the ratio (λ) of a central similarity applied to $\mathcal{C}$ (from O) covering $\mathcal{C}'$ must be an upper bound for the Banach–Mazur distance $d_{BM}(\mathcal{C}, \mathcal{C}')$. Since $\mathcal{C}$ and $\mathcal{C}'$ play symmetric roles, the inequality in (1) follows.

Turning to the proof of (2), assume that $\mathcal{C}, \mathcal{C}' \in \mathfrak{B}$ are both symmetric. As before, we may identify $\mathcal{C}$ with an affine copy contained in $\mathcal{C}'$ having maximal volume. We first note that $\mathcal{C}$ and $\mathcal{C}'$ have the same center. Indeed, otherwise we reflect $\mathcal{C}$ to the center of $\mathcal{C}'$, and the convex hull of $\mathcal{C}$ and its reflected image would contain an affine copy of $\mathcal{C}$ of greater volume than the volume of $\mathcal{C}$, a contradiction.

Let O be the common center of $\mathcal{C}$ and $\mathcal{C}'$. As before, let $X \in \partial \mathcal{C}'$ and $C \in [O, X] \cap \partial \mathcal{C}$. Let $X' \in \partial \mathcal{C}'$ and $C' \in \partial \mathcal{C}$ be the respective antipodals with respect to O. Finally, let $\mathcal{H}$ and $\mathcal{H}'$ be parallel hyperplanes supporting $\mathcal{C}$ at C and $\mathcal{C}'$ at C', respectively. The previous construction applies to $X, X', \mathcal{H}, \mathcal{H}'$, and $\mathcal{C}$, and we have

$$n \geq \rho = \frac{d(X, X')}{d(C, C')} = \frac{d(X, O)}{d(C, O)}.$$

Varying $X \in \partial \mathcal{C}'$, we obtain

$$d_{BM}(\mathcal{C}, \mathcal{C}') \leq \sup_{X \in \partial \mathcal{C}'} \frac{d(X, O)}{d(C, O)} \leq n.$$

Part (2) of the theorem follows.

For the proof of (3) and (4), assume that C is an ellipsoid. Performing a suitable affine transformation, we may assume that $C = \bar{B}(O)$ is the *closed unit ball* inscribed in C' as an affine copy of C with maximal volume. Let $C, C' \in \partial C$ and $X, X' \in \partial C'$ as before.

Due to the geometry of the (unit) sphere ∂C, the points of support of all the hyperplanes that contain X and support C comprise a (codimension one) small sphere of ∂C if $X \neq C$, and it reduces to the singleton $\{C\}$ if $X = C$. We let $\mathcal{H}_0 \subset \mathcal{X}$ denote the hyperplane that cuts out this small sphere from ∂C if $X \neq C$; and, in the limiting case $X = C$, we let $\mathcal{H}_0$ denote the hyperplane tangent to C at C. We define $\mathcal{H}_0'$ similarly, using X' and C'. Since $O \in [X, X']$, the hyperplanes $\mathcal{H}_0$ and $\mathcal{H}_0'$ are parallel. We now claim that

$$\frac{d(X, X')}{d(\mathcal{H}_0, \mathcal{H}_0')} \leq n. \tag{3.3.4}$$

To prove this we assume that $X \neq C$ and $X' \neq C'$ since the limiting cases follow easily. Assume that the opposite inequality holds in (3.3.4). Translate the hyperplane $\mathcal{H}_0$ toward X, resp. $\mathcal{H}_0'$ toward X', to a new (parallel) hyperplane $\mathcal{H}$, resp. $\mathcal{H}'$, such that we still have

$$\frac{d(X, X')}{d(\mathcal{H}, \mathcal{H}')} > n. \tag{3.3.5}$$

Let C_0 denote the part of C between the two hyperplanes $\mathcal{H}$ and $\mathcal{H}'$, and let $\mathcal{D} \subset C$, resp. $\mathcal{D}' \subset C$, be the intersection of C with the half-space given by $\mathcal{H}$ that contains X, resp. $\mathcal{H}'$ that contains X'. (See Figure 3.3.4.)

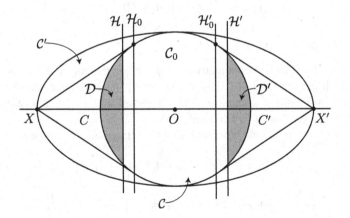

Fig. 3.3.4

For the configuration $\mathcal{C}_0$, X, X', $\mathcal{H}$, and $\mathcal{H}'$, construct the continuous one-parameter family $\{\psi_r\}_{r\in(0,1]} \subset \mathrm{Aff}(\mathcal{X})$, $\psi_1 = I$, that satisfies the analogue of (3.3.1):

$$\psi_r(\mathcal{C}_0) \subset [X, \mathcal{C}_0, X'] \subset \mathcal{C}', \quad r \in (0, 1].$$

By (3.3.5), there exists $r_0 \in (0, 1)$, such that ψ_r is volume increasing for $r \in (r_0, 1)$.

By construction, the spherical cuts $\mathcal{D}$ and $\mathcal{D}'$ have positive distance from $\partial\mathcal{C}'$. Since ψ_r converges to $\psi_1 = I$ as $r \to 1$ uniformly on compact sets, we see that there exists $r_1 \in (r_0, 1)$ such that, for $r \in (r_1, 1]$, we have $\psi_r(\mathcal{D}) \subset \mathcal{C}'$ and $\psi_r(\mathcal{D}') \subset \mathcal{C}'$. Putting everything together, we see that $\psi_r(\mathcal{C}) = \psi_r(\mathcal{C}_0) \cup \psi_r(\mathcal{D}) \cup \psi_r(\mathcal{D}') \subset \mathcal{C}', r \in (r_1, 1]$. Since $\psi_r, r \in (r_1, 1)$, is volume increasing, this contradicts to the maximality of the volume of $\mathcal{C}$ in $\mathcal{C}'$. The inequality in (3.3.4) follows.

By elementary geometry, we have

$$d(X, O)d(\mathcal{H}_0, O) = d(C, O)^2 = 1 \quad \text{and} \quad d(X', O)d(\mathcal{H}'_0, O) = d(C', O)^2 = 1.$$

With these, (3.3.4) becomes

$$n \geq \frac{d(X, X')}{d(\mathcal{H}_0, \mathcal{H}'_0)} = \frac{d(X, O) + d(X', O)}{d(\mathcal{H}_0, O) + d(\mathcal{H}'_0, O)}$$

$$= \frac{d(X, O) + d(X', O)}{1/d(X, O) + 1/d(X', O)} = d(X, O)d(X', O).$$

As before, using $d(X', O) \geq 1$, we have

$$d_{BM}(\mathcal{C}, \mathcal{C}') \leq \sup_{X\in\partial\mathcal{C}'} d(X, O) \leq n.$$

Part (3) of the theorem follows.

Finally, assume that $\mathcal{C}'$ is symmetric. We first claim that the center O' of $\mathcal{C}'$ must coincide with the center O of $\mathcal{C}$. Indeed, as before, otherwise we can reflect $\mathcal{C}$ to O' and the convex hull of $\mathcal{C}$ and its reflected unit ball would contain an ellipsoid of greater volume than that of $\mathcal{C}$, a contradiction.

Thus, in the computation above we have $d(X, O) = d(X', O)$, and we obtain

$$d_{BM}(\mathcal{C}, \mathcal{C}')^2 \leq \sup_{X\in\partial\mathcal{C}'} d(X, O)^2 \leq n.$$

Part (4) of the theorem follows. This completes the proof.

3.4 The Centroidal Minkowski Measure and Stability

Let $C \in \mathfrak{B}$. Recall that the *centroid* $g(C)$ of C, the *center of mass* of a uniform mass distribution on C, is an interior point of C. (See Problem 9.)

To avoid confusion with the already heavily used letters c and m (in various font styles), we employ the letter g based on the Archimedean identification of $g(C)$ with the *center of gravity* (in uniform gravitational field).

In this section we discuss the *centroidal Minkowski measure* $\mathfrak{m}^g = \mathfrak{m}^g_C$ defined as the maximum distortion at the centroid: $\mathfrak{m}^g_C = \mathfrak{m}_C(g(C))$, where $\mathfrak{m} = \mathfrak{m}_C : \mathrm{int}\, C \to \mathbb{R}$ is the maximum distortion as in (2.1.3). We prove the *original* Minkowski–Radon inequality for $\mathfrak{m}^g_C$, and Groemer's stability estimate for the corresponding lower and upper bounds.

By the definition of the Minkowski measure, we have $\mathfrak{m}^*_C \leq \mathfrak{m}^g_C$, $C \in \mathfrak{B}$. The following example shows that in general these two measures are different.

Example 3.4.1. Let

$$C = \{(x, y) \in \mathbb{R}^2 \mid x^2 + y^2 \leq 1,\ y \geq 0\}$$

be the unit half-disk in $\mathbb{R}^2$. A simple computation shows that the maximum distortion $\mathfrak{m}_C$ attains its minimum at the (unique) critical point at $O^* = (0, \sqrt{2}-1)$. (See also [Hammer 2].) (In fact, the critical point must be on the vertical axis, say $(0, y)$, $0 < y < 1$, and the only competing ratios are the boundary points $(0, 0)$, $(0, 1)$, and $(-1, 0)^o = (1 - y^2, 2y)/(1 + y^2)$.) Thus, we have $\mathfrak{m}^*_C = \sqrt{2}$.

The *centroid* $g(C) = (0, 4/3\pi)$ is *different* from O^*, and we have

$$\mathfrak{m}^g_C = \frac{9\pi^2 + 16}{9\pi^2 - 16} > \sqrt{2} = \mathfrak{m}^*_C.$$

(Note that the difference is only ≈ 0.025187295.)

Theorem 3.4.2 ([Minkowski 2, Radon]). *For $C \in \mathfrak{B}$, we have*

$$1 \leq \mathfrak{m}^g_C \leq n. \tag{3.4.1}$$

The lower bound is attained if and only if C is symmetric with respect to $g(C)$. The upper bound is attained if and only if C is a convex cone.

Proof. Since $1 \leq \mathfrak{m}^*_C \leq \mathfrak{m}^g_C$, $C \in \mathfrak{B}$, the lower estimate in (3.4.1) and the fact that it is attained only by symmetric convex bodies is clear from the analogous statement for the Minkowski measure $\mathfrak{m}^*$ discussed in Section 2.1.

Turning to the upper bound in (3.4.1), we first note that, for any convex cone $K \in \mathfrak{B}$, we have $\mathfrak{m}^g_K = n$. (This is an elementary exercise in integration; see Problem 10.)

Now let $C \in \mathfrak{B}$. Setting the centroid of C as the base point in (3.2.2), we have

$$\mathrm{m}_C^g = \max_{N \in \mathcal{S}} \frac{h_C(N)}{h_C(-N)},$$

where the support function is with respect to $g(C)$. Choose a unit vector $N_0 \in \mathcal{S}$ at which the maximum is attained

$$\mathrm{m}_C^g = \frac{h_C(N_0)}{h_C(-N_0)}. \tag{3.4.2}$$

As usual, let $\mathcal{H}'$ and $\mathcal{H}''$ be parallel hyperplanes supporting C in the orthogonal directions $\pm N_0$, and $\mathcal{H}$ yet another parallel hyperplane through $g(C)$. Fix $V \in C \cap \mathcal{H}'$. This configuration defines a (unique) convex cone $\mathcal{K} = [\mathcal{K}_0, V] \in \mathfrak{B}$ with $\mathcal{K}_0 \subset \mathcal{H}''$ such that $\mathcal{K} \cap \mathcal{H} = C \cap \mathcal{H}$. Let $\mathcal{G}' \supset \mathcal{H}'$ and $\mathcal{G}'' \supset \mathcal{H}''$ denote the two closed half-spaces with common boundary hyperplane $\mathcal{H} = \partial \mathcal{G}' = \partial \mathcal{G}''$. Since $\mathcal{K}$ is a cone, we have $\mathrm{m}_{\mathcal{K}}^g = n$. To derive the upper estimate $\mathrm{m}_C^g \leq \mathrm{m}_{\mathcal{K}}^g = n$ in (3.4.1) it remains to show that $g(\mathcal{K}) \in \mathcal{G}''$ (since $g(C) \in \mathcal{H}$).

For $\mathcal{A} \subset \mathcal{X}$, we denote $\mathcal{A}' = \mathcal{A} \cap \mathcal{G}'$ and $\mathcal{A}'' = \mathcal{A} \cap \mathcal{G}''$. By construction, we have

$$\mathcal{K}' \subset C' \quad \text{and} \quad C'' \subset \mathcal{K}''. \tag{3.4.3}$$

In order to locate the centroid $g(\mathcal{K})$ we endow $\mathcal{X}$ with a coordinate system $x = (x_1, \ldots, x_n)$, placing the origin at $g(C)$ and setting $N_0 = (0, \ldots, 0, 1)$. With this, we have $\mathcal{G}' = \{x_n \geq 0\}$ and $\mathcal{G}'' = \{x_n \leq 0\}$. Therefore $g(\mathcal{K}) \in \mathcal{G}''$ is equivalent to

$$g_n(\mathcal{K}) \leq 0, \tag{3.4.4}$$

where the numerical subscript indicates the respective coordinate of the point.

Letting $a' = h_C(N_0) > 0$ and $a'' = -h_C(-N_0) < 0$, for $a'' \leq t \leq a'$, we let $\mathcal{H}_t = \{x_n = t\}$, the hyperplane parallel to $\mathcal{H}$ and intersecting the nth axis at $x_n = t$. Clearly, we have $\mathcal{H}_{a'} = \mathcal{H}'$, $\mathcal{H}_{a''} = \mathcal{H}''$, and $\mathcal{H}_0 = \mathcal{H}$. By the definition of the centroid, we have

$$g_n(C) = \frac{\int_{a''}^{a'} t \, \mathrm{vol}(C \cap \mathcal{H}_t) dt}{\mathrm{vol}(C)} = 0,$$

where we used the volume in the respective dimensions without explicit subscripts.

Using the inclusions in (3.4.3), we now estimate

$$g_n(C) \, \mathrm{vol}(C) = \int_{a''}^{a'} t \, \mathrm{vol}(C \cap \mathcal{H}_t) dt$$

$$= \int_{a''}^{0} t \, \mathrm{vol}(C'' \cap \mathcal{H}_t) dt + \int_{0}^{a'} t \, \mathrm{vol}(C' \cap \mathcal{H}_t) dt$$

$$\geq \int_{a''}^{0} t\,\mathrm{vol}(\mathcal{K}'' \cap \mathcal{H}_t)dt + \int_{0}^{a'} t\,\mathrm{vol}(\mathcal{K}' \cap \mathcal{H}_t)dt$$

$$= \int_{a''}^{a'} t\,\mathrm{vol}(\mathcal{K} \cap \mathcal{H}_t)dt = g_n(\mathcal{K})\,\mathrm{vol}\,(\mathcal{K}).$$

Since $g_n(\mathcal{C}) = 0$ we obtain (3.4.4). The upper estimate in (3.4.1) follows.

Finally, note that if $\mathfrak{m}_{\mathcal{C}}^g = n$ then $g_n(\mathcal{C}) = g_n(\mathcal{K}) = 0$, and equality holds everywhere. In particular, equalities hold in the inclusions (3.4.3), and $\mathcal{C} = \mathcal{K}$ is a cone. Theorem 3.4.2 follows.

Remark. For the original proofs, see [Minkowski 2] ($n = 2, 3$) and [Radon] ($n \geq 2$). For further proofs and study, see [Neumann, Ehrhart, Estermann, Süss 1, Hammer 1, Klee 2, Leichtweiss 1, Birch, Yaglom–Boltyanskii] and also [Grünbaum 2]. The proof in [Bonnesen-Fenchel, 34] is outlined in Problem 11. The proof above was tailored to mesh well with the discussion on the stability estimates below by [Groemer 1].

For a stability estimate of the lower bound in (3.4.1) we have the following:

Theorem 3.4.3 ([Groemer 1]). *Let $0 \leq \epsilon \leq n-1$. If a convex body $\mathcal{C} \in \mathfrak{B}$ satisfies*

$$\mathfrak{m}_{\mathcal{C}}^g \leq 1 + \epsilon$$

then we have

$$d_H(\mathcal{C}, \tilde{\mathcal{C}}) \leq \frac{D_{\mathcal{C}}}{2} \frac{n}{n+1}\epsilon, \tag{3.4.5}$$

where $\tilde{\mathcal{C}} \in \mathcal{B}$ is the Minkowski symmetral of $\mathcal{C}$ with respect to $g(\mathcal{C})$.

Proof. This is a simple adjustment of the proof of Theorem 3.2.1. The base point now is the centroid $g(\mathcal{C})$ placed at the origin. The supremum of the support function in (3.2.6) can be directly estimated from above by $D_{\mathcal{C}}\,n/(n+1)$ using Theorem 3.4.2.

Remark. Stability of the lower bound for $\mathfrak{m}_{\mathcal{C}}^g$ in terms of the Banach–Mazur distance is immediate, since, by (3.2.11), we have

$$d_{BM}(\mathcal{C}, \tilde{\mathcal{C}}) \leq \mathfrak{m}_{\mathcal{C}}^* \leq \mathfrak{m}_{\mathcal{C}}^g, \quad \mathcal{C} \in \mathfrak{B}.$$

Stability for the upper bound in (3.4.1) is more involved. As expected, a suitable distance function is provided by symmetric difference metric $d_\triangle$. (See Section 1.1/B; especially Remark 2 at the end for comparisons of the Hausdorff and the symmetric difference metrics.)

Theorem 3.4.4 ([Groemer 1]). *Let $\epsilon \geq 0$. If a convex body $\mathcal{C} \in \mathfrak{B}$ satisfies*

$$\mathfrak{m}_{\mathcal{C}}^g \geq n - \epsilon$$

then there exists a convex cone $\mathcal{K} \in \mathfrak{B}$ such that

$$d_\Delta(\mathcal{C}, \mathcal{K}) \le \frac{4}{n+1} \, \text{vol} \, (\mathcal{C}) \, \epsilon. \tag{3.4.6}$$

Proof. We begin with a few adjustments. Since $\mathcal{C} \Delta \mathcal{K} \subset \mathcal{C}$, in view of (3.4.6), for any cone $\mathcal{K} \subset \mathcal{C}$, we may assume

$$0 \le \epsilon < \frac{n+1}{4}. \tag{3.4.7}$$

In addition, decreasing the value of ϵ if needed, we may also assume that $m_\mathcal{C}^g = n - \epsilon$.
Choose $N_0 \in \mathcal{S}$ such that

$$m_\mathcal{C}^g = \frac{h_\mathcal{C}(N_0)}{h_\mathcal{C}(-N_0)} = n - \epsilon,$$

where the support function $h_\mathcal{C}$ (here and below) is with respect to $g(\mathcal{C})$.
Finally, performing a suitable similarity $S_{\lambda, g(\mathcal{C})}$, we may assume

$$h_\mathcal{C}(N_0) = n - \epsilon \quad \text{and} \quad h_\mathcal{C}(-N_0) = 1. \tag{3.4.8}$$

We now construct the cone $\mathcal{K} \in \mathfrak{B}$ approximating $\mathcal{C}$ as in the proof of Theorem 3.4.2, and retain the corresponding notations: $\mathcal{K} = [\mathcal{K}_0, V]$, $V \in \mathcal{C} \cap \mathcal{H}'$ and $\mathcal{K}_0 \subset \mathcal{H}''$, such that $\mathcal{K} \cap \mathcal{H} = \mathcal{C} \cap \mathcal{H}$. Using the inclusions in (3.4.3), the symmetric difference of $\mathcal{C}$ and $\mathcal{K}$ decomposes as

$$\mathcal{C} \Delta \mathcal{K} = (\mathcal{C}' \setminus \mathcal{K}') \cup (\mathcal{K}'' \setminus \mathcal{C}'')$$

with disjoint interiors in the union. With these, the symmetric difference distance in (3.4.6) rewrites as

$$d_\Delta(\mathcal{C}, \mathcal{K}) = (\text{vol} \, (\mathcal{C}') - \text{vol} \, (\mathcal{K}')) - (\text{vol} \, (\mathcal{C}'') - \text{vol} \, (\mathcal{K}'')). \tag{3.4.9}$$

For future reference, observe that $\mathcal{K}'$ is a cone and $\mathcal{K}''$ is a truncated cone.
As before, we endow $\mathcal{X}$ with a coordinate system (but) *placing the origin at* $g(\mathcal{K})$, and setting $N_0 = (0, \ldots, 0, 1)$. By (3.4.8), the height of the cone $\mathcal{K}$ is equal to $n + 1 - \epsilon$. Taking proportions, we have

$$h_\mathcal{K}(N_0) = n \left(1 - \frac{\epsilon}{n+1}\right) \quad \text{and} \quad h_\mathcal{K}(-N_0) = 1 - \frac{\epsilon}{n+1}. \tag{3.4.10}$$

We also have

$$g_n(\mathcal{C}) = h_\mathcal{K}(N_0) - h_\mathcal{C}(N_0) = \frac{\epsilon}{n+1} \ge 0.$$

By the definition of the centroid, we then have

$$\text{vol}\,(\mathcal{C}')g_n(\mathcal{C}') + \text{vol}\,(\mathcal{C}'')g_n(\mathcal{C}'') = \text{vol}\,(\mathcal{C})\frac{\epsilon}{n+1}. \qquad (3.4.11)$$

We need to compare the ingredients here with those of the cone. By Theorem 3.4.2, we have $\mathsf{m}(g(\mathcal{C}')) \leq n = \mathsf{m}(g(\mathcal{K}'))$ (as $\mathcal{K}'$ is a cone), and this gives

$$g_n(\mathcal{C}') \geq g_n(\mathcal{K}'). \qquad (3.4.12)$$

The analogous (and expected) inequality

$$g_n(\mathcal{C}'') \geq g_n(\mathcal{K}'') \qquad (3.4.13)$$

is also true but more involved to prove. We defer the proof till the end, and proceed with the main line of the argument.

Replacing all centroidal coordinates in (3.4.11) via (3.4.12) and (3.4.13), we obtain

$$\text{vol}\,(\mathcal{C}')g_n(\mathcal{K}') - \text{vol}\,(\mathcal{C}'')|g_n(\mathcal{K}'')| \leq \text{vol}\,(\mathcal{C})\frac{\epsilon}{n+1}, \qquad (3.4.14)$$

where the absolute value is because $g_n(\mathcal{K}'') \leq 0$ (as $g(\mathcal{K}) = 0$).

Since the centroid of the cone $\mathcal{K}$ is the origin, we also have

$$\text{vol}\,(\mathcal{K}')g_n(\mathcal{K}') - \text{vol}\,(\mathcal{K}'')|g_n(\mathcal{K}'')| = 0,$$

Using this in (3.4.14), we obtain

$$\big(\text{vol}\,(\mathcal{C}') - \text{vol}\,(\mathcal{K}')\big)\,g_n(\mathcal{K}') - \big(\text{vol}\,(\mathcal{C}'') - \text{vol}\,(\mathcal{K}'')\big)\,|g_n(\mathcal{K}'')| \leq \text{vol}\,(\mathcal{C})\frac{\epsilon}{n+1}.$$
$$(3.4.15)$$

The centroidal nth coordinates here are easy to estimate. Since $\mathcal{K}'$ is a cone, by (3.4.10), we have

$$g_n(\mathcal{K}') = \frac{1}{n+1}\left(h_{\mathcal{K}}(N_0) - g_n(\mathcal{C})\right) + g_n(\mathcal{C}) \geq \frac{n-\epsilon}{n+1} > \frac{1}{4}.$$

Since $\mathcal{K}''$ is a truncated cone of height 1, the distance of the centroid $g(\mathcal{K}'')$ from the base (in $\mathcal{H}''$) is $\leq 1/2$. Thus, we have

$$|g_n(\mathcal{K}'')| \geq h_{\mathcal{K}}(-N_0) - \frac{1}{2} = \frac{1}{2} - \frac{\epsilon}{n+1} > \frac{1}{4}.$$

Using these in (3.4.15), the main estimate (3.4.6) follows via (3.4.9).

To complete the proof of the theorem, it remains to prove (3.4.13).

The convex bodies C'' and K'' are between the supporting hyperplanes $\mathcal{H}$ and $\mathcal{H}''$. As in the proof of Theorem 3.4.2, we let $\mathcal{H}_t = \{x_n = t\}$ denote the hyperplane parallel to $\mathcal{H}$ and intersecting the nth axis at $x_n = t$. Then $\mathcal{H}_t$ intersects C'' (and K'') if and only if $a'' \leq t \leq a'$, where $\mathcal{H} = \mathcal{H}_{a'}$ and $\mathcal{H}'' = \mathcal{H}_{a''}$ with $a' = g_n(C) = \epsilon/(n+1) \geq 0$ and $a'' = -h_K(-N_0) = -(1 - \epsilon/(n+1)) < 0$.

We now perform a *rotational symmetrization* with respect to the nth axis. (See [Bonnesen-Fenchel, 41].) This amounts to replacing $C'' \cap \mathcal{H}_t$, $a'' \leq t \leq a'$, with the metric ball $\mathcal{B}_{r(t)}(0, \ldots, t) \subset \mathcal{H}_t$ of the same volume: $\mathrm{vol}(C'' \cap \mathcal{H}_t) = \kappa_{n-1} r(t)^{n-1}$, where κ_n is the volume of the n-dimensional unit ball. Clearly, the nth coordinate of the centroid $g(C'')$ stays the same. Now the crux is that the *symmetrized set stays convex*, that is, $r : [a'', a'] \to \mathbb{R}$ is concave. This is a simple consequence of convexity of C'' and the Brunn–Minkowski inequality. (See the next section or Problem 12.) Performing the same symmetrization for K'', the symmetrized set is still a truncated cone containing the symmetrized C''.

Now, to prove (3.4.13), we therefore may assume that $C'' \subset K''$ are rotational symmetric with respect to the nth axis.

Denote by $\hat{\mathcal{H}}$ the hyperplane parallel to $\mathcal{H}$ and containing $g(C'')$. Consider the truncated cone (or cylinder) $\hat{K} \subset K$ with bases $\hat{K} \cap \mathcal{H} = K \cap \mathcal{H}$ and $\hat{K} \cap \mathcal{H}'' \subset K \cap \mathcal{H}''$ satisfying $\hat{K} \cap \hat{\mathcal{H}} = C'' \cap \hat{\mathcal{H}}$. An argument similar to the proof of Theorem 3.4.2 shows that $g_n(\hat{K}) \leq g_n(C'')$. For (3.4.13), it remains to show that

$$g_n(\hat{K}) \geq g_n(K''). \tag{3.4.16}$$

This is a comparison of the centroids of two truncated cones. The integration leading to (3.4.16) is entirely elementary (see Problem 13).

3.5   The Brunn–Minkowski Inequality and Its Stability

The Brunn–Minkowski inequality relates volumes of convex bodies under convex combinations. It is the cornerstone of the Brunn–Minkowski theory which provides a powerful arsenal to tackle a host of geometric problems about volume, surface area, width, etc. For our purposes, we state the Brunn–Minkowski theorem for convex bodies only.

Theorem 3.5.1. *Let $C_0, C_1 \in \mathfrak{B}$. Then, for $\alpha \in [0, 1]$, we have*

$$\mathrm{vol}\,((1 - \alpha)C_0 + \alpha C_1)^{1/n} \geq (1 - \alpha)\,\mathrm{vol}(C_0)^{1/n} + \alpha\,\mathrm{vol}(C_1)^{1/n}. \tag{3.5.1}$$

Moreover, equality holds for some $\alpha \in (0, 1)$ if and only if C_0 and C_1 are homothetic.

The Brunn–Minkowski theorem has a long and prominent history fully expounded in the survey article of [Gardner]. Along with the early history of this important result, [Bonnesen-Fenchel, 48] presents a classical proof which can be traced back to the original ideas of Brunn but uses the analytic formulation of Minkowski as simplified by [Kneser–Süss].

The Brunn–Minkowski theorem holds in general for compact sets $\mathcal{A}_0, \mathcal{A}_1 \in \mathfrak{C}$ using the Lebesgue measure on $\mathcal{X}$ in place of the (n-dimensional) volume. The once again classical proof of this employs an inductive argument in disjoint unions of open parallelepipeds, and uses the fact that, with respect to the Lebesgue measure, these unions can approximate any compact set with arbitrary precision. This proof is posed in Problem 12 (with generous hints).

In his monograph [Eggleston 1, 5.5] chooses this as his first proof of the Brunn –Minkowski inequality (also repeated in [Berger, 11.8.8.1, pp. 371–372]). (For the equality, see also [Bonnesen-Fenchel, 48] and [Hadwiger, p. 187].)

The second and more delicate proof in [Eggleston 1, 5.5, pp. 99–100] applies to convex bodies but it also recovers the equality case. The method employed there goes back to Blaschke who used an infinite cyclical Steiner symmetrization procedure to reduce the convex body to a rotational symmetric one as in Groemer's proof of Theorem 3.4.4. (See also Problem 16.)

Finally, note that [Schneider 2, 7.2] presents three proofs: the classical Kneser-Süss proof noted above (see also the discussion below), another using the *mass transportation* method of [Knothe] and recovering the equality case as well, and a more recent proof (for not necessarily convex sets) based on the Prékopa–Leindler inequality.

The purpose of this section is to present a detailed proof of [Groemer 5] on the *stability* of the Brunn–Minkowski inequality. As a byproduct of the proof, we will also recover the original Brunn–Minkowski theorem above. This will be used in Section 3.8/A to derive the classical lower bound of the Rogers–Shephard volume ratio.

Groemer's proof is delicate and complex, and uses the setting of the proof of [Kneser–Süss].

To state Groemer's theorem we need a suitable distance function on $\mathfrak{B}$. Given $\mathcal{C} \in \mathfrak{B}$, there is a dilatation (a central similarity followed by a translation) $\phi \in \text{Dil}$ (Section 1.1/C) such that the homothetic copy $\mathcal{C}' = \phi(\mathcal{C})$ has volume $\text{vol}\,\mathcal{C}' = 1$ and centroid $g(\mathcal{C}') = 0$, the origin. We define

$$d_H^{\text{Dil}}(\mathcal{C}_0, \mathcal{C}_1) = d_H(\mathcal{C}_0', \mathcal{C}_1'),$$

where d_H is the Hausdorff distance function, and $\mathcal{C}_0'$ and $\mathcal{C}_1'$ are respective homothetic copies of $\mathcal{C}_0$ and $\mathcal{C}_1$ satisfying $\text{vol}\,\mathcal{C}_0' = \text{vol}\,\mathcal{C}_1' = 1$ and $g(\mathcal{C}_0') = g(\mathcal{C}_1') = 0$.

Theorem 3.5.2 ([Groemer 5]). *Let* $\mathcal{C}_0, \mathcal{C}_1 \in \mathfrak{B}$ *be convex bodies with* $v_0 = (\text{vol}\,\mathcal{C}_0)^{1/n}$ *and* $v_1 = (\text{vol}\,\mathcal{C}_1)^{1/n}$, *and set* $D = \max\,(D_{\mathcal{C}_0}/v_0, D_{\mathcal{C}_1}/v_1)$. *If, for* $\epsilon \geq 0$ *and* $\alpha \in (0, 1)$, *we have*

$$\text{vol}\,((1 - \alpha)\mathcal{C}_0 + \alpha\mathcal{C}_1)^{1/n} \leq (1 + \epsilon)^{1/n} \left((1 - \alpha)(\text{vol}\,\mathcal{C}_0)^{1/n} + \alpha(\text{vol}\,\mathcal{C}_1)^{1/n}\right)$$

$$(3.5.2)$$

then

$$h_H^{\mathrm{Dil}}(C_0, C_1) \leq \lambda_n \left(\frac{(1-\alpha)v_0 + \alpha v_1}{\sqrt{\alpha(1-\alpha)v_0 v_1}} + 2 \right) D \epsilon^{1/(n+1)}, \qquad (3.5.3)$$

where

$$\lambda_n = 3^{1-1/n} 2^{1+1/(n+1)} n^{1-1/(n+1)} < 6n.$$

Remark 1. We first note that Theorem 3.5.2 implies Theorem 3.5.1. Indeed, if in (3.5.1) we had the opposite inequality then (3.5.2) would hold with $\epsilon = 0$. Therefore we would have $d_H^{\mathrm{Dil}}(C_0, C_1) = 0$, that is, homothety of C_0 and C_1 with equality in (3.5.1); a contradiction.

Remark 2. The non-linear ϵ-dependence on the right-hand side of the inequality in (3.5.2) can be changed to an additive ϵ-dependence at the expense of a somewhat larger constant in (3.5.3); see [Groemer 5, Theorem 3]. Theorem 3.5.2, however, captures the essence of stability, so that we will not pursue this technical point any further.

We begin the proof of Theorem 3.5.2 by a simple reduction. Using the substitutions $C_0/v_0 \mapsto C_0$, $C_1/v_1 \mapsto C_1$, $\alpha v_1 / ((1-\alpha)v_0 + \alpha v_1) \mapsto \alpha$ (and consequently $(1-\alpha)v_0 / ((1-\alpha)v_0 + \alpha v_1) \mapsto 1 - \alpha$), and *changing the notation*, (3.5.2) reduces to

$$\mathrm{vol}\,((1-\alpha)C_0 + \alpha C_1) \leq 1 + \epsilon, \quad \mathrm{vol}\,C_0 = \mathrm{vol}\,C_1 = 1, \quad \alpha \in (0, 1), \qquad (3.5.4)$$

while (3.5.3) gives

$$h_H^{\mathrm{Dil}}(C_0, C_1) \leq \lambda_n \left(\frac{1}{\sqrt{\alpha(1-\alpha)}} + 2 \right) D \epsilon^{1/(n+1)}, \qquad (3.5.5)$$

where $D = \max(D_{C_0}, D_{C_1})$. (Note that, under h_H^{Dil}, the new convex bodies C_0 and C_1 need only be translated to a common centroid (which we choose to be the origin).)

To prove Theorem 3.5.2, we will assume that (3.5.4) holds. The proof of the upper estimate in (3.5.5) will be long and technical and will be accomplished in the rest of this section.

We begin with the main construction (of [Kneser–Süss]) first applied to a single convex body $C \in \mathfrak{B}$ with $\mathrm{vol}\,C = 1$. Let $N \in \mathcal{S}$ be a fixed unit vector and endow $\mathcal{X}$ with a coordinate system $x = (x_1, \ldots, x_n)$ such that $N = (0, \ldots, 0, 1)$. For $a \in \mathbb{R}$, we set $\mathcal{G}_a = \mathcal{G}_{(0,\ldots,0,a)}(N) = \{x \in \mathcal{X} \mid x_n \leq a\}$ and $\mathcal{H}_a = \partial \mathcal{G}_a = \{x \in \mathcal{X} \mid x_n = a\}$.

For the given $C \in \mathfrak{B}$, using the support function h_C of C, we have $h_C(N) = \sup\{a \in \mathbb{R} \mid C \cap \mathcal{H}_a \neq \emptyset\}$ and $-h_C(-N) = \inf\{a \in \mathbb{R} \mid C \cap \mathcal{H}_a \neq \emptyset\}$.

Since $\mathrm{vol}\,C = 1$, for every $t \in [0, 1]$, there exists a unique $a(t) \in [-h_C(-N), h_C(N)]$ such that

$$\mathrm{vol}\,(C \cap \mathcal{G}_{a(t)}) = t. \qquad (3.5.6)$$

Note that $a : [0, 1] \to \mathbb{R}$ is a continuous function, and we have

$$a(0) = -h_C(-N) \quad \text{and} \quad a(1) = h_C(N).$$

Finally, let $v : [0, 1] \to \mathbb{R}$ be the (continuous) function defined by

$$v(t) = \text{vol}_{n-1}(C \cap \mathcal{H}_{a(t)}), \quad t \in [0, 1].$$

Now let $C_0, C_1 \in \mathfrak{B}$ with $\text{vol}\, C_0 = \text{vol}\, C_1 = 1$. We apply the main construction to C_0 and C_1, and obtain the (continuous) functions $a_0, v_0 : [0, 1] \to \mathbb{R}$ and $a_1, v_1 : [0, 1] \to \mathbb{R}$, where the subscripts indicate the respective convex body.

In view of the expression of the Hausdorff distance in (1.2.2) in terms of the respective support functions, it is good to keep in mind that the ultimate aim of the proof is to give an upper bound for

$$|h_{C_1}(N) - h_{C_0}(N)| = |a_1(1) - a_0(1)|.$$

Using (3.5.6) we see that a_0 and a_1 are strictly increasing, and, for $i = 0, 1$, we have

$$\text{vol}\, (C_i \cap \mathcal{G}_{a_i(t)}) = \int_{a_i(0)}^{a_i(t)} \text{vol}_{n-1}(C_i \cap \mathcal{H}_\tau)\, d\tau = \int_{a_i(0)}^{a_i(t)} v_i(a_i^{-1}(\tau))\, d\tau = t.$$

Differentiating with respect to the variable t, we obtain

$$\frac{da_i}{dt} = \frac{1}{v_i(t)}, \quad t \in (0, 1), \ i = 0, 1.$$

We combine these into a single function $s = (1 - \alpha)a_0 + \alpha a_1 : [0, 1] \to \mathbb{R}$ and obtain

$$\frac{ds}{dt} = \frac{1 - \alpha}{v_0(t)} + \frac{\alpha}{v_1(t)}. \tag{3.5.7}$$

Forming affine combinations, by the definition of s, we have

$$(1 - \alpha)(C_0 \cap \mathcal{H}_{a_0(t)}) + \alpha(C_1 \cap \mathcal{H}_{a_1(t)}) \subset ((1 - \alpha)C_0 + \alpha C_1) \cap \mathcal{H}_{s(t)}, \quad t \in [0, 1].$$

Taking the respective volumes, we obtain

$$\text{vol}_{n-1}(((1 - \alpha)C_0 + \alpha C_1) \cap \mathcal{H}_{s(t)})$$
$$\geq \text{vol}_{n-1}((1 - \alpha)(C_0 \cap \mathcal{H}_{a_0(t)}) + \alpha(C_1 \cap \mathcal{H}_{a_1(t)}))$$
$$\geq \left((1 - \alpha)v_0(t)^{1/(n-1)} + \alpha v_1(t)^{1/(n-1)}\right)^{n-1}.$$

In the last inequality, we used the Brunn–Minkowski inequality (3.5.1) in dimension $n - 1$ (by projecting the slices $C_i \cap \mathcal{H}_a$, $a \in \mathbb{R}$, $i = 0, 1$, to a single hyperplane).

We now assume that the entire proof is under the umbrella of a general induction step $n - 1 \Rightarrow n$, so that the inequality above follows from the induction hypothesis. During the proof of Theorem 3.5.2 we will use this induction hypothesis one more time. Once the entire proof is finished, Theorem 3.5.2 will automatically imply the Brunn–Minkowski theorem in dimension n, and this will also complete the general induction step $n - 1 \Rightarrow n$.

Integrating and using (3.5.7), we have

$$
\operatorname{vol}\left((1 - \alpha)\mathcal{C}_0 + \alpha\mathcal{C}_1\right) = \int_{s(0)}^{s(1)} \operatorname{vol}_{n-1}\left(((1 - \alpha)\mathcal{C}_0 + \alpha\mathcal{C}_1) \cap \mathcal{H}_s\right) ds
$$

$$
= \int_0^1 \operatorname{vol}_{n-1}\left(((1 - \alpha)\mathcal{C}_0 + \alpha\mathcal{C}_1) \cap \mathcal{H}_{s(t)}\right) \frac{ds}{dt} dt
$$

$$
\geq \int_0^1 \left((1-\alpha)v_0^{1/(n-1)} + \alpha v_1^{1/(n-1)}\right)^{n-1} \left(\frac{1 - \alpha}{v_0(t)} + \frac{\alpha}{v_1(t)}\right) dt.
$$

Jensen's inequality asserts that the final integral here is ≥ 1. Due to the conditions $\operatorname{vol}\mathcal{C}_0 = \operatorname{vol}\mathcal{C}_1 = 1$, this means that we obtained the Brunn–Minkowski inequality *in dimension n*. With this the main induction step $n - 1 \Rightarrow n$ is complete and the Brunn–Minkowski inequality follows. This is the classical proof of [Kneser–Süss] (also repeated in [Bonnesen-Fenchel, 48]).

Alternatively, following [Groemer 5] (and disregarding Jensen's inequality), we may stay within the general induction step $n - 1 \Rightarrow n$ and, aiming higher, refine our approach to obtain more delicate estimates.

To do this we briefly return to a single convex body $\mathcal{C} \in \mathfrak{B}$, $\operatorname{vol}\mathcal{C} = 1$, and derive a lower bound for $v(t)$, $t \in [0, 1]$.

Lemma 3.5.3. *Let $\mathcal{C} \in \mathfrak{B}$ with $\operatorname{vol}\mathcal{C} = 1$ and diameter $D_\mathcal{C}$. Then the function v defined above has the lower estimate*

$$
v(t) \geq \left(\frac{2}{3}\right)^{1-1/n} \frac{1}{D_\mathcal{C}} \min\left(t^{1-1/n}, (1 - t)^{1-1/n}\right), \quad t \in [0, 1]. \tag{3.5.8}
$$

Proof. As in Theorem 3.4.4, we first perform a rotational symmetrization with respect to the x_n-axis (by replacing each slice $\mathcal{C} \cap \mathcal{H}_{a(t)}$, $t \in [0, 1]$, with the $(n - 1)$-dimensional closed metric ball in $\mathcal{H}_{a(t)}$ of volume $v(t)$ and center on $\mathbb{R} \cdot N$). By construction, $a(t)$ and $v(t)$, $t \in [0, 1]$, stay the same.

The symmetrized body stays *convex*. As above (and Theorem 3.4.4), this is a consequence of the Brunn–Minkowski inequality (3.5.1) *in dimension $n - 1$*. Since we are within the general induction step $n-1 \Rightarrow n$, the induction hypothesis applies here.

Finally, rotational symmetrization *does not increase the diameter* of the convex body. (For a proof of this using Blaschke's symmetrization via Steiner's, see

Problem 16. Incidentally, as a byproduct of this problem, convexity of the symmetrized body also follows but we preferred to use the already existing induction hypothesis.)

Thus, from now on we may assume that C is rotational symmetric with respect to the x_n-axis. We denote by $r(t)$ the radius of the closed metric ball $C \cap \mathcal{H}_{a(t)}, t \in [0, 1]$. By continuity of r, we have $r(t_0) = \max_{[0,1]} r$ for some $t_0 \in [0, 1]$.

Clearly, we have

$$v(t_0) \, D_C \geq \operatorname{vol} C = 1. \tag{3.5.9}$$

We set

$$c_n = 1 - \frac{2}{3n}. \tag{3.5.10}$$

Case I. Let $t \in [0, 1]$ be such that $r(t) > c_n r(t_0)$. By (3.5.9) and (3.5.10), we have

$$
v(t) \geq c_n^{n-1} v(t_0) \geq \frac{c_n^{n-1}}{D_C} = \left(1 - \frac{2}{3n}\right)^{n-1} \frac{1}{D_C}
$$

$$
\geq \left(1 - \frac{2}{3}\right)^{1-1/n} \frac{1}{D_C} = \left(\frac{2}{3}\right)^{1-1/n} \frac{1}{D_C} \left(\frac{1}{2}\right)^{1-1/n}
$$

$$
\geq \left(\frac{2}{3}\right)^{1-1/n} \frac{1}{D_C} \min\left(t^{1-1/n}, (1-t)^{1-1/n}\right),
$$

where (in the second line) we used the Bernoulli inequality $(1 - x/n)^n \geq 1 - x$, $x \in [0, 1]$. (The latter is the consequence of convexity of the function $x \mapsto (1-x/n)^n$ on $[0, 1]$; therefore its graph is above its tangent line at $x = 0$.) The lemma follows in this case.

Case II. Let $t \in [0, 1]$ be such that $r(t) \leq c_n r(t_0)$. We have the alternatives

$$a(t) < a(t_0) \quad \text{or} \quad a(t) > a(t_0). \tag{3.5.11}$$

Due to the construction, these two cases (and the lower bound in (3.5.8)) are symmetric with respect to the substitution $t \leftrightarrow 1 - t, t \in [0, 1]$. Thus, from now on, we may assume that for the given $t \in [0, 1]$ the first inequality holds in (3.5.11).

Let $\mathcal{K}_0$ be the convex cone with base $C \cap \mathcal{H}_{a(t_0)}$ such that $\mathcal{K}_0 \cap \mathcal{H}_{a(t)} = C \cap \mathcal{H}_{a(t)}$. (Compare this with the proof of Theorem 3.4.2.) Let $\mathcal{K} \subset \mathcal{K}_0$ be the convex cone truncated from $\mathcal{K}_0$ by the hyperplane $\mathcal{H}_{a(t)}$. By convexity of C, we have

$$C \cap \mathcal{G}_{a(t)} \subset \mathcal{K}.$$

Taking volumes, by (3.5.6), we obtain

$$t \leq \operatorname{vol} \mathcal{K}. \tag{3.5.12}$$

We estimate the volume of the cone $\mathcal{K}$ using elementary geometry. Let h, resp. h_0, denote the height of $\mathcal{K}$, resp. $\mathcal{K}_0$. Taking into account the various proportions as well as (3.5.9), we calculate

$$\text{vol}\,\mathcal{K} = h\frac{v(t)}{n} = \frac{r(t)}{r(t_0) - r(t)}\frac{v(t)}{n}(h_0 - h)$$

$$\leq \frac{r(t)/r(t_0)}{1 - r(t)/r(t_0)}\frac{v(t)}{n}D_C$$

$$\leq \frac{1}{n(1 - c_n)}\left(\frac{v(t)}{v(t_0)}\right)^{1/(n-1)}v(t)D_C$$

$$\leq \frac{3}{2}(v(t)D_C)^{n/(n-1)}.$$

Combining this with (3.5.12), we arrive at

$$v(t) \geq \left(\frac{2}{3}\right)^{1-1/n}\frac{1}{D_C}t^{1-1/n}.$$

The lemma follows.

We now bring in the main assumption in (3.5.4). By our previous lower estimate, this gives

$$\int_0^1 \left((1-\alpha)v_0^{1/(n-1)} + \alpha v_1^{1/(n-1)}\right)^{n-1}\left(\frac{1-\alpha}{v_0(t)} + \frac{\alpha}{v_1(t)}\right)dt \leq 1 + \epsilon. \qquad (3.5.13)$$

The rest of the proof consists of a careful estimation of this integral.

First, we need a strong version of the inequality for the (weighted) arithmetic/geometric means:

Lemma 3.5.4. *For $a_0, a_1 > 0$ and $\alpha \in (0, 1)$, we have*

$$(1-\alpha)a_0 + \alpha a_1 - a_0^{1-\alpha}a_1^\alpha \geq \frac{\alpha(1-\alpha)}{2\max(a_0, a_1)}(a_0 - a_1)^2. \qquad (3.5.14)$$

Proof. We may assume that $a_0 \geq a_1$, so that $\max(a_0, a_1) = a_0$. Letting $c = (a_0 - a_1)/a_0 \in [0, 1)$ and using the binomial expansion, we calculate

$$(1-\alpha)a_0 + \alpha a_1 - a_0^{1-\alpha}a_1^\alpha = a_0\left(1 - \alpha c - (1 - c)^\alpha\right)$$

$$= a_0\left(1 - \alpha c - \left(1 + \sum_{k=0}^\infty \frac{(-\alpha)(1-\alpha)\dots(k-\alpha)}{(k+1)!}c^{k+1}\right)\right)$$

$$\geq a_0\left(1 - \alpha c - \left(1 - \alpha c - \frac{\alpha(1-\alpha)}{2}c^2\right)\right) = \frac{\alpha(1-\alpha)}{2a_0}(a_0 - a_1)^2.$$

The lemma follows.

We substitute $a_0 = 1/b_0$ and $a_1 = 1/b_1$ in (3.5.14), and obtain

$$b_0^{1-\alpha} b_1^\alpha \left(\frac{1-\alpha}{b_0} + \frac{\alpha}{b_1} \right) \geq 1 + b_0^{1-\alpha} b_1^\alpha \frac{\alpha(1-\alpha)}{2} \left(\frac{1}{b_0} - \frac{1}{b_1} \right)^2 \min(b_0, b_1).$$

$$(3.5.15)$$

Returning to the main line, to estimate the integral in (3.5.13), we first note

$$\left((1-\alpha) v_0^{1/(n-1)} + \alpha v_1^{1/(n-1)} \right)^{n-1} \geq v_1^{1-\alpha} v_2^\alpha. \tag{3.5.16}$$

(Indeed, $(1-\alpha) + \alpha x \geq x^\alpha$, $x \geq 0$, since the function $x \mapsto x^\alpha - 1$, $x \geq 0$, is convex so that its graph is below its tangent line at $x = 1$.) Now making use of (3.5.16) and (3.5.15) (with $v_0 = b_0$ and $v_1 = b_1$) reduces the inequality in (3.5.13) to the simpler form

$$\frac{\alpha(1-\alpha)}{2} \int_0^1 \left(\frac{1}{v_1} - \frac{1}{v_0} \right)^2 v_0^{1-\alpha} v_1^\alpha \, v \, dt \leq \epsilon, \tag{3.5.17}$$

where $v = \min(v_0, v_1)$.

For the next step we employ (3.5.8) in our setting:

$$v_i(t) \geq \left(\frac{2}{3} \right)^{1-1/n} \frac{1}{D} \min \left(t^{1-1/n}, (1-t)^{1-1/n} \right), \quad t \in [0, 1], \ i = 0, 1. \tag{3.5.18}$$

(Recall that $D = \max(D_{C_0}, D_{C_1})$.) We claim that, for $0 \leq \epsilon \leq n/2$, (3.5.17) and (3.5.18) imply

$$\int_0^1 \left| \frac{1}{v_1} - \frac{1}{v_0} \right| dt \leq \frac{\lambda_n}{2} \left(\frac{1}{\sqrt{\alpha(1-\alpha)}} + 2 \right) D \epsilon^{1/(n+1)}, \tag{3.5.19}$$

where λ_n is given in Theorem 3.5.2.

To show this we let

$$\delta = \frac{1}{2} \left(\frac{2}{n} \right)^{n/(n+1)} \epsilon^{n/(n+1)} \in [0, 1/2], \tag{3.5.20}$$

split the integral in (3.5.19) into three parts, and estimate

$$\int_0^1 \left| \frac{1}{v_1} - \frac{1}{v_0} \right| dt \leq \int_0^\delta \frac{2}{v} dt + \int_\delta^{1-\delta} \left| \frac{1}{v_1} - \frac{1}{v_0} \right| dt + \int_{1-\delta}^1 \frac{2}{v} dt$$

$$\leq \int_0^\delta \frac{2}{v} dt + \int_{1-\delta}^1 \frac{2}{v} dt \tag{3.5.21}$$

$$
+ \left(\int_\delta^{1-\delta} \left(\frac{1}{v_1} - \frac{1}{v_0} \right)^2 v_0^{1-\alpha} v_1^\alpha dt \right)^{1/2} \left(\int_0^1 \frac{dt}{v_0^{1-\alpha} v_1^\alpha} \right)^{1/2} ,
$$

where we used Hölder's inequality.

By making use of (3.5.18), we have $v(t) \geq (2/3)^{1-1/n}(1/D)t^{1-1/n}$ if $t \in [0, \delta]$, and similarly $v(t) \geq (2/3)^{1-1/n}(1/D)(1-t)^{1-1/n}$ if $t \in [1-\delta, 1]$. Thus

$$
\int_0^\delta \frac{2}{v} \, dt + \int_{1-\delta}^1 \frac{2}{v} \, dt \leq \left(\frac{3}{2} \right)^{1-1/n} 2D \left(\int_0^\delta t^{1/n-1} dt + \int_{1-\delta}^1 (1-t)^{1/n-1} dt \right)
$$

$$
\leq \left(\frac{3}{2} \right)^{1-1/n} 4Dn\delta^{1/n}.
$$

Again by (3.5.18), for $\delta \leq t \leq 1-\delta$, we have $v(t) \geq (2/3)^{1-1/n}(1/D)\delta^{1-1/n}$. Using this and (3.5.17), we have

$$
\int_\delta^{1-\delta} \left(\frac{1}{v_1} - \frac{1}{v_0} \right)^2 v_0^{1-\alpha} v_1^\alpha dt \leq \left(\frac{3}{2} \right)^{1-1/n} D\delta^{1/n-1} \int_0^1 \left(\frac{1}{v_1} - \frac{1}{v_0} \right)^2 v_0^{1-\alpha} v_1^\alpha v \, dt
$$

$$
\leq \left(\frac{3}{2} \right)^{1-1/n} D\delta^{1/n-1} \frac{2\epsilon}{\alpha(1-\alpha)}.
$$

Finally, we have $v_0^{1-\alpha} v_1^\alpha \geq (2/3)^{1-1/n}(1/D)t^{1-1/n}$ if $0 \leq t \leq 1/2$, and similarly $v_0^{1-\alpha} v_1^\alpha \geq (2/3)^{1-1/n}(1/D)(1-t)^{1-1/n}$ if $1/2 \leq t \leq 1$. Using these, we have

$$
\int_0^1 \frac{dt}{v_0^{1-\alpha} v_1^\alpha} \leq \left(\frac{3}{2} \right)^{1-1/n} D \left(\int_0^{1/2} t^{1/n-1} dt + \int_{1/2}^1 (1-t)^{1/n-1} dt \right)
$$

$$
\leq \left(\frac{3}{2} \right)^{1-1/n} 2^{1-1/n} nD.
$$

Putting all these together in (3.5.21), we finally arrive at

$$
\int_0^1 \left| \frac{1}{v_1} - \frac{1}{v_0} \right| dt \leq \left(\frac{3}{2} \right)^{1-1/n} 4Dn\delta^{1/n}
$$

$$
+ \left(\frac{3}{2} \right)^{1-1/n} D \left(\frac{2\epsilon}{\alpha(1-\alpha)} 2^{1-1/n} n\delta^{1/n-1} \right)^{1/2}
$$

$$
= \lambda_n D\epsilon^{1/(n+1)} + \frac{\lambda_n}{2\sqrt{\alpha(1-\alpha)}} D\epsilon^{1/(n+1)}.
$$

The estimate in (3.5.19) follows.

We now return to the geometric setting. The importance of the integral in (3.5.19) is clear from the formula

$$\left| a_1(t) - a_0(t) - (a_1(0) - a_0(0)) \right| = \left| \int_0^t \left(\frac{da_1}{d\tau} - \frac{da_0}{d\tau} \right) d\tau \right| \leq \int_0^1 \left| \frac{1}{v_1} - \frac{1}{v_0} \right| dt.$$
(3.5.22)

Performing suitable translations, from now on we may assume that $g(\mathcal{C}_0)=g(\mathcal{C}_1)=0$. Thus

$$\int_0^1 a_0(t)dt = \int_0^1 a_1(t)dt = 0,$$

and, by (3.5.22), we have

$$\begin{aligned}
|a_1(0) - a_0(0)| &= \left| \int_0^1 (a_1(t) - a_0(t) - (a_1(0) - a_0(0)))dt \right| \\
&\leq \int_0^1 |a_1(t) - a_0(t) - (a_1(0) - a_0(0))| \, dt \qquad (3.5.23) \\
&\leq \int_0^1 \left| \frac{1}{v_1} - \frac{1}{v_0} \right| dt.
\end{aligned}$$

Using (3.5.19) along with (3.5.22) and (3.5.23), we finally obtain

$$|a_1(1) - a_0(1)| \leq 2 \int_0^1 \left| \frac{1}{v_1} - \frac{1}{v_0} \right| dt \leq \lambda_n \left(\frac{1}{\sqrt{\alpha(1-\alpha)}} + 2 \right) D \epsilon^{1/(n+1)}$$
(3.5.24)

Now, as noted at the beginning of the proof, we have $a_i(1) = h_{\mathcal{C}_i}(N)$, $i = 0, 1$, the value of the support function of $\mathcal{C}_i$ in the direction N. Since this direction in $\mathcal{S}$ was arbitrary, (1.2.2) along with (3.5.24) give

$$d_H(\mathcal{C}_0, \mathcal{C}_1) = \sup_{\mathcal{S}} |h_{\mathcal{C}_1} - h_{\mathcal{C}_0}| \leq \lambda_n \left(\frac{1}{\sqrt{\alpha(1-\alpha)}} + 2 \right) D \epsilon^{1/(n+1)}.$$

Finally, it remains to remove the condition $\epsilon \leq n/2$. But, for $\epsilon > n/2$, we have

$$\lambda_n \left(\frac{1}{\sqrt{\alpha(1-\alpha)}} + 2 \right) D \epsilon^{1/(n+1)} > 2D,$$

so that the Hausdorff-distance estimate above is automatic.

Theorem 4.5.2 follows.

3.6   The General Concept of Measures of Symmetry

By now we have developed a sufficient variety of geometric and analytic constructions in convexity to tackle the basic question of this book: How to quantify symmetry for convex sets?

In this section we introduce the general concept of measure of symmetry due to Branko Grünbaum. To motivate this, recall that in Proposition 3.2.2 (and also in the stability estimate in Theorem 3.2.1) there is no restriction as to how close the Minkowski measure $\mathfrak{m}_{\mathcal{C}}^*$ of a convex body $\mathcal{C} \in \mathfrak{B}$ should be to its minimum 1. Thus, beyond stability, the inequality in (3.2.11) asserts in general that $\mathfrak{m}_{\mathcal{C}}^*$ is an upper bound to how far a convex body is from being symmetric; it provides a *measure of symmetry* (or asymmetry) for convex bodies.

This prompts the following definition: A continuous affine invariant function $\mathfrak{f} : \mathfrak{B} \rightarrow \mathbb{R}$ is called a *measure of symmetry* if $\inf_{\mathfrak{B}} \mathfrak{f}$ is attained precisely on symmetric convex bodies (that is, for $\mathcal{C} \in \mathfrak{B}$, we have $\mathfrak{f}(\mathcal{C}) = \inf_{\mathfrak{B}} \mathfrak{f}$ if and only if $\mathcal{C}$ is (centrally) symmetric).

Based on the intuitive concept that simplices are the least symmetric of the convex bodies, one may be tempted to include in the definition that $\sup_{\mathfrak{B}} \mathfrak{f}$ is attained precisely on simplices. This is indeed the case for several (classical) measures of symmetry, but not all. For example, the *centroidal Minkowski measure* $\mathfrak{m}^g$ (as a measure of symmetry to be discussed below) attains its supremum on *all* convex cones (Theorem 3.4.2).

By affine invariance, a measure of symmetry $\mathfrak{f}$ factors through the quotient map $\mathfrak{B} \rightarrow \mathfrak{B}/\text{Aff}(\mathcal{X})$ yielding a continuous function $\mathfrak{f}_0 : \mathfrak{B}/\text{Aff}(\mathcal{X}) \rightarrow \mathbb{R}$. In particular, $\mathfrak{f}$ is bounded, and the extremal values $\inf_{\mathfrak{B}} \mathfrak{f}$ and $\sup_{\mathfrak{B}} \mathfrak{f}$ are attained.

Conversely, if $\mathfrak{f}_0 : \mathfrak{B}/\text{Aff}(\mathcal{X}) \rightarrow \mathbb{R}$ is *any* continuous function whose infimum is attained precisely at (the Aff $(\mathcal{X})$-equivalence classes of) symmetric convex bodies, then precomposition of $\mathfrak{f}_0$ with the quotient map $\mathfrak{B} \rightarrow \mathfrak{B}/\text{Aff}(\mathcal{X})$ is a measure of symmetry.

A measure of symmetry can be composed with a homeomorphism (of the image to another closed interval) to obtain a new measure of symmetry. In particular, in the definition, the choice of the infimum as well as the actual values of the infimum and the supremum are irrelevant.

Of interest are those measures of symmetry that arise through (convex) geometric and measure theoretic constructions.

Remark. In his seminal paper [Grünbaum 2] initiated a systematic treatment of measures of symmetry. Apart from his specific range restriction (the interval [0, 1]), the definition above is identical to Grünbaum's affine invariant measures of symmetry.

As expected, the archetype of measures of symmetry is the Minkowski measure:

Proposition 3.6.1. *The Minkowski measure* $\mathfrak{m}^* : \mathfrak{B} \to \mathbb{R}$, $\mathfrak{m}^*(\mathcal{C}) = \mathfrak{m}_{\mathcal{C}}^*$, $\mathcal{C} \in \mathfrak{B}$, *is a measure of symmetry.*

Proof. Affine invariance and the fact that the lowest level-set characterizes the symmetric convex bodies follow from the definition of $\mathfrak{m}^*$ (Section 2.1). It remains to establish sequential continuity of $\mathfrak{m}^*$.

Let $\mathcal{C} \in \mathfrak{B}$, and $\{\mathcal{C}_k\}_{k\geq 1} \subset \mathfrak{B}$ a sequence such that $\lim_{k\to\infty} \mathcal{C}_k = \mathcal{C}$ with respect to the Banach–Mazur distance. We set

$$d_{BM}(\mathcal{C}_k, \mathcal{C}) = 1 + \delta_k \text{ with } \lim_{k\to\infty} \delta_k = 0. \tag{3.6.1}$$

Due to affine invariance of $\mathfrak{m}^*$ (and d_{BM}), we may assume that $\mathcal{C} \subset \mathcal{B}$. Moreover, by the same reason, each $\mathcal{C}_k$, $k \geq 1$, can be replaced by an affine equivalent copy to obtain

$$\mathcal{C}_k \subset \mathcal{C} \subset (1 + \delta_k)\mathcal{C}_k + Z_k, \ k \geq 1. \tag{3.6.2}$$

(For the right-hand side, note that $\mathrm{Dil}(\mathcal{X}) \lhd \mathrm{Aff}(\mathcal{X})$; Section 1.1/C.) We first claim that $|Z_k| \leq \delta_k$, $k \geq 1$. Indeed, using (3.6.2), we have

$$\mathcal{C}_k - Z_k \subset (1 + \delta_k)\mathcal{C}_k \subset \mathcal{C}_k + \delta_k\mathcal{C} \subset \mathcal{C}_k + \delta_k\mathcal{B},$$

and the claim follows.

Returning to (3.6.2), we then have

$$\mathcal{C}_k \subset \mathcal{C} \subset (1 + \delta_k)\mathcal{C}_k + Z_k \subset \mathcal{C}_k + \delta_k\mathcal{C} + \delta_k\mathcal{B} \subset \mathcal{C}_k + 2\delta_k\mathcal{B}.$$

In particular, we obtain that the Hausdorff distance

$$d_H(\mathcal{C}_k, \mathcal{C}) \leq 2\delta_k, \ k \geq 1.$$

Thus, with the choices made, we have $\lim_{k\to\infty} \mathcal{C}_k = \mathcal{C}$ with respect to the *Hausdorff metric*. In particular, for any compact subset $\mathcal{C}_0 \subset \mathrm{int}\,\mathcal{C}$, there exists $k_0 \geq 1$ such that we have $\mathcal{C}_0 \subset \mathrm{int}\,\mathcal{C}_k$ for $k \geq k_0$, and the maximum distortions $\mathfrak{m}_{\mathcal{C}_k}$ converge to $\mathfrak{m}_{\mathcal{C}}$ *uniformly on* $\mathcal{C}_0$. Sequential continuity of $\mathfrak{m}^*$ now follows by Lemma 2.1.10.

Remark. The following general statement can be extracted from the proof above. Let $\mathcal{C} \in \mathfrak{B}$, and assume that $\mathcal{C}$ is contained in the unit ball: $\mathcal{C} \subset \mathcal{B}$. Then, for any $\mathcal{C}' \in \mathfrak{B}$ there exists $\psi \in \mathrm{Aff}(\mathcal{X})$ such that

$$d_H(\mathcal{C}, \psi(\mathcal{C}')) \leq 2\left(d_{BM}(\mathcal{C}, \mathcal{C}') - 1\right).$$

Up to this point we were discussing upper bounds for the Banach–Mazur distance. For a lower bound, we have

$$(1 \leq) \max\left(\frac{m_{\mathcal{C}}^*}{m_{\mathcal{C}'}^*}, \frac{m_{\mathcal{C}'}^*}{m_{\mathcal{C}}^*}\right) \leq d_{BM}(\mathcal{C}, \mathcal{C}'), \quad \mathcal{C}, \mathcal{C}' \in \mathfrak{B}. \tag{3.6.3}$$

This gives

$$|\ln m_{\mathcal{C}}^* - \ln m_{\mathcal{C}'}^*| \leq \ln d_{BM}(\mathcal{C}, \mathcal{C}'), \quad \mathcal{C}, \mathcal{C}' \in \mathfrak{B}.$$

In other words, with respect to the Banach–Mazur distance $\ln d_{BM}$, the function $\ln m^*$ is Lipschitz with Lipschitz constant equal to 1.

In particular, we have

$$|m_{\mathcal{C}}^* - m_{\mathcal{C}'}^*| \leq n(d_{BM}(\mathcal{C}, \mathcal{C}') - 1).$$

(Indeed, by simple calculus, we have $(1/n)|x - y| \leq |\ln(x) - \ln(y)|$, $1 \leq x, y \leq n$, and $\ln(z) \leq z - 1$, $z > 0$.) This gives another proof of continuity of m^*.

To show (3.6.3), let $\delta \geq 0$ and $\mathcal{C}, \mathcal{C}' \in \mathfrak{B}$ such that $d_{BM}(\mathcal{C}, \mathcal{C}') \leq 1 + \delta$. By the definition of the Banach–Mazur distance, we have

$$\mathcal{C}' \subset \phi(\mathcal{C}) \subset (1 + \delta)\mathcal{C}' + Y, \tag{3.6.4}$$

for some $\phi \in \mathrm{Aff}(\mathcal{X})$ and $Y \in \mathcal{X}$. Due to affine invariance of m^* (and d_{BM}), we may assume that ϕ is the *identity*.

Let $\lambda > 0$ such that $m_{\mathcal{C}}^* \leq \lambda$. By (3.2.12) in Lemma 3.2.3, we have

$$\mathcal{C} + X \subset -\lambda \mathcal{C}, \tag{3.6.5}$$

for some $X \in \mathcal{X}$. Combining (3.6.4) and (3.6.5), we obtain

$$\mathcal{C}' + X + \lambda Y \subset \mathcal{C} + X + \lambda Y \subset -\lambda(\mathcal{C} - Y) \subset -\lambda(1 + \delta)\mathcal{C}'.$$

Once again, (3.2.12) (this time applied to $\mathcal{C}'$) gives $m_{\mathcal{C}'}^* \leq \lambda(1 + \delta)$. We obtain

$$m_{\mathcal{C}'}^* \leq m_{\mathcal{C}}^* \cdot d_{BM}(\mathcal{C}, \mathcal{C}'). \tag{3.6.6}$$

Switching the roles of $\mathcal{C}$ and $\mathcal{C}'$, we arrive at (3.6.3).

The inequality in (3.6.3) implies

$$m_{\mathcal{C}}^* \leq d_{BM}(\mathcal{C}, \mathcal{C}_0), \quad \mathcal{C} \in \mathfrak{B},$$

for any symmetric convex body $\mathcal{C}_0 \in \mathfrak{B}$ since $m_{\mathcal{C}_0}^* = 1$. In addition, by (3.2.11), for the Minkowski symmetral $\tilde{\mathcal{C}} = (\mathcal{C} - \mathcal{C})/2$, equality holds. We conclude that the minimum distance of a convex body $\mathcal{C}$ from any symmetric convex body is realized by the Minkowski symmetral of $\mathcal{C}$.

The centroid $g(C)$ of a convex body $C \subset \mathcal{X}$, is an *affine invariant point* in the sense that, for any $\phi \in \text{Aff}(\mathcal{X})$, we have $g(\phi(C)) = \phi(g(C))$. (This follows immediately by the respective affine change of variables in the integrals defining the centroid.)

We obtain that the centroidal Minkowski measure $\mathfrak{m}^g : \mathfrak{B} \to \mathbb{R}$, $\mathfrak{m}^g(C) = \mathfrak{m}_C^g$, $C \in \mathfrak{B}$, is an affine invariant function. Moreover, we have the following:

Proposition 3.6.2. *The centroidal Minkowski measure* $\mathfrak{m}^g : \mathfrak{B} \to \mathbb{R}$ *is a measure of symmetry.*

Proof. We need to prove sequential continuity of $\mathfrak{m}^g$. We let $C \in \mathfrak{B}$, and $\{C_k\}_{k \geq 1} \subset \mathfrak{B}$ a sequence such that $\lim_{k \to \infty} C_k = C$ with respect to the Banach–Mazur distance. We need to show

$$\lim_{k \to \infty} \mathfrak{m}^g(C_k) = \mathfrak{m}^g(C). \tag{3.6.7}$$

(For notational convenience, convex bodies will be displayed in subscripts and in functional arguments interchangeably.)

As in the proof of Proposition 3.6.1, replacing the participating convex bodies by suitable affine copies, we obtain $\lim_{k \to \infty} C_k = C$ with respect to the Hausdorff metric.

Simple integration now gives $\lim_{k \to \infty} g(C_k) = g(C)$. As noted previously, for any compact subset $C_0 \subset \text{int } C$, there exists $k_0 \geq 1$ such that we have $C_0 \subset \text{int } C_k$ for $k \geq k_0$, and the maximum distortions $\mathfrak{m}_{C_k}$ converge to $\mathfrak{m}_C$ uniformly on C_0. Now (3.6.7) and therefore Proposition 3.6.2 follow.

Remark. The centroid $g : \mathfrak{B} \to \mathcal{X}$ is an archetype of an *affine invariant point* which, by definition, is a continuous map $p : \mathfrak{B} \to \mathcal{X}$ (with respect to the Hausdorff distance d_H on $\mathfrak{B}$) and satisfies $p(\phi(C)) = \phi(p(C))$, $\phi \in \text{Aff}(\mathcal{X})$, $C \in \mathfrak{B}$.

As [Grünbaum 2] noted, similar to the construction of the centroidal Minkowski measure $\mathfrak{m}^g(C) = \mathfrak{m}(g(C))$, $C \in \mathfrak{B}$, evaluating the maximum distortion function $\mathfrak{m} : \text{int } C \to \mathbb{R}$ on (proper) affine invariant points of C one obtains a whole host of new affine measures of symmetry. (Here *proper* means that $p(C) \in \text{int } C$, $C \in \mathfrak{B}$.) Grünbaum called these *derived* measures of symmetry.

Examples of affine invariant points yielding new derived measures of symmetry abound, for example: (1) the *surface-area centroid*, that is, the center of uniform mass distribution on the *boundary* [Schneider 2, 5.4]; (2) the center O of John's ellipsoid; (3) the center O' of the *Löwner ellipsoid*. (See [Grünbaum 2].)

In fact [Meyer–Schütt–Werner 2, Meyer–Schütt–Werner 3] proved that the space of affine invariant points is *infinite dimensional*.

More recently [Meyer–Schütt–Werner 1] introduced the concept of *dual affine invariant points* as follows. Given an affine invariant point $p : \mathfrak{B} \to \mathcal{X}$, a dual $q : \mathfrak{B} \to \mathcal{X}$ of p is an affine invariant point if it satisfies $q(C^{p(C)}) = p(C)$, $C \in \mathfrak{B}$. Here C^O is the *dual* of C with respect to $O \in \text{int } C$ (Section 3.1).

They proved that the dual affine invariant point may not exist, but if it does then it is unique. A prime example of a dual pair is the centers O and O' of the John and Löwner ellipsoids. (Section 3.2.) In addition, the dual of the centroid g is the *Santaló point* $s : \mathfrak{B} \to \mathcal{X}$ defined, for $C \in \mathfrak{B}$, as the unique point $s(C) \in \operatorname{int} C$ satisfying $\operatorname{vol}(C^{s(C)}) = \max_{O \in \operatorname{int} C} \operatorname{vol}(C^O)$.

3.7 Winternitz Measures

Let $C \in \mathfrak{B}$. Given $O \in \operatorname{int} C$ and $N \in \mathcal{S}$, consider the half-space $\mathcal{G}_O(N) \subset \mathcal{X}$ with boundary hyperplane $\mathcal{H}_O(N)$ containing O and having *outward normal vector* N:

$$\mathcal{G}_O(N) = \{X \in \mathcal{X} \mid \langle X, N \rangle \le \langle O, N \rangle\}.$$

(See Example 1.1.1.) We define

$$\mathfrak{w}_C(O) = \sup_{N \in \mathcal{S}} \frac{\operatorname{vol}(C \cap \mathcal{G}_O(N))}{\operatorname{vol}(C \cap \mathcal{G}_O(-N))}. \tag{3.7.1}$$

A standard continuity argument implies that the supremum is attained. Varying $O \in \operatorname{int} C$, we obtain the continuous function $\mathfrak{w}_C : \operatorname{int} C \to \mathbb{R}$ with boundary behavior $\mathfrak{w}_C(O) \to \infty$ as $d(O, \partial C) \to 0$. The infimum of $\mathfrak{w}_C$ is therefore attained on a compact subset $C^* \subset \operatorname{int} C$, called the (Winternitz) *critical set* of C. By definition, we have $\mathfrak{w}_C \ge 1$.

Proposition 3.7.1 ([Blaschke 3]). *Let $C \in \mathfrak{B}$ and $O^* \in C^*$ a critical point. Assume that the supremum $\mathfrak{w}^* = \mathfrak{w}(O^*)$ of the volume ratios in (3.7.1) is attained at $N_0 \in \mathcal{S}$. Then O^* is the centroid of the intersection $C \cap \mathcal{H}_{O^*}(N_0)$.*

Proof. For simplicity, we may assume that O^* is at the origin. Setting $\mathcal{H}_0 = \mathcal{H}_0(N_0)$ and $C_0 = C \cap \mathcal{H}_0$, we need to show that $g(C_0) = 0$.

Let $N \in \mathcal{H}_0 \cap \mathcal{S}$ and $N_\alpha = \cos\alpha \cdot N_0 + \sin\alpha \cdot N$, $0 \le \alpha \le \pi/2$. To compare $C \cap \mathcal{G}_0(N_0)$ with the "rotated" $C \cap \mathcal{G}_0(N_\alpha)$, we write

$$C \cap \mathcal{G}_0(N_\alpha) = ((C \cap \mathcal{G}_0(N_0)) \cup C_\alpha^+) \setminus C_\alpha^-,$$

where

$$C_\alpha^+ = C \cap \mathcal{G}_0(N_\alpha) \setminus \mathcal{G}_0(N_0),$$
$$C_\alpha^- = C \cap \mathcal{G}_0(N_0) \setminus \mathcal{G}_0(N_\alpha).$$

Hence, we obtain

$$\operatorname{vol}(C \cap \mathcal{G}_0(N_\alpha)) = \operatorname{vol}(C \cap \mathcal{G}_0(N_0)) + \operatorname{vol} C_\alpha^+ - \operatorname{vol} C_\alpha^-.$$

Since $N_0 \in \mathcal{S}$ is maximal, we have

$$\operatorname{vol} C_\alpha^+ - \operatorname{vol} C_\alpha^- \le 0. \tag{3.7.2}$$

To estimate the volume difference we introduce a coordinate system $x = (x_1, \ldots, x_n)$ in $\mathcal{X}$ such that $N_0 = (0, \ldots, 0, 1)$ and $N = (0, \ldots, 0, -1, 0)$. (Note the sign change.) Letting $\mathcal{H}_t = \{x_{n-1} = t\} = \mathcal{H}_{(0,\ldots,0,t,0)}(N)$ we have

$$\operatorname{vol} \mathcal{C}_\alpha^+ - \operatorname{vol} \mathcal{C}_\alpha^- = \int_0^{a'} \operatorname{vol} (\mathcal{C}_\alpha^+ \cap \mathcal{H}_t) \, dt - \int_{a''}^0 \operatorname{vol} (\mathcal{C}_\alpha^- \cap \mathcal{H}_t) \, dt,$$

with some fixed bounds $a'' < 0 < a'$. We approximate each (signed) intersection $\mathcal{H}_t \cap \mathcal{C}_\alpha^\pm$ by the $((n-1)$-dimensional) cylinder with base $\mathcal{C}_0 \cap \mathcal{H}_t$ and height $t \tan \alpha$. Since the boundary of $\mathcal{C}$ is Lipschitz continuous (Section 1.1), we obtain

$$\operatorname{vol} \mathcal{C}_\alpha^+ - \operatorname{vol} \mathcal{C}_\alpha^- = \tan \alpha \int_{a''}^{a'} t \operatorname{vol} (\mathcal{C}_0 \cap \mathcal{H}_t) \, dt + O(\alpha^2), \quad \alpha \to 0. \tag{3.7.3}$$

The integral is $g_{n-1}(\mathcal{C}_0)$, the $(n-1)$st coordinate of the centroid of $\mathcal{C}_0$. Comparing (3.7.2)–(3.7.3), we arrive at

$$g_{n-1}(\mathcal{C}_0) \le 0.$$

Changing the sign of N, we obtain the opposite inequality, so that $g(\mathcal{C}_0) \in \mathcal{H}_0(N)$. Now, varying $N \in \mathcal{S} \cap \mathcal{H}_0$, we get $g(\mathcal{C}_0) = 0$. The proposition follows.

Remark. For the original proof, see [Blaschke 3].

Proposition 3.7.2. *Let* $\mathcal{C} \in \mathfrak{B}$. *The function* $\mathfrak{w}_\mathcal{C} : \operatorname{int} \mathcal{C} \to \mathbb{R}$ *is quasi-convex. It attains its infimum at a unique interior point of* $\mathcal{C}$, *so that the critical set* $\mathcal{C}^*$ *is a singleton.*

Proof. Let $O_0, O_1 \in \operatorname{int} \mathcal{C}$. Setting $O_\lambda = (1 - \lambda)O_0 + \lambda O_1, \lambda \in [0, 1]$, we claim

$$\mathfrak{w}_\mathcal{C}(O_\lambda) \le \max(\mathfrak{w}_\mathcal{C}(O_0), \mathfrak{w}_\mathcal{C}(O_1)). \tag{3.7.4}$$

Observe that quasi-convexity (convexity of the level-sets) is a direct consequence of this.

To prove the claim, let $\lambda \in (0, 1)$ and assume that the supremum in (3.7.1) defining $\mathfrak{w}_\mathcal{C}(O_\lambda)$ is attained at $N_0 \in \mathcal{S}$. By definition, the hyperplane $\mathcal{H}_{O_\lambda}(N_0)$ splits $\mathcal{C}$ in the volume ratio $\mathfrak{w}_\mathcal{C}(O_\lambda)$. Since O_λ is in the interior of the line segment $[O_0, O_1]$, one of the parallel hyperplanes $\mathcal{H}_{O_0}(N_0)$ or $\mathcal{H}_{O_1}(N_0)$ splits $\mathcal{C}$ in a volume ratio $\ge \mathfrak{w}_\mathcal{C}(O_\lambda)$. Thus, (3.7.4) and the first statement of the proposition follow.

As a byproduct, we also see that the inequality in (3.7.4) is strict if $[O_0, O_1] \not\subset \mathcal{H}_{O_\lambda}(N_0)$.

For the second statement first note that, as a consequence of quasi-convexity of $\mathfrak{w}_\mathcal{C}$ just proved, the critical set $\mathcal{C}^* \subset \operatorname{int} \mathcal{C}$ is (compact and) convex.

Assume, on the contrary, that there exist distinct critical points $O_0^*, O_1^* \in \mathcal{C}^*$. Consider the line segment $[O_0^*, O_1^*] \subset \mathcal{C}^*$ parametrized as usual by $\lambda \mapsto O_\lambda^* = (1 - \lambda)O_0^* + \lambda O_1^*, \lambda \in [0, 1]$. Then, for any $\lambda \in (0, 1)$ and any $N_0 \in \mathcal{S}$ that realizes the supremum in (3.7.1) for $\mathfrak{w}_C(O_\lambda^*)$, we have $[O_0^*, O_1^*] \subset \mathcal{H}_{O_\lambda^*}(N_0)$ since otherwise strict inequality would hold in (3.7.4) contradicting minimality of the critical set. On the other hand, by Proposition 3.7.1, the only critical point in the intersection $C \cap \mathcal{H}_{O_\lambda^*}(N_0)$ is the centroid. This is a contradiction.

Remark. For other proofs, see [Süss 2] ($n = 2$) and the unpublished paper of Hammer (Volumes cut from convex bodies by planes, Preprint 1960, unpublished) ($n \geq 3$).

Using Proposition 3.7.2, the following geometric picture emerges. Given $C \in \mathfrak{B}$, for $r > \mathfrak{w}_C^*$, the level-set $C_r = \{O \in \operatorname{int} C \mid \mathfrak{w}_C(O) \leq r\}$ is a convex body in $\operatorname{int} C$, and the boundaries in the monotonic family $\{C_r\}_{r > \mathfrak{w}_C^*}$ give rise to a foliation of $\operatorname{int} C \setminus C^*, C^* = \{O^*\}$, with topological spheres. Moreover, for $O \in \partial C_r, r > \mathfrak{w}_C^*$, the half-space $\mathcal{G}_O(N)$ that realizes the supremum in (3.7.1) contains C_r, in particular, its boundary hyperplane $\mathcal{H}_O(N)$ supports C_r.

As noted in [Grünbaum 3, Grünbaum 4], another property of the critical set $C^* = \{O^*\}$ is that the supremum in (3.7.1) defining $\mathfrak{w}_C(O^*)$ is assumed at least $n+1$ times. In other words, there exist at least $n + 1$ hyperplanes containing O^* and splitting the volume of C in the ratio $\mathfrak{w}_C^* = \mathfrak{w}_C(O^*)$.

In analogy with the Minkowski measure, we now define the *Winternitz measure* of C as

$$\mathfrak{w}_C^* = \mathfrak{w}_C(O^*) = \inf_{O \in \operatorname{int} C} \mathfrak{w}_C(O), \tag{3.7.5}$$

where $C^* = \{O^*\}$ is the critical set, and the *centroidal Winternitz measure* of C as

$$\mathfrak{w}_C^g = \mathfrak{w}_C(g(C)), \ C \in \mathfrak{B},$$

where $g(C) \in \operatorname{int} C$ is the centroid of C.

Clearly, we have

$$1 \leq \mathfrak{w}_C^* \leq \mathfrak{w}_C^g, \ C \in \mathfrak{B}.$$

For an upper estimate of the Winternitz measures, we have the following:

Theorem 3.7.3. *For $C \in \mathfrak{B}$, we have*

$$\mathfrak{w}_C^g \leq \left(1 + \frac{1}{n}\right)^n - 1. \tag{3.7.6}$$

Equality holds if and only if C is a convex cone.

Proof. For brevity, let $e_n = (1 + 1/n)^n$. Let $C \in \mathfrak{B}$ be such that $\mathfrak{w}_C^g \geq e_n - 1$. As usual, setting centroid at the origin, $g(C) = 0$, there exists $N_0 \in \mathcal{S}$ such that

$$\frac{\operatorname{vol}(C \cap \mathcal{G}_0(N_0))}{\operatorname{vol}(C \cap \mathcal{G}_0(-N_0))} \geq e_n - 1. \tag{3.7.7}$$

In the light of the proof of Theorem 3.4.2, we let $\mathcal{G}' = \mathcal{G}_0(N_0)$, $\mathcal{G}'' = \mathcal{G}_0(-N_0)$, and $\mathcal{H} = \mathcal{H}_0(N_0) = \partial\mathcal{G}' = \partial\mathcal{G}''$, and endow $\mathcal{X}$ with a coordinate system $x = (x_1, \ldots, x_n)$ such that $N_0 = (0, \ldots, 0, 1)$. With this, we have $\mathcal{G}' = \{x_n \leq 0\}$, $\mathcal{G}'' = \{x_n \geq 0\}$, and $\mathcal{H} = \{x_n = 0\}$. (Note the opposite inequalities.) Finally, for $\mathcal{A} \subset \mathcal{X}$, we denote $\mathcal{A}' = \mathcal{A} \cap \mathcal{G}'$ and $\mathcal{A}'' = \mathcal{A} \cap \mathcal{G}''$.

As in the proof of Theorem 3.4.4, we perform a rotational symmetrization with respect to the nth axis. Adjusting the notation, we observe that (3.7.6) and (3.7.7) remain valid. We may therefore assume that $\mathcal{C}$ is rotationally symmetric with respect to the nth axis.

We now define the cone $\mathcal{K} \in \mathfrak{B}$ "matching" $\mathcal{C}$ as follows. First, let $\mathcal{K}'' \in \mathfrak{B}$ be the cone with base $\mathcal{C} \cap \mathcal{H}$ and vertex $(0, \ldots, 0, v) = v \cdot N_0$, $v > 0$, such that $\mathrm{vol}\,\mathcal{K}'' = \mathrm{vol}\,\mathcal{C}''$. Second, we extend $\mathcal{K}''$ along its generators to obtain $\mathcal{K}$ such that the (affine span of the) base of $\mathcal{K}$ is parallel to $\mathcal{H}$ and $\mathrm{vol}\,\mathcal{K} = \mathrm{vol}\,\mathcal{C}$. By construction, $\mathcal{K}'$ is a truncated cone with $\mathrm{vol}\,\mathcal{K}'' = \mathrm{vol}\,\mathcal{C}''$, and (3.7.7) gives

$$\frac{\mathrm{vol}\,\mathcal{K}'}{\mathrm{vol}\,\mathcal{K}''} \geq e_n - 1. \tag{3.7.8}$$

In addition, convexity of $\mathcal{C}$ implies

$$g_n(\mathcal{K}'') \geq g_n(\mathcal{C}'') \quad \text{and} \quad g_n(\mathcal{K}') \geq g_n(\mathcal{C}'),$$

where the subscript indicates the nth coordinate. These give $g(\mathcal{K}) \in \mathcal{G}''$. In particular, we have

$$\mathcal{G}_0(N_0) \subset \mathcal{G}_{g(\mathcal{K})}(N_0) \quad \text{and} \quad \mathcal{G}_{g(\mathcal{K})}(-N_0) \subset \mathcal{G}_0(-N_0).$$

Taking intersections with $\mathcal{K}$, by (3.7.8), we obtain

$$\frac{\mathrm{vol}\,(\mathcal{K} \cap \mathcal{G}_{g(\mathcal{K})}(N_0))}{\mathrm{vol}\,(\mathcal{K} \cap \mathcal{G}_{g(\mathcal{K})}(-N_0))} \geq \frac{\mathrm{vol}\,\mathcal{K}'}{\mathrm{vol}\,\mathcal{K}''} \geq e_n - 1.$$

The left-hand side is the volume ratio of a cone with a hyperplane section through the centroid and parallel to the base. By elementary reasoning, this ratio is $e_n - 1$. Thus, in all estimates in the proof above, equalities hold. The theorem follows.

Remark 1. As noted in [Blaschke 2, pp. 54–55], it was A. Winternitz who observed first that a line through the centroid cuts a planar convex body into an area-ratio between 4/5 and 5/4. In addition he also noted that the triangles were the only extremal bodies. Winternitz's results have been rediscovered many times. For a detailed history of this problem, see [Grünbaum 3] and Hammer (Volumes cut from convex bodies by planes, preprint 1960, unpublished). The proof above is due to [Grünbaum 3].

Remark 2. A stability estimate of the upper bound for $\mathfrak{w}^g$ was obtained by [Groemer 1] as follows: Given $\mathcal{C} \in \mathfrak{B}$ and $\epsilon \geq 0$, there exists a universal constant λ_n, such that

$$\mathfrak{w}_{\mathcal{C}}^g \geq \left(1 + \frac{1}{n}\right) - 1 - \epsilon$$

implies that $\mathcal{C}$ contains a convex cone $\mathcal{K} \subset \mathcal{C}$ satisfying

$$d_\Delta(\mathcal{C}, \mathcal{K}) \leq \lambda_n \mathrm{vol}\,(\mathcal{C})\, \epsilon^{1/2n^2}.$$

(Here d_Δ is the symmetric difference metric; see Section 1.1/B.)

Summarizing, for $\mathcal{C} \in \mathfrak{B}$, we have

$$1 \leq \mathfrak{w}_{\mathcal{C}}^* \leq \mathfrak{w}_{\mathcal{C}}^g \leq \left(1 + \frac{1}{n}\right) - 1. \tag{3.7.9}$$

By Theorem 3.7.3 just proved, the upper bound is attained by $\mathfrak{w}_{\mathcal{C}}^g$, that is $\mathfrak{w}_{\mathcal{C}}^g = (1 + 1/n)^n - 1$, if and only if $\mathcal{C}$ is a convex cone.

We now claim that the (same) upper bound is attained in (3.7.9) by $\mathfrak{w}_{\mathcal{C}}^*$ if and only if $\mathcal{C}$ is a simplex.

Indeed, assume that $\mathfrak{w}_{\mathcal{C}}^* = \mathfrak{w}(O^*) = e_n - 1$, $e_n = (1 + 1/n)^n$. Then, we also have $\mathfrak{w}_{\mathcal{C}}^g = e_n - 1$. By (the proof of) Theorem 3.7.3, $\mathcal{C}$ is a convex cone. Setting the centroid $g(\mathcal{C})$ at the origin, the base of the cone $\mathcal{C}$ is orthogonal to a unit vector $N_0 \in \mathcal{S}$ that realizes equality in (3.7.7). In addition, the vertex V_0 (of $\mathcal{C}$ as a cone) is contained in $\mathrm{int}\,\mathcal{G}_0(-N_0)$.

Note that, by assumption, the critical point O^* is the centroid: $O^* = g(\mathcal{C}) = 0$.

Now let $O \in \mathrm{int}\,\mathcal{C} \cap \mathrm{int}\,\mathcal{G}_0(N_0)$. Let $N \in \mathcal{S}$ realize the supremum for $\mathfrak{w}(O)$ in (3.7.1). Since $\mathfrak{w}_{\mathcal{C}}(O) > \mathfrak{w}_{\mathcal{C}}^*$, we have $0 \in \mathcal{G}_0(N)$. Letting O approach to 0 along a line segment, the limit of a convergent subsequence of the corresponding normal vectors $N \in \mathcal{S}$ is a unit vector $N_1 \neq N_0$ such that equality holds in (3.7.7) for N_0 replaced by N_1. Applying once again the proof of Theorem 3.7.3, we obtain that $\mathcal{C}$ is a convex cone with base orthogonal to N_1. As before, the corresponding vertex V_1 is in $\mathrm{int}\,\mathcal{G}_0(-N_1)$.

Comparing the two cone structures of $\mathcal{C}$, we see that first vertex V_0 is contained in the base of the second cone structure, and the second vertex V_1 is contained in the base of the first cone structure. We now use induction to define $N_k \in \mathcal{S}$, $k = 0, \ldots, n$ (as a limit of extremal normals on a line segment $[O, 0]$ with $O \in \mathrm{int}\,\mathcal{C} \cap \mathrm{int}\,\mathcal{G}_0(N_0) \cap \ldots \cap \mathrm{int}\,\mathcal{G}_0(N_{k-1})$). We obtain that $\mathcal{C}$ has $n + 1$ distinct cone structures, so that it must be a simplex.

The lower bound in (3.7.9) is attained *precisely* on symmetric convex bodies (simultaneously by $\mathfrak{w}_{\mathcal{C}}^*$ and $\mathfrak{w}_{\mathcal{C}}^g$). The "if" part is obvious. The "only if" part has a long and circuitous history. The planar case $n = 2$ is easy, and proofs of the three-dimensional case have been published by [Funk 1, 2] and [Kubota]. The general case $n \geq 3$ has been proved by [Petty], [Schneider 3], and [Falconer]. See also [Groemer 3, Theorems 5.6.8–5.6.10] (for star-bodies), and also for the corresponding stability of the lower bound.

Finally, note that the proofs of Propositions 3.6.1 and 3.6.2 suitably modified give that $\mathfrak{w}^* : \mathfrak{B} \to \mathbb{R}$, $\mathfrak{w}^*(\mathcal{C}) = \mathfrak{w}_{\mathcal{C}}^*$, and $\mathfrak{w}^g : \mathfrak{B} \to \mathbb{R}$, $\mathfrak{w}^g(\mathcal{C}) = \mathfrak{w}_{\mathcal{C}}^g$, $\mathcal{C} \in \mathfrak{B}$, are measures of symmetry.

Remark 1. The Winternitz and Minkowski measures have simple comparisons

$$(\mathfrak{w}^*)^{1/n} \leq \mathfrak{m}^* \quad \text{and} \quad (\mathfrak{w}^g)^{1/n} \leq \mathfrak{m}^g. \tag{3.7.10}$$

These follow from the estimate

$$\frac{1}{\mathfrak{m}(O)^n} \leq \mathfrak{w}(O) \leq \mathfrak{m}(O)^n, \quad O \in \text{int}\,\mathcal{C}, \ \mathcal{C} \in \mathfrak{B}. \tag{3.7.11}$$

To prove this, let $N \in \mathcal{S}$ a unit vector, and $\mathcal{C}^\pm = \mathcal{C} \cap \mathcal{G}_O(\pm N)$. Assuming that O is the origin, the remark after Lemma 3.2.3 gives $-\mathcal{C} \subset \mathfrak{m}(0)\,\mathcal{C}$. Applying this to the two parts $\mathcal{C}^\pm$, we obtain $-\mathcal{C}^\pm \subset \mathfrak{m}(0)\mathcal{C}^\mp$. Taking volumes, we have $\text{vol}\,(\mathcal{C}^\pm) \leq \mathfrak{m}(0)^n \,\text{vol}\,(\mathcal{C}^\mp)$. These give

$$\frac{1}{\mathfrak{m}(0)^n} \leq \frac{\text{vol}\,(\mathcal{C}^+)}{\text{vol}\,(\mathcal{C}^-)} \leq \mathfrak{m}(0)^n.$$

Taking the supremum in $N \in \mathcal{S}$, (3.7.11) and hence (3.7.10) follow.

Remark 2. Analogously to the Minkowski measure, several variants of the Winternitz measure can be defined; see [Grünbaum 2]. For the *surface Winternitz measure*, the lower bound is again attained on symmetric convex bodies. (See [Groemer 3, Theorem 5.5.17] along with further results.)

3.8 Other Measures of Symmetry

As noted in the preface, this book is not a comprehensive survey on measures of symmetry. Nevertheless, we will discuss two additional measures of symmetry not only because of their beautiful geometry, but also because they provide additional insights to our previous studies. The *Rogers–Shephard volume ratio* sheds new light to the nature of the Minkowski symmetral and links up with the Brunn–Minkowski inequality, and Qi Guo's *L^p-Minkowski measure* opens a path to generalizations of the classical Minkowski measure to L^p-setting. In our rudimentary treatments of these measures we focus on the main points only.

A. The Rogers–Shephard Volume Ratio. Let $\mathfrak{rs} : \mathfrak{B} \to \mathbb{R}$ be defined by

$$\mathfrak{rs}(C) = \frac{\text{vol}\,(C - C)}{\text{vol}\,C}, \quad C \in \mathfrak{B},$$

where $C - C = 2\tilde{C} \in \mathfrak{B}$ is the *difference body* of C. ($\tilde{C}$ is the Minkowski symmetral of C as in Proposition 3.2.2.) Then we have

$$2^n \le \mathfrak{rs}(C) \le \binom{2n}{n}, \quad C \in \mathfrak{B}. \tag{3.8.1}$$

Moreover, equality holds on the left-hand side if and only if C is symmetric, and on the right-hand side if and only if C is a simplex.

The inequality on the left-hand side of (3.8.1) is a direct consequence of the Brunn–Minkowski inequality (3.5.1) (with $C_0 = C$, $C_1 = -C$, and $\alpha = 1/2$). Equality is clearly attained for symmetric C (since $C - C = 2C$). Conversely, also by the Brunn–Minkowski theorem, if equality is attained then C and $-C$ are homothetic. Therefore, they must be equal, so that C is symmetric.

The inequality on the right-hand side of (3.8.1) is known as the [Rogers–Shephard] inequality. Following their original and illuminating approach, we now derive this, and give a brief indication why equality is attained for simplices only. (See also [Eggleston 1, 5.6]. For a more recent proof, see [Schneider 2, 10.1].)

Let $C \in \mathfrak{B}$. The idea of Rogers–Shephard is to "spread out" translates of C in the product space $\mathcal{X} \times \mathcal{X}$ to form a convex body $\mathcal{D}$, and to calculate the volume of $\mathcal{D}$ in two different ways; by vertical and horizontal slicing along the two factors $\mathcal{X}$.

We let

$$\mathcal{D} = \{(X, Y) \in \mathcal{X} \times \mathcal{X} \mid X \in C, X + Y \in C\} = \bigcup_{X \in C} \{X\} \times (C - X).$$

Clearly, $\mathcal{D}$ is a convex body in $\mathcal{X} \times \mathcal{X}$; in particular, $\dim \mathcal{D} = 2 \dim \mathcal{X} = 2n$.

For any set $\mathcal{A} \subset \mathcal{X} \times \mathcal{X}$, we denote by $\mathcal{A}' \subset \mathcal{X}$, resp. $\mathcal{A}'' \subset \mathcal{X}$, the projection of $\mathcal{A}$ to the first, resp. second, factor of $\mathcal{X} \times \mathcal{X}$. We have $\mathcal{D}' = C$ and $\mathcal{D}'' = C - C$; both convex bodies in $\mathcal{X}$.

The vertical slicing of $\mathcal{D}$ is simple since by definition we have $\mathcal{D} \cap (\{X\} \times \mathcal{X}) = \{X\} \times (C - X)$, $X \in C$. We then integrate

$$\text{vol}_{2n}(\mathcal{D}) = \int_C \text{vol}_n(C - X)\, dX = \int_C \text{vol}_n C\, dX = (\text{vol}_n C)^2, \tag{3.8.2}$$

where the subscripts indicate the respective volume dimensions.

Turning to the horizontal slicing of $\mathcal{D}$, for $Y \in \mathcal{D}'' = C - C$, we clearly have

$$\mathcal{D} \cap (\mathcal{X} \times \{Y\}) = (C \cap (C - Y)) \times \{Y\}.$$

In particular, for $Y = 0$, we have $\mathcal{D} \cap (\mathcal{X} \times \{0\}) = C \times \{0\}$.

From now on (without loss of generality) we assume that the origin 0 is an interior point of C. For $0 \neq Y \in \mathcal{D}''$, we give a lower estimate of the horizontal slice $\mathcal{D} \cap (\mathcal{X} \times \{Y\})$ as follows. By convexity of $\mathcal{D}''$, there is a unique $C \in \partial \mathcal{D}''$ such that $Y \in [0, C]$. We let $Y = \lambda C$, $\lambda \in (0, 1]$. Since $C \in \mathcal{D}'' = C - C$, there exists $X \in \mathcal{D}' = C$ such that $(X, C) \in \mathcal{D}$. We have

$$\mathcal{D} \supset [(X, C), \mathcal{D} \cap (\mathcal{X} \times \{0\})] = [(X, (1/\lambda)Y), C \times \{0\}]. \tag{3.8.3}$$

The convex hull on the right-hand side is a cone with vertex (X, C) and base $C \times \{0\}$. Taking the horizontal slice of both sides in (3.8.3) by $\mathcal{X} \times \{Y\}$ and taking proportions, we obtain

$$\mathcal{D} \cap (\mathcal{X} \times \{Y\}) = (C \cap (C - Y)) \times \{Y\} \supset (1 - \lambda)C \times \{Y\}. \tag{3.8.4}$$

We can now integrate:

$$
\begin{aligned}
\operatorname{vol}_{2n}(\mathcal{D}) &= \int_{\mathcal{D}''} \operatorname{vol}_n(\mathcal{D} \cap (\mathcal{X} \times \{Y\}))\, dY \\
&\geq \int_{\mathcal{D}''} (1 - \lambda)^n \operatorname{vol}_n(C \times \{Y\})\, dY \\
&= \int_0^1 (1 - \lambda)^n \operatorname{vol}_n C \, d(\lambda^n \operatorname{vol}_n(\mathcal{D}'')) \\
&= \operatorname{vol}_n C \cdot \operatorname{vol}_n(C - C) \int_0^1 n (1 - \lambda)^n \lambda^{n-1}\, d\lambda,
\end{aligned}
$$

where the last but one integral is with respect to the level-sets of λ as a function on $\mathcal{D}''$.

Finally, by elementary integration (using repeated integration by parts, say), we obtain

$$\int_0^1 n (1 - \lambda)^n \lambda^{n-1}\, d\lambda = \binom{2n}{n}^{-1}.$$

Substituting this to the computation above and using (3.8.2), the Rogers–Shephard upper estimate in (3.8.1) follows.

Assume now that equality holds. Then equality also holds in (3.8.4) and we have

$$C \cap (C - Y) = (1 - \lambda)C, \quad Y \in C - C.$$

This means that any non-trivial intersection of C with a translate of C must be homothetic with C. This property characterizes the simplices. (For details, see [Rogers–Shephard], also [Böröczky 2], and Chapter 2.1 of the article of H. Heil and H. Martini in [Gruber–Wills].)

By a considerable refinement of the argument above, [Böröczky 2] went much further and established a stability version of this argument. He showed

$$\mathfrak{rs}(\mathcal{C}) \geq (1 - \epsilon) \binom{2n}{n} \quad \Rightarrow \quad d_{BM}(\mathcal{C}, \Delta) \leq 1 + n^{50n^2}\epsilon, \quad \mathcal{C} \in \mathfrak{B}.$$

Remark. A stability estimate for the lower bound in (3.8.1) is an immediate consequence of Theorem 3.5.2 (with $C_0 = \mathcal{C}, C_1 = -\mathcal{C}$, and $\alpha = 1/2$). If $\mathcal{C} \in \mathfrak{B}$ with $\mathrm{vol}\,\mathcal{C} = 1$, then we have

$$2^n \leq \mathfrak{rs}(\mathcal{C}) \leq (1 + \epsilon)^{1/n} 2^n \quad \Rightarrow \quad h_H^{\mathrm{Dil}}(\mathcal{C}, -\mathcal{C}) \leq 4\lambda_n D_{\mathcal{C}} \epsilon^{1/(n+1)}.$$

Finally, note that, in terms of the Minkowski measure, we also have

$$\mathfrak{rs}(\mathcal{C}) = (1 + \epsilon)\, 2^n \quad \Rightarrow \quad 1 + c\epsilon \leq \mathfrak{m}_{\mathcal{C}}^* \leq 1 + c'\,\epsilon,$$

where $c, c' > 0$ depend only on n. (Compare this with Theorem 3.2.1 and Proposition 3.2.2.) The lower bound is obvious and the upper bound is due to [Diskant].

B. Guo's L^p–Minkowski Measure. Let $\mathcal{C} \in \mathfrak{B}$. Recall from Section 3.2 that the Minkowski measure $\mathfrak{m}_{\mathcal{C}}^*$ can be expressed in terms of the support function of $\mathcal{C}$ as

$$\mathfrak{m}_{\mathcal{C}}^* = \inf_{O \in \mathrm{int}\,\mathcal{C}} \max_{N \in \mathcal{S}} \frac{h_{\mathcal{C},O}(N)}{h_{\mathcal{C},O}(-N)}. \tag{3.8.5}$$

Here we used the base point dependent support function $h_{\mathcal{C},O} : \mathcal{X} \to \mathbb{R}$, $O \in \mathrm{int}\,\mathcal{C}$, given by

$$h_{\mathcal{C},O}(X) = \sup_{C \in \mathcal{C}} \langle C - O, X \rangle = h_{\mathcal{C}}(X) - \langle O, X \rangle, \ X \in \mathcal{X}.$$

The basic idea of [Guo 2] is to view the maximum in (3.8.5) as the L^∞-*norm* of the function $\rho_{\mathcal{C},O} : \mathcal{S} \to \mathbb{R}$ given by

$$\rho_{\mathcal{C},O}(N) = \frac{h_{\mathcal{C},O}(-N)}{h_{\mathcal{C},O}(N)}, \ N \in \mathcal{S},$$

and define a corresponding L^p-Minkowski measure for $1 \leq p < \infty$. (Note the sign change in the argument; a latter technical convenience.)

For the L^p-norm one needs a suitable (probability) measure $m_{\mathcal{C},O}$ on $\mathcal{S}$ (depending on the base point $O \in \mathrm{int}\,\mathcal{C}$). For this one can use the *surface area measure* $S(\mathcal{C}, \cdot) = S(\mathcal{C}, \ldots, \mathcal{C}, \cdot)$ of $\mathcal{C}$ on the unit sphere $\mathcal{S} \subset \mathcal{X}$, where $S(\mathcal{C}_1, \ldots, \mathcal{C}_{n-1}, \cdot)$ is the *mixed area measure* of $\mathcal{C}_1, \ldots, \mathcal{C}_{n-1} \in \mathfrak{B}$ on $\mathcal{S}$. (See [Schneider 2, 5.1].) With this, the probability measure $m_{\mathcal{C},O}$ on $\mathcal{S}$ is defined by

$$m_{C,O}(\omega) = \frac{\int_\omega h_{C,O}(N)\, dS(C, N)}{\int_S h_{C,O}(N)\, dS(C, N)}, \quad \omega \subset S.$$

Finally, for $1 \leq p \leq \infty$, we let

$$m_{p,C}(O) = \begin{cases} \left(\int_S \rho_{C,O}(N)^p \, dm_{C,O}(N) \right)^{1/p} & \text{if } 1 \leq p < \infty \\ \sup_{N \in S} \rho_{C,O}(N) & \text{if } p = \infty. \end{cases} \tag{3.8.6}$$

Now, Guo's *L^p-Minkowski measure* is defined as

$$m_{p,C}^* = \inf_{O \in \text{int} C} m_{p,C}(O), \quad 1 \leq p \leq \infty. \tag{3.8.7}$$

By (3.2.2), we have $m_C = m_{\infty,C}$, so that $m_C^* = m_{\infty,C}^*$. (The notation for m_p^* in [Guo 2] is as$_p$ reflecting a prevailing view of this as a measure of "asymmetry.")

The L^p-Minkowski measures $m_p^* : \mathfrak{B} \to \mathbb{R}$, $m_p^*(C) = m_{p,C}^*$, $C \in \mathfrak{B}$, $1 \leq p \leq \infty$, form an *increasing* sequence of affine invariant measures [Guo 2, Theorem 3]. The bounds are the same as for the Minkowski measure:

$$1 \leq m_p^* \leq n, \quad 1 \leq p \leq \infty.$$

In addition, for $C \in \mathfrak{B}$ and $1 \leq p \leq \infty$, the lower bound $m_{p,C}^* = 1$ is attained if and only if C is symmetric, and the upper bound $m_{p,C}^* = n$ is attained if and only if C is a simplex.

We now discuss the critical set at which the infimum in (3.8.7) is attained. It is clear from the definition that $m_{1,C}$ ($p = 1$) is constant on int C. At the other extreme, $m_{\infty,C}^* = m_C^*$ ($p = \infty$) is attained precisely on the critical set C^* of C, a compact convex set (Section 2.1). In contrast, as shown in [Guo 2, Theorem 4], for $1 < p < \infty$, the level-sets of $m_{p,C}$ are *strictly convex* so that the infimum in (3.8.7) is attained at a single point.

Derived L^p-measures can be defined using the function $m_{p,C}$ in a natural way; for example, we have the centroidal L^p-Minkowski measure $m_{p,C}^g = m_{p,C}(g(C))$, $C \in \mathfrak{B}$.

Using the volume ratio (3.7.1) instead of $\rho_{C,O}$ in (3.8.6), one can also define the L^p-Winternitz measure $\mathfrak{w}_{p,C}^*$, $1 \leq p \leq \infty$, and the respective derived measures, such as the centroidal L^p-Winternitz measure $\mathfrak{w}_{p,C}^g$, $1 \leq p \leq \infty$.

3.9 The Circumradius and Inradius in Minkowski Space and Stability

As an application of the Minkowski measure, in this section we briefly return to our earlier study of the ratios R_C/D_C and d_C/r_C, $C \in \mathfrak{B}$, in Section 1.5. Our goal here is to derive stability estimates for upper bounds of these ratios. The Jung–Steinhagen universal estimates (1.5.3) and (1.5.4) are not suitable for stability, however, since the upper bounds are not attained by well-defined extremal classes of convex bodies. (See Remark 2 after Theorem 1.5.1.)

In these estimates we assumed that $\mathcal{X}$ was Euclidean. We now relax this condition and consider the problem of giving universal upper bounds for the ratios R_C/D_C and d_C/r_C, $C \in \mathfrak{B}$, for all *Minkowski structures* on $\mathcal{X}$. We will show that in Minkowski spaces $\mathcal{X}$ we have the following

$$\frac{R_C}{D_C} \leq \frac{n}{n+1} \quad \text{and} \quad \frac{d_C}{r_C} \leq n+1, \quad C \in \mathfrak{B}. \tag{3.9.1}$$

The first inequality is due to [Bohnenblust] in 1938. Both inequalities have been proved by [Leichtweiss 1] in 1955, and a few years later independent proofs have been given by [Eggleston 2]. Following [Schneider 1] (and the original approach of Eggleston) we will give short proofs of both inequalities.

The upper bounds are sharp and attained on any simplex Δ whose difference body $\Delta - \Delta = \bar{B} \subset \mathcal{X}$ is the unit ball. Conversely, if equality holds for a convex body $C \in \mathfrak{B}$ in either of the inequalities in (3.9.1) then C is still a simplex with some specific properties as described in [Leichtweiss 1, Satz 2 and Satz 3]. In particular, and in contrast to (1.5.3) and (1.5.4), we see that the upper bounds in (3.9.1) are better suited for stability as they are attained by simplices only.

We will actually derive the following sharper estimates:

$$\frac{R_C}{D_C} \leq \frac{m_C^*}{m_C^* + 1} \quad \text{and} \quad \frac{d_C}{r_C} \leq m_C^* + 1, \quad C \in \mathfrak{B}.$$

By (2.1.7), we have $(1 \leq) m^* \leq n$, so that these immediately imply (3.9.1).

The stability estimates for the upper bounds in (3.9.1) are contained in the following:

Theorem 3.9.1 ([Schneider 1]). *Let $\mathcal{X}$ be a Minkowski space of dimension n and $0 < \epsilon < 1/n$. If $C \in \mathfrak{B}$ satisfies one of the conditions*

$$\frac{R_C}{D_C} > \frac{n-\epsilon}{n-\epsilon+1} \quad \text{or} \quad \frac{d_C}{r_C} > n-\epsilon+1, \tag{3.9.2}$$

then there exists a simplex $\Delta \in \mathfrak{B}$ such that

$$d_{BM}(C, \Delta) < 1 + \frac{n+1}{1-n\epsilon}\epsilon. \tag{3.9.3}$$

The crux in the approach of [Schneider 1] to derive (3.9.3) is to express all ingredients in terms of a function $\rho : \mathfrak{B} \times \mathfrak{B} \to \mathbb{R}$ defined by

$$\rho(C, C') = \min\{\lambda > 0 \mid C + X \subset \lambda C' \text{ for some } X \in \mathcal{X}\}, \quad C, C' \in \mathfrak{B}. \tag{3.9.4}$$

Remark 1. Lemma 3.2.3 immediately gives

$$m_C^* = \rho(C, -C), \quad C \in \mathfrak{B}. \tag{3.9.5}$$

Remark 2. The reciprocal of $\rho(C, C')$ is called the *inradius of C relative to C'*:

$$r(C, C') = \max\{\lambda \geq 0 \mid \lambda C + X \subset C' \text{ for some } X \in \mathcal{X}\}, \; C, C' \in \mathfrak{B}.$$

For the properties of the relative inradius and further developments, see [Schneider 2, 3.1].

A crucial property of ρ (not mentioned in [Schneider 1]; see [Toth 11]) is *submultiplicativity*:

Lemma 3.9.2. *We have*

$$\rho(C, C'') \leq \rho(C, C') \cdot \rho(C', C''), \quad C, C', C'' \in \mathfrak{B}. \tag{3.9.6}$$

Proof. Let $\lambda \geq \rho(C, C')$ and $\lambda' \geq \rho(C', C'')$, so that we have

$$C + X \subset \lambda C' \quad \text{and} \quad C' + X' \subset \lambda' C'', \text{ for some } X, X' \in \mathcal{X}.$$

Combining these, we obtain

$$C + X + \lambda X' \subset \lambda C' + \lambda X' \subset \lambda \lambda' C''.$$

Thus, we have $\lambda \lambda' \geq \rho(C, C'')$. The lemma follows.

The metric invariants of a convex body studied in Section 1.5 can be expressed in terms of the function ρ as follows:

Proposition 3.9.3. *Let $C \in \mathfrak{B}$. We have*

$$D_C = 2\rho(\tilde{C}, \bar{B}), \tag{3.9.7}$$

$$d_C = \frac{2}{\rho(\bar{B}, \tilde{C})}, \tag{3.9.8}$$

$$R_C = \rho(C, \bar{B}), \tag{3.9.9}$$

$$r_C = \frac{1}{\rho(\bar{B}, C)}, \tag{3.9.10}$$

where $\tilde{C} = (C - C)/2$ is the Minkowski symmetral.

Proof. A typical element of the *difference body* $2\tilde{C} = C - C$ is the difference of a pair of elements in C. Taking norms, we see that $|X - X'| \leq 2\lambda$, for all $X, X' \in C$, if and only if $\tilde{C} \subset \lambda \bar{B}$. Equation (3.9.7) follows.

To prove (3.9.8) we will make use of the support function $h_C : \mathcal{X}^* \to \mathbb{R}$ of C defined by $h_C(\phi) = \sup_{C \in C} \phi(C)$, $\phi \in \mathcal{X}^*$. (See the remark after Corollary 1.2.4.) We have

$$\frac{2}{\lambda} \le \frac{h_C(\phi) + h_C(-\phi)}{h_{\tilde{B}}(\phi)} \text{ for all } \phi \in \mathcal{X}^* \Leftrightarrow \bar{B} \subset \lambda \tilde{C} \Leftrightarrow \rho(\bar{B}, \tilde{C}) \le \lambda,$$

where the last equivalence is because $\bar{B}$ and $\tilde{C}$ are both symmetric (with respect to the origin). Since $d_C = \inf\{h_C(\phi) + h_C(-\phi) \mid \phi \in \mathcal{X}^*, |\phi| = 1\}$ (Section 1.5, where $|\phi| = \max_{C \in S} \phi(C) = \max_{C \in B} \phi(C) = h_{\tilde{B}}(\phi)$, this implies (3.9.8).

Finally, (3.9.9) and (3.9.10) follow from the definitions of the circumradius and inradius (Section 1.5).

In view of (3.9.7)–(3.9.10), the two inequalities in (3.9.1) can be written in more symmetric forms as

$$\frac{R_C}{D_C} = \frac{\rho(C, \bar{B})}{2\rho(\tilde{C}, \bar{B})} \le \frac{n}{n+1} \quad \text{and} \quad \frac{d_C}{r_C} = \frac{2\rho(\bar{B}, C)}{\rho(\bar{B}, \tilde{C})} \le n + 1, \ C \in \mathfrak{B}. \tag{3.9.11}$$

The crux in establishing these estimates (and the respective stability) is the following intermediate step:

Proposition 3.9.4. *Let $C \in \mathfrak{B}$. We have*

$$\rho(C, \tilde{C}) = \frac{2m_C^*}{m_C^* + 1} \tag{3.9.12}$$

$$\rho(\tilde{C}, C) = \frac{m_C^* + 1}{2}. \tag{3.9.13}$$

Proof. We mimic the proof of Proposition 3.2.2. In general, for $\lambda > 0$ and $X \in \mathcal{X}$, we clearly have

$$C + X \subset -\lambda C \iff C + \frac{1}{\lambda + 1} X \subset \frac{2\lambda}{\lambda + 1} \tilde{C}. \tag{3.9.14}$$

For brevity, let $k^* = \rho(C, \tilde{C})$. By (3.9.5), we have $C + X \subset -m^* C$, for some $X \in \mathcal{X}$. Applying (3.9.14) [and the definition of ρ in (3.9.4)], we obtain $k^* \le 2m^*/(m^* + 1)$. Conversely, by the definition of k^*, we have $C + X \subset k^* \tilde{C}$, for some $X \in \mathcal{X}$. Applying (3.9.14) (this time backwards), we obtain $m^* \le k^*/(2 - k^*)$, or equivalently, $k^* \ge 2m^*/(m^* + 1)$. Equation (3.9.12) follows.

Turning to (3.9.13), let $k_* = \rho(\tilde{C}, C)$. In general, for $\lambda > 0$ and $X \in \mathcal{X}$, we clearly have

$$C + X \subset -\lambda C \iff 2\tilde{C} + X \subset -(\lambda + 1)C. \tag{3.9.15}$$

As before, this gives $k_* = (m^* + 1)/2$. Equation (3.9.13) follows.

After these preparations we are now ready to prove the inequalities in (3.9.11). Let $\mathcal{C} \in \mathfrak{B}$. By sub-multiplicativity of ρ in (3.9.6), we have $\rho(\mathcal{C}, \bar{\mathcal{B}}) \leq \rho(\mathcal{C}, \tilde{\mathcal{C}})\rho(\tilde{\mathcal{C}}, \bar{\mathcal{B}})$. Dividing and using (3.9.12), we have

$$\frac{R_{\mathcal{C}}}{D_{\mathcal{C}}} = \frac{\rho(\mathcal{C}, \bar{\mathcal{B}})}{2\rho(\tilde{\mathcal{C}}, \bar{\mathcal{B}})} \leq \frac{\rho(\mathcal{C}, \tilde{\mathcal{C}})}{2} = \frac{\mathrm{m}^*_{\mathcal{C}}}{\mathrm{m}^*_{\mathcal{C}} + 1} \leq \frac{n}{n + 1}. \tag{3.9.16}$$

where we used the Minkowski–Radon inequality $(1 \leq) \mathrm{m}^*_{\mathcal{C}} \leq n$ in (2.1.7). The first inequality in (3.9.11) follows.

In a similar vein, we have $\rho(\bar{\mathcal{B}}, \mathcal{C}) \leq \rho(\bar{\mathcal{B}}, \tilde{\mathcal{C}})\rho(\tilde{\mathcal{C}}, \mathcal{C})$. Dividing and using (3.9.13), we have

$$\frac{2\rho(\bar{\mathcal{B}}, \mathcal{C})}{\rho(\bar{\mathcal{B}}, \tilde{\mathcal{C}})} \leq 2\rho(\tilde{\mathcal{C}}, \mathcal{C}) \leq \mathrm{m}^*_{\mathcal{C}} + 1 \leq n + 1. \tag{3.9.17}$$

The second inequality in (3.9.11) also follows.

Proof of Theorem 3.9.1. Let $0 < \epsilon < 1/n$, and assume that $\mathcal{C} \in \mathfrak{B}$ satisfies (3.9.2). Monotonicity of the bounds in (3.9.16) and (3.9.17) *in the variable* $\mathrm{m}^*_{\mathcal{C}}$ gives $\mathrm{m}^*_{\mathcal{C}} > n - \epsilon$. Theorem 3.2.4 now applies yielding (3.9.3).

Remark. [Schneider 1] also proved the inequality

$$\frac{\rho(\mathcal{C}, \mathcal{C}')}{\rho(\tilde{\mathcal{C}}, \tilde{\mathcal{C}}')} \leq n, \quad \mathcal{C}, \mathcal{C}' \in \mathfrak{B}_{\mathcal{X}}. \tag{3.9.18}$$

Once again, (3.9.18) is a consequence of sub-multiplicativity of ρ and (3.9.12) and (3.9.13) as

$$\frac{\rho(\mathcal{C}, \mathcal{C}')}{\rho(\tilde{\mathcal{C}}, \tilde{\mathcal{C}}')} \leq \rho(\mathcal{C}, \tilde{\mathcal{C}})\rho(\tilde{\mathcal{C}}', \mathcal{C}') \leq \frac{2\mathrm{m}^*_{\mathcal{C}}}{\mathrm{m}^*_{\mathcal{C}} + 1} \frac{\mathrm{m}^*_{\mathcal{C}'} + 1}{2} \leq \frac{n}{n + 1}(n + 1) \leq n. \tag{3.9.19}$$

If the upper bound in (3.9.18) is attained then (3.9.19) immediately implies that $\mathrm{m}^*_{\mathcal{C}} = \mathrm{m}^*_{\mathcal{C}'} = n$, so that $\mathcal{C}$ and $\mathcal{C}'$ are both simplices. In addition, as shown by [Schneider 1], $\mathcal{C}'$ must be homothetic to $-\mathcal{C}$; in fact, this characterizes the upper bound n.

For a stability estimate of the upper bound in (3.9.18), assume

$$n - \frac{n}{n + 1}\epsilon \leq \frac{\rho(\mathcal{C}, \mathcal{C}')}{\rho(\tilde{\mathcal{C}}, \tilde{\mathcal{C}}')}.$$

Combining this with (3.9.19) and $\mathrm{m}^*_{\mathcal{C}}, \mathrm{m}^*_{\mathcal{C}'} \leq n$, a simple estimation gives $\mathrm{m}^*_{\mathcal{C}} \geq n - n\epsilon$ and $\mathrm{m}^*_{\mathcal{C}'} \geq n - \epsilon$. Now Theorem 3.2.4 asserts the existence of simplices $\Delta, \Delta' \in \mathfrak{B}$, such that

$$d_{BM}(\mathcal{C}, \Delta) \leq 1 + \frac{(n+1)n}{1 - n^2\epsilon}\epsilon \quad \text{and} \quad d_{BM}(\mathcal{C}', \Delta') \leq 1 + \frac{n+1}{1 - n\epsilon}\epsilon.$$

A delicate analysis in [Schneider 1] gives much more: For $0 \leq \epsilon < 1/(n(5n^2 + 1))$, there exists a simplex Δ_0 with centroid at the origin such that

$$(1 - n(5n^2 + 1)\epsilon)\Delta_0 \subset \phi(\mathcal{C}) \subset \Delta_0 \quad \text{and} \quad (1 - 2n\epsilon)\Delta_0' \subset \phi'(\mathcal{C}') \subset \Delta_0$$

for some $\phi, \phi' \in \text{Aff}(\mathcal{X})$. (In fact, ϕ and ϕ' can be chosen to be (positive) homotheties.)

Exercises and Further Problems

1. Show that, under the musical equivalences (Section 3.1), affine diameters correspond to affine diameters. More precisely, given $\mathcal{C} \in \mathfrak{B}$ and $O \in \text{int}\,\mathcal{C}$, let $[C, C^o] \subset \mathcal{C}$ be an affine diameter (through O) with parallel hyperplanes at the endpoints as the level-sets $f^{-1}(0)$ and $f^{-1}(1)$ of an affine functional $f \in \text{aff}_{\mathcal{C}}$ normalized for $\mathcal{C}$. Show that $[f^{\sharp}, (f^{\sharp})^o] \subset \mathcal{C}^O$ (through O) is an affine diameter of the dual $\mathcal{C}^O$ with parallel hyperplanes at the endpoints given by $(C^{\flat})^{-1}(0)$ and $(C^{\flat})^{-1}(1)$ of the affine functional $C^{\flat} \in \text{aff}_{\mathcal{C}^O}$ normalized for $\mathcal{C}^O$.
2. Assume that $\mathcal{C} \in \mathfrak{B}$ is *not symmetric*. Show that the only symmetric convex body in the 1-parameter family $\{(1 - \lambda)\mathcal{C} - \lambda\mathcal{C}\}_{\lambda \in [0,1]}$ is $\tilde{\mathcal{C}}$ (corresponding to $\lambda = 1/2$).
3.* Let $\mathcal{C} \in \mathfrak{B}$ and $O^* \in \mathcal{C}^*$ a critical point. Reflect $\mathcal{C}$ to O^* to obtain $\mathcal{C}' \in \mathfrak{B}$. Show that, for the *symmetric* convex body $\mathcal{C}_0 = \mathcal{C} \cap \mathcal{C}' \in \mathfrak{B}$, we have $d_{BM}(\mathcal{C}, \mathcal{C}_0) \leq \mathfrak{m}_{\mathcal{C}}^*$. (This construction is due to Qi Guo. It gives an alternative method for estimating the Banach–Mazur distance of a convex body $\mathcal{C} \in \mathfrak{B}$ from a symmetric one (not the Minkowski symmetral of $\mathcal{C}$); compare with Proposition 3.2.2.)
4. Show that an affine invariant continuous function of $\mathfrak{C}_{\mathcal{X}}$ (with respect to the Hausdorff metric) is constant; therefore, any continuous function on the quotient $\mathfrak{C}_{\mathcal{X}}/\text{Aff}(\mathcal{X})$ is constant. Hence, the quotient is not metrizable.
5.* Given $\mathcal{C} \in \mathfrak{B}$, show the existence and uniqueness of the John ellipsoid $\mathcal{E} \subset \mathcal{C}$ (an ellipsoid of maximal volume contained in $\mathcal{C}$) using the following steps: (1) Let $\mathcal{E} \subset \mathcal{X}$ be an ellipsoid. Show that $\mathcal{E}$ is the image of the unit ball $\bar{\mathcal{B}} \in \mathcal{X}$ under an affine transformation $\phi \in \text{Aff}(\mathcal{X})$ of the form $\phi(X) = AX + Z, X \in \mathcal{X}$, with $A \in GL(\mathcal{X})$ positive definite and $Z \in \mathcal{X}$. (Use polar decomposition in $GL(\mathcal{X})$.) Also note that $\text{vol}(\mathcal{E}) = \det(A)\text{vol}(\bar{\mathcal{B}})$. (2) Let $P(\mathcal{X}) \subset GL(\mathcal{X})$ denote the convex cone of all positive definite linear transformations. Show that the subset

$$\{(A, Z) \in P(\mathcal{X}) \times \mathcal{X} \mid A\bar{\mathcal{B}} + Z \subset \mathcal{C}\}$$

is compact and convex. For existence, maximize the volume over this set. For uniqueness: (3) Let $\mathcal{E} = A\bar{B} + Z$ and $\mathcal{E}' = A'\bar{B} + Z', A, A' \in P(\mathcal{X}), Z, Z' \in \mathcal{X}$, be two ellipsoids in $\mathcal{C}$ of maximal volume. Show that $A = A'$. (4) Continuing (3), show that $Z = Z'$.

6.* (1) Show that $d_{BM}(\mathcal{E}, \Delta) = n$ if $\mathcal{E}$ is an ellipsoid and Δ is a simplex. (2) Let $Q = \{(x_1, \ldots, x_n) \in \mathbb{R}^n \mid |x_i| \leq 1, i = 1, \ldots, n\}$ be the unit cube in $\mathbb{R}^n$. Show that the John's ellipsoid is the unit ball $\bar{B}$, so that $d_{BM}(\bar{B}, \mathcal{C}) = \sqrt{n}$.

7. Show that

$$d_{BM}(\mathcal{C}, \bar{B}) = \inf_{\phi \in \text{Aff}(\mathcal{X})} \frac{R_{\phi(\mathcal{C})}}{r_{\phi(\mathcal{C})}}.$$

8.* Prove "superminimality" of the Minkowski measure:

$$\mathfrak{m}^*_{\mathcal{C}' + \mathcal{C}''} \leq \max(\mathfrak{m}^*_{\mathcal{C}'}, \mathfrak{m}^*_{\mathcal{C}''}), \quad \mathcal{C}', \mathcal{C}'' \in \mathfrak{B}.$$

9.* Let $\mathcal{C} \in \mathfrak{B}$. Show that the centroid $g(\mathcal{C})$ of $\mathcal{C}$ is contained in the interior of $\mathcal{C}$.

10. Show that, for any convex cone $\mathcal{K} \in \mathfrak{B}$, we have $\mathfrak{m}^g_{\mathcal{K}} = n$.

11. Prove the Minkowski–Radon inequality $\mathfrak{m}^g_{\mathcal{C}} \leq n, \mathcal{C} \in \mathfrak{B}$, using the following steps: (1) Use an approximation argument to show that it is enough to prove the inequality for polytopes; (2) Let $\mathcal{C}$ be a polytope and assume that the centroid $g(\mathcal{C})$ is at the origin. By (3.2.2) one needs to show that $B(N)/(n+1) \leq h_{\mathcal{C}}(N) \leq B(N)n/(n+1)$, where $B(N) = h_{\mathcal{C}}(N) + h_{\mathcal{C}}(-N)$ is the distance between the two supporting hyperplanes $\mathcal{H}'$ and $\mathcal{H}''$ of $\mathcal{C}$ orthogonal to $N \in \mathcal{S}$. Choose a point $V \in \mathcal{C} \cap \mathcal{H}'$ and decompose $\mathcal{C}$ into finitely many pyramids with common vertex V and bases, the faces of $\mathcal{C}$ disjoint from V. Use Problem 10 to show that the distance of the centroid of any participating pyramid to $\mathcal{H}'$ is $\leq B(N)n/(n+1)$, and, therefore, to $\mathcal{H}''$ is $\geq B(N)/n + 1$).

12.* Prove the Brunn–Minkowski inequality: For $\mathcal{A}_1, \mathcal{A}_2 \in \mathfrak{C}$, the function $f_{\mathcal{A}_1, \mathcal{A}_2}$: $[0, 1] \to \mathbb{R}, f_{\mathcal{A}_1, \mathcal{A}_2}(\lambda) = \text{vol}(\lambda \mathcal{A}_1 + (1 - \lambda)\mathcal{A}_2)^{1/n}, \lambda \in [0, 1]$, is concave.

13.* Prove (3.4.16).

14. Use sub-multiplicativity of ρ to prove the following: $r_{\mathcal{C}}/R_{\mathcal{C}'} \leq \rho(\mathcal{C}, \mathcal{C}') \leq R_{\mathcal{C}}/r_{\mathcal{C}'}, \mathcal{C}, \mathcal{C}' \in \mathfrak{B}$.

15.* Define $d : \mathfrak{B} \times \mathfrak{B} \to \mathbb{R}$ by $d(\mathcal{C}, \mathcal{C}') = \rho(\mathcal{C}, \mathcal{C}')\rho(\mathcal{C}', \mathcal{C}), \mathcal{C}, \mathcal{C}' \in \mathfrak{B}$, where ρ is defined in (3.9.4). Show that (1) $d(\mathcal{C}, \mathcal{C}') \geq 1$, and equality holds if and only if $\mathcal{C}$ and $\mathcal{C}'$ are positively homothetic, that is, $\mathcal{C}' = S_{\lambda, X}(\mathcal{C})$ for some $\lambda > 0$ and $X \in \mathcal{X}$. (2) Show that d is sub-multiplicative. Conclude that $\ln(d)$ is a metric on the quotient of $\mathfrak{B}$ by the relation of positive homothety. (3) Show that $d_{BM} \leq d$. (4) Calculate: $d(\mathcal{C}, \tilde{\mathcal{C}}) = \mathfrak{m}^*_{\mathcal{C}}$ (in particular, recover (3.2.11)), $d(\mathcal{C}, \bar{B}) = R_{\mathcal{C}}/r_{\mathcal{C}}$ (in particular, d is not bounded), and $d(\tilde{\mathcal{C}}, \bar{B}) = D_{\mathcal{C}}/d_{\mathcal{C}}$, $\mathcal{C} \in \mathfrak{B}$.

16.* Prove that rotational symmetrization does not increase the diameter using the following steps: (1) First, define the *Steiner symmetrization* $\mathcal{C}[\mathcal{H}]$ of a convex body $\mathcal{C} \in \mathfrak{B}$ with respect to a hyperplane $\mathcal{H} \subset \mathcal{X}$ as follows. For $C \in \mathcal{C}$, let ℓ_C denote the line through C and perpendicular to $\mathcal{H}$. Then, for each $C \in \mathcal{C}$, translate the chord $C \cap \ell_C$ (or point) within ℓ_C to a symmetric position with respect to $\mathcal{H}$ (that is, the midpoint of the translated chord (or the point itself) is on $\mathcal{H}$). The union $C[\mathcal{H}]$ of the translated chords is the *Steiner symmetral* of C. Clearly, $C[\mathcal{H}]$ is symmetric with respect to $\mathcal{H}$. Prove the following: (a) $C[\mathcal{H}] \in \mathfrak{B}$. (b) $\mathrm{vol}\, C[\mathcal{H}] = \mathrm{vol}\, C$. (c) Under Steiner symmetrization, the surface area does not increase: $\mathrm{vol}_{n-1} \partial(C[\mathcal{H}]) \leq \mathrm{vol}_{n-1} \partial C$, and equality holds if and only if C is symmetric with respect to a hyperplane parallel to $\mathcal{H}$. (See the end of Section 1.2.) (d) $D_{C[\mathcal{H}]} \leq D_C$. (2) Let $\mathcal{H}_0, \ldots, \mathcal{H}_{n-2} \subset \mathcal{X}$ be $n - 1$ hyperplanes containing the line $\mathbb{R} \cdot N$ such that their mutual (dihedral) angles are irrational multiples of π. Apply the composition of Steiner's symmetrizations: $C \mapsto C[\mathcal{H}_0] \ldots [\mathcal{H}_{n-2}]$, and repeat this process cyclically to obtain the sequence $\{C_i\}_{i \geq 0}$, $C_0 = C$, $C_{i+1} = C_i[\mathcal{H}_{i(\mathrm{mod}(n-1))}]$, $i \geq 0$. (Observe that C_{i+1} is symmetric with respect to $\mathcal{H}_{i(\mathrm{mod}(n-1))}$, $i \geq 0$.) Use Blaschke's selection theorem (Section 1.1/B) along with (1/b-c) to show that $\{C_i\}_{i \geq 0}$ subconverges to a convex body C_∞ which is symmetric with respect to the hyperplanes $\mathcal{H}_1, \ldots, \mathcal{H}_{n-1}$. (3) Use the irrationality condition on the mutual angles of the hyperplanes to prove Blaschke's result that C_∞ is rotationally symmetrix with axis $\mathbb{R} \cdot N$. Conclude that C_∞ is the convex body obtained from C by rotational symmetrization. (4) Use (1/d) to deduce $D_{C_\infty} \leq D_C$.

Chapter 4
Mean Minkowski Measures

4.1 Mean Minkowski Measures: Arithmetic Properties

In this final chapter we introduce a sequence of new measures of symmetry $\{\sigma_k\}_{k\geq1}$. For a convex body $\mathcal{C}$, the kth term $\sigma_k = \sigma_{\mathcal{C},k}$, $k \geq 1$, is a function on the interior of $\mathcal{C}$. For an interior point O of $\mathcal{C}$, $\sigma_k(O)$ measures how far are the k-dimensional affine slices of $\mathcal{C}$ (across O) from a k-simplex (viewed from O). The minimum value of $\sigma_k(O)$ is 1 corresponding to a k-dimensional simplicial slice of $\mathcal{C}$. For $k \geq 2$, the maximum value of $\sigma_k(O)$ corresponds to symmetric $\mathcal{C}$ with respect to O. In this section we derive a host of arithmetic properties of the sequence $\{\sigma_k\}_{k\geq1}$, and show various connections with the maximal distortion $\mathfrak{m}_{\mathcal{C}}$, and the Minkowski measure $\mathfrak{m}_{\mathcal{C}}^*$.

As usual, we work in a Euclidean space $\mathcal{X}$ of dimension n. Let $\mathcal{C} \in \mathfrak{B} = \mathfrak{B}_{\mathcal{X}}$ and $O \in \text{int}\,\mathcal{C}$. Given $k \geq 1$, a multi-set $\{C_0, \ldots, C_k\} \subset \partial\mathcal{C}$ (repetition allowed) is called a k-configuration of $\mathcal{C}$ (with respect to O) if $O \in [C_0, \ldots, C_k]$. The set of all k-configurations of $\mathcal{C}$ is denoted by $\mathfrak{C}_k(O) = \mathfrak{C}_{\mathcal{C},k}(O)$.

We define the kth mean Minkowski measure $\sigma_k = \sigma_{\mathcal{C},k} : \text{int}\,\mathcal{C} \to \mathbb{R}$ by

$$\sigma_k(O) = \inf_{\{C_0,\ldots,C_k\}\in\mathfrak{C}_k(O)} \sum_{i=0}^{k} \frac{1}{\Lambda(C_i,O) + 1}, \quad O \in \text{int}\,\mathcal{C}, \tag{4.1.1}$$

where $\Lambda : \partial\mathcal{L} \times \text{int}\,\mathcal{C} \to \mathbb{R}$ is the (interior) distortion ratio (Section 2.1). (The subscript $\mathcal{C}$ will usually be suppressed when no ambiguity is present.)

Since Λ is continuous (Proposition 2.1.1) and $\partial\mathcal{C}$ is compact, the infimum in (4.1.1) is attained. A k-configuration at which $\sigma_k(O)$ attains its minimum is called *minimizing* or *minimal*, for short.

© Springer International Publishing Switzerland 2015
G. Toth, *Measures of Symmetry for Convex Sets and Stability*,
Universitext, DOI 10.1007/978-3-319-23733-6_4

Our primarily focus will be the *functions* σ_k, $k \geq 1$, but occasionally we will make note of the centroidal measures $\sigma_k^g = \sigma_{C,k}^g = \sigma_{C,k}(g(C))$, $k \geq 1$, and the suprema $\sigma_k^* = \sigma_{C,k}^* = \sup_{O \in \text{int}\,C} \sigma_{C,k}(O)$, $k \geq 1$. (In Section 4.4 we will show that the suprema are attained.)

A 1-configuration is an antipodal pair $\{C, C^o\} \subset \partial C$ and, since $\Lambda(C^o, O) = 1/\Lambda(C, O)$, we have

$$\frac{1}{\Lambda(C^o, O) + 1} = \frac{1}{1/\Lambda(C, O) + 1} = 1 - \frac{1}{\Lambda(C, O) + 1}. \qquad (4.1.2)$$

We obtain that $\sigma_1 = \sigma_1^* = \sigma_1^g = 1$ identically on $\text{int}\,C$.

Remark. For a short discussion on the measures of symmetry

$$\inf_{\{C_0, \ldots, C_k\} \in \mathfrak{C}_k(C, O)} \sum_{i=0}^{k} \Lambda(C_i, O) \quad \text{and} \quad \inf_{\{C_0, \ldots, C_k\} \in \mathfrak{C}_k(C, O)} \prod_{i=0}^{k} \Lambda(C_i, O)$$

(at least for $k = n$), see [Grünbaum 2, 6.1].

Any k-configuration, $k \geq 1$, (with respect to an interior point O) can always be *extended* to a $(k + l)$-configuration, $l \geq 1$, by adding l copies of a boundary point of C at which $\Lambda(., O)$ attains its maximum distortion $\mathfrak{m}(O) = \max_{C \in \partial C} \Lambda(C, O)$ (Section 2.1). Thus we have the following *sub-arithmeticity*:

$$\sigma_{k+l}(O) \leq \sigma_k(O) + \frac{l}{\mathfrak{m}(O) + 1}, \qquad O \in \text{int}\,C, \; k, l \geq 1. \qquad (4.1.3)$$

We claim that equality holds for $k = n$ and $l \geq 1$, that is, the sequence $\{\sigma_k\}_{k \geq 1}$ is *arithmetic* with difference $1/(\mathfrak{m} + 1)$ from the nth term onwards.

Indeed, let $\{C_0, \ldots, C_{n+l}\} \in \mathfrak{C}_{n+l}(O)$ be a *minimal* $(n + l)$-configuration. Since $O \in [C_0, \ldots, C_{n+l}]$, by Carathéodory's theorem (Section 1.3), a subset of $(n + 1)$ points (or less) $\{C_0, \ldots, C_n\}$, say, contains O in its convex hull, so that $\{C_0, \ldots, C_n\} \in \mathfrak{C}_n(O)$ is an n-configuration. We thus have

$$\sigma_{n+l}(O) = \sum_{i=0}^{n+l} \frac{1}{\Lambda(C_i, O) + 1}$$

$$= \sum_{i=0}^{n} \frac{1}{\Lambda(C_i, O) + 1} + \sum_{j=1}^{l} \frac{1}{\Lambda(C_{n+j}, O) + 1}$$

$$\geq \sigma_n(O) + \frac{l}{\mathfrak{m}(O) + 1}.$$

By (4.1.3), the opposite inequality also holds, so that arithmeticity of the subsequence $\{\sigma_k\}_{k \geq n}$ follows.

Note that, as a byproduct, we obtain that a subconfiguration of a minimal configuration is also minimal.

We also see that

$$\lim_{k \to \infty} \frac{\sigma_{n+k}}{k} = \lim_{k \to \infty} \frac{\sigma_k}{k} = \frac{1}{m+1}.$$

In view of the arithmeticity just established, the most important member of the sequence $\{\sigma_k\}_{k \geq 1}$ is $\sigma_n = \sigma_{C,n}$, $\dim C = n$. We call this *the mean Minkowski measure* of C. We will usually suppress the numerical index and write $\sigma = \sigma_C$.

For $1 \leq k \leq n$, we also have

$$\sigma_{C,k}(O) = \inf_{(O \in) \mathcal{E} \subset \mathcal{X}, \dim \mathcal{E} = k} \sigma_{C \cap \mathcal{E}}(O), \qquad (4.1.4)$$

where the infimum is over affine subspaces $\mathcal{E} \subset \mathcal{X}$ (of dimension k) containing O.

Theorem 4.1.1. *Let $C \in \mathfrak{B}$. For $k \geq 1$, we have*

$$1 \leq \sigma_k \leq \frac{k+1}{2}. \qquad (4.1.5)$$

Assuming $k \geq 2$, $\sigma_k(O) = (k+1)/2$ for some $O \in \operatorname{int} C$ if and only if C is symmetric with respect to O. If, for some $k \geq 1$, $\sigma_k(O) = 1$ at $O \in \operatorname{int} C$, then $k \leq n$ and C has a k-dimensional simplicial intersection across O, that is, there exists a k-dimensional affine subspace $\mathcal{E} \subset \mathcal{X}$, $O \in \mathcal{E}$, such that $C \cap \mathcal{E}$ is a k-simplex. Conversely, if C has a simplicial intersection with a k-dimensional affine subspace $\mathcal{E}$ then $\sigma_k = 1$ identically on $\operatorname{int} C \cap \mathcal{E}$.

We begin with a lemma which will be useful in several instances in the future.

Lemma 4.1.2. *Let $C \in \mathfrak{B}$ and $O \in \operatorname{int} C$. Assume that $\{C_0, \dots, C_k\} \in \mathfrak{C}_k(O)$ such that $\Delta = [C_0, \dots, C_k]$ is a k-simplex with $O \in \operatorname{int} \Delta$ (relative interior). We write $O = \sum_{i=0}^{k} \lambda_i C_i$, $\sum_{i=0}^{k} \lambda_i = 1$, $\{\lambda_0, \dots, \lambda_k\} \subset [0, 1]$. Then, for $i = 0, \dots, k$, we have*

$$\lambda_i \leq \frac{1}{\Lambda_C(C_i, O) + 1}. \qquad (4.1.6)$$

Equality holds if and only if the ith face $\Delta_i = [C_0, \dots, \widehat{C_i}, \dots, C_k]$ antipodal to C_i is contained in ∂C.

Proof. For $i = 0, \dots, k$, let $\bar{C}_i^o$, denote the antipodal of C_i in Δ with respect to O. (See Figure 4.1.1.)

Clearly, $\bar{C}_i^o \in [C_i^o, O]$ so that $\Lambda_\Delta(C_i, O) \geq \Lambda_C(C_i, O)$. By Example 2.1.7, $\Lambda_\Delta(C_i, O) = (1 - \lambda_i)/\lambda_i$ and (4.1.6) follows.

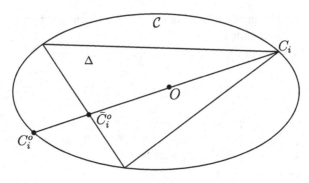

Fig. 4.1.1

For a given $i = 0, \ldots, k$, equality holds in (4.1.6) if and only if $C_i^o = \bar{C}_i^o$. Since $\bar{C}_i^o \in \text{int } \Delta_i$ and the vertices $C_0, \ldots, \widehat{C}_i, \ldots, C_k$ of Δ_i are on the boundary of $\mathcal{C}$, this holds if and only if any hyperplane supporting $\mathcal{C}$ at C_i^o contains Δ_i. Thus, $\Delta_i \subset \partial \mathcal{C}$ and the lemma follows.

Proof of Theorem 4.1.1. We first derive the upper bound in (4.1.5) as it is simpler. Given $O \in \text{int } \mathcal{C}$, using sub-arithmeticity in (4.1.3), we have

$$\sigma_k(O) \le \sigma_1(O) + \frac{k-1}{\mathfrak{m}(O)+1} \le 1 + \frac{k-1}{2} = \frac{k+1}{2},$$

where we used that $\mathfrak{m}(O) = \max_{\partial \mathcal{C}} \Lambda(., O) \ge 1$. The upper bound in (1.4.5) follows.

Let $k \ge 2$. If equality holds throughout then $\mathfrak{m}(O) = 1$ so that $\Lambda(., O) = 1$ identically on $\partial \mathcal{C}$. This means that $\mathcal{C}$ is symmetric with respect to O. The converse is obvious: If $\mathcal{C}$ is symmetric with respect to O, then, once again, $\Lambda(., O) = 1$ identically on $\partial \mathcal{C}$, and, for *any* configuration $\{C_0, \ldots, C_k\} \in \mathfrak{C}_k(O)$, $k \ge 1$, the sum in the infimum in (4.1.1) is equal $(k+1)/2$.

We now derive the lower bound in (4.1.5). Let $O \in \text{int } \mathcal{C}$, and $\{C_0, \ldots, C_k\} \in \mathfrak{C}_k(O)$ be *any* k-configuration. By Carathéodory's Theorem 1.3.1, there is a subconfiguration, say $\{C_0, \ldots, C_m\} \in \mathfrak{C}_m(O)$, $m \le k$, such that $\Delta = [C_0, \ldots, C_m]$ is an m-simplex. (See also the second part of the proof of Corollary 1.3.3.) Moreover, assuming that m is minimal we have $O \in \text{int } \Delta$. As in Lemma 4.1.2, we write $O = \sum_{i=0}^{m} \lambda_i C_i$, $\sum_{i=0}^{m} \lambda_i = 1$, $\{\lambda_0, \ldots, \lambda_m\} \subset [0, 1]$. By (4.1.6), we have

$$\sum_{i=0}^{k} \frac{1}{\Lambda_{\mathcal{C}}(C_i, O)+1} \ge \sum_{i=0}^{m} \frac{1}{\Lambda_{\mathcal{C}}(C_i, O)+1} \ge \sum_{i=0}^{m} \lambda_i = 1. \qquad (4.1.7)$$

Thus, the lower bound $\sigma_k(\mathcal{C}, O) \ge 1$ is proved.

Finally, let $k \ge 1$, and assume that $\sigma_k(O) = 1$ for some $O \in \text{int } \mathcal{C}$. Let $\{C_0, \ldots, C_k\} \in \mathfrak{C}_k(O)$ be a *minimal* k-configuration:

$$\sigma_k(O) = \sum_{i=0}^{k} \frac{1}{\Lambda(C_i, O) + 1} = 1.$$

Comparing this with (4.1.7), we see that equalities hold there, in particular, $m = k$, $\Delta = [C_0, \ldots, C_k]$ is a k-simplex with O in its (relative) interior, and equality holds in (4.1.6) for all $i = 0, \ldots, k$. Lemma 4.1.2 now finishes the proof.

Remark. The sequence of mean Minkowski measures $\{\sigma_k\}_{k\geq 1}$ was introduced in [Toth 5] with a different proof of Theorem 4.1.1.

The various estimates above can be summarized as follows. Combining the trivial lower estimate for (4.1.1) with the lower estimate in (4.1.5), and sub-arithmeticity in (4.1.3) (as in the proof above), for $k \geq 1$, we obtain

$$\max\left(1, \frac{k+1}{m(O)+1}\right) \leq \sigma_k(O) \leq 1 + \frac{k-1}{m(O)+1}, \qquad O \in \text{int}\, C. \qquad (4.1.8)$$

In addition, once again by (4.1.3) and the discussion above, we have $\sigma_{k+1}(O) - \sigma_k(O) \leq 1/(m(O) + 1)$ with equality for $k \geq n$. We conclude that the sequence $\{(k, \sigma_k(O))\}_{k\geq 1}$ is contained in the strip (with slope $1/(m(O) + 1)$) bounded by two parallel lines as shown in Figure 4.1.2.

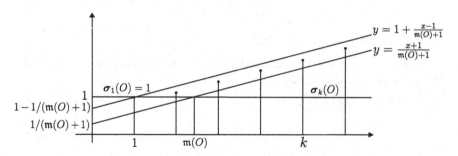

Fig. 4.1.2

Remark. If $\{C_0, \ldots, C_k\} \in \mathfrak{C}_k(O)$ is a k-configuration then, taking antipodals, a simple computation shows that $\{C_0^o, \ldots, C_k^o\} \in \mathfrak{C}_k(O)$ is also a k-configuration. Thus the antipodal map gives rise to an involution $^o : \mathfrak{C}_k(O) \to \mathfrak{C}_k(O)$. Using (4.1.2), we obtain

$$k + 1 = \sum_{i=0}^{k} \frac{1}{\Lambda(C_i, O) + 1} + \sum_{i=0}^{k} \frac{1}{\Lambda(C_i^o, O) + 1} \geq 2\sigma_k(O).$$

Thus the upper bound in (4.1.5) follows again. (For another application of this, see Problem 1.)

As a simple application of Theorem 4.1.1, we recover the characterization of the upper bound of the Minkowski measure in (2.1.7):

Corollary 4.1.3. *Let* $C \in \mathfrak{B}$. *If* $\mathfrak{m}_C^* = n$ *then* C *is a simplex.*

Proof. Assuming $\mathfrak{m}^* = n$, by Corollary 2.4.13, C^* consists of a single point O^*, say, and it is in the interior of the convex hull of $\mathcal{M}^*$. Therefore, by Carathéodory's theorem (Section 1.3), there exists $\{C_0, \dots, C_n\} \subset \mathcal{M}^*$ with $O^* \in \text{int}\,[C_0, \dots, C_n]$, and we have

$$\sum_{i=0}^{n} \frac{1}{\Lambda(C_i, O^*) + 1} = \sum_{i=0}^{n} \frac{1}{\mathfrak{m}^* + 1} = 1.$$

Since $\{C_0, \dots, C_n\}$ is an n-configuration, Theorem 4.1.1 applies. We obtain that C is an n-simplex.

Remark. We note here that the long computational part in the proof of Lemma 2.4.7 can be bypassed using the lower bound in (4.1.5). Indeed, if $\{C_0, \dots, C_k\} \subset \partial C$ is the set of vertices of a k-simplex and $0 \in \text{int}\,[C_0, \dots, C_k]$ with $\Lambda(C_i, 0) > m$, $i = 0, \dots, k$, then $\{C_0, \dots, C_k\}$ is a k-configuration of C, and we have

$$1 \leq \sigma_k(0) \leq \sum_{i=0}^{k} \frac{1}{\Lambda(C_i, 0) + 1} < \frac{k+1}{m+1}.$$

Thus, $k > m$ follows.

We now return to the main line and show that the sequence $\{\sigma_k\}_{k \geq 1}$ is *super-additive* in the following sense:

Theorem 4.1.4. *For* $k, l \geq 1$, *we have*

$$\sigma_{k+l} - \sigma_{k+1} \geq \sigma_l - \sigma_1, \quad \sigma_1 = 1. \tag{4.1.9}$$

Proof. Let $O \in \text{int}\,C$ and consider a *minimal* $(k+l)$-configuration $\{C_0, \dots, C_{k+l}\} \in \mathfrak{C}_{k+l}(O)$. We write

$$O = \sum_{i=0}^{k+l} \lambda_i C_i, \quad \sum_{i=0}^{k+l} \lambda_i = 1, \quad \{\lambda_0, \dots, \lambda_{k+l}\} \subset [0,1]. \tag{4.1.10}$$

By definition, the partial sum $\sum_{j=1}^{l} \lambda_{k+j} \geq 0$. We now split the proof into two cases.

I. If this partial sum is zero then $O \in [C_0, \dots, C_k]$ so that $\{C_0, \dots, C_k\} \in \mathfrak{C}_k(O)$.
   As noted above, this subconfiguration must be minimal and $\Lambda(C_{k+j}, O) = \mathfrak{m}(O)$, $j = 1, \dots, l$. Using sub-arithmeticity in (4.1.3) repeatedly, we now calculate

$$\sigma_{k+l}(O) = \sum_{i=0}^{k} \frac{1}{\Lambda(C_i, O) + 1} + \frac{l}{\mathfrak{m}(O) + 1}$$

$$= \sigma_k(O) + \frac{l}{\mathfrak{m}(O) + 1}$$

$$\geq \sigma_{k+1}(O) + \frac{l-1}{\mathfrak{m}(O) + 1}$$

$$\geq \sigma_{k+1}(O) + \sigma_l - \sigma_1(O).$$

Thus, (4.1.9) follows in this case.

II. Assume now that the partial sum above is positive. We let

$$O' = \sum_{j=1}^{l} \frac{\lambda_{k+j}}{\lambda_{k+1} + \ldots + \lambda_{k+l}} C_{k+j}.$$

First, assume that $O' = O$. Then $\{C_{k+1}, \ldots, C_{k+l}\} \in \mathfrak{C}_{l-1}(O)$ and, as before, this subconfiguration must be minimal and $\Lambda(C_i, O) = \mathfrak{m}(O)$, $i = 1, \ldots, k$. Once again, using (4.1.3), we calculate

$$\sigma_{k+l}(O) = \frac{k+1}{\mathfrak{m}(O) + 1} + \sum_{j=1}^{l} \frac{1}{\Lambda(C_{k+j}, O) + 1}$$

$$= \frac{k+1}{\mathfrak{m}(O) + 1} + \sigma_{l-1}(O)$$

$$\geq \sigma_{k+1}(O) - \sigma_1(O) + \frac{1}{\mathfrak{m}(O) + 1} + \sigma_{l-1}(O)$$

$$\geq \sigma_{k+1}(O) + \sigma_l(O) - \sigma_1(O).$$

Thus, (4.1.9) follows in this case.

Second, assume that $O' \neq O$. Let $C' \in \partial \mathcal{C}$ be the intersection of the ray emanating from O and passing through O' with the boundary of $\mathcal{C}$. Using the definition of O', the decomposition of O in (4.1.10) can be written as

$$O = \sum_{i=0}^{k} \lambda_i C_i + (\lambda_{k+1} + \ldots \lambda_{k+l}) O'.$$

Since $O' \in [O, C']$, $C' \in \partial \mathcal{C}$, we obtain $\{C_0, \ldots, C_k, C'\} \in \mathfrak{C}_{k+1}$. On the other hand, $O \in [O', C'^o]$ and $O' \in [C_{k+1}, \ldots, C_{k+l}]$ so that $\{C_{k+1}, \ldots, C_{k+l}, C'^o\} \in \mathfrak{C}_l(O)$. Using these, we calculate

$$\sigma_{k+l}(O) = \sum_{i=0}^{k} \frac{1}{\Lambda(C_i, O) + 1} + \sum_{j=1}^{l} \frac{1}{\Lambda(C_{k+j}, O) + 1}$$

$$= \sum_{i=0}^{k} \frac{1}{\Lambda(C_i, O) + 1} + \frac{1}{\Lambda(C') + 1}$$

$$+ \sum_{j=1}^{l} \frac{1}{\Lambda(C_{k+j}, O) + 1} + \frac{1}{\Lambda(C'^o) + 1} - 1$$

$$\geq \sigma_{k+1}(O) + \sigma_l(O) - \sigma_1(O).$$

Thus, (4.1.9) follows. The proof of the theorem is complete.

As an immediate consequence of (4.1.9) (and the lower bound in (4.1.5)), for $k \geq 1$, we have

$$\sigma_{k+2} - \sigma_{k+1} \geq \sigma_2 - \sigma_1 \geq \sigma_2 - 1 \geq 0.$$

Hence the sequence $\{\sigma_k\}_{k \geq 1}$ is *non-decreasing*: $\sigma_{k+1} \geq \sigma_k, k \geq 1$.

Theorem 4.1.5. *For $O \in$ int C, the length $d(O)$ of the initial string of 1's in $\{\sigma_k(O)\}_{k \geq 1}$ is the dimension of the maximal simplicial slice of C across O. We have*

$$d(O) \leq m(O). \tag{4.1.11}$$

If equality holds then the sequence $\{\sigma_k(O)\}_{k \geq 1}$ is arithmetic from the $d(O)$th term onwards:

$$\sigma_k(O) = \frac{k+1}{m(O) + 1}, \quad k \geq d(O).$$

Proof. The first statement is a direct consequence of Theorem 4.1.1. The inequality in (4.1.11) and the last statement follow from (4.1.8) and sub-arithmeticity in (4.1.3):

$$\max\left(1, \frac{k+1}{m(O) + 1}\right) \leq \sigma_k(O) \leq \sigma_{d(O)}(O) + \frac{k - d(O)}{m(O) + 1}$$

$$= \frac{k + 1 + m(O) - d(O)}{m(O) + 1}, \quad k \geq d(O).$$

Note that this can also be read off from Figure 4.1.2 using

$$\sigma_{k+1}(O) - \sigma_k(O) \leq \frac{1}{m(O) + 1}, \quad k \geq 1. \tag{4.1.12}$$

We finish this section by a long example which will play an important part in Appendix A.

Example 4.1.6. Let $\mathcal{H}$ be a Euclidean vector space, dim $\mathcal{H} = h$, and $S^2(\mathcal{H})$ the space of symmetric linear endomorphisms of $\mathcal{H}$. On $S^2(\mathcal{H})$ the natural scalar product is defined by

$$\langle C, C' \rangle = \text{trace}\,(C \cdot C'), \quad C, C' \in S^2(\mathcal{H}).$$

With this $S^2(\mathcal{H})$ is a Euclidean vector space. Let $S_0^2(\mathcal{H})$ be the linear subspace of *traceless* endomorphisms:

$$S_0^2(\mathcal{H}) = \{C \in S^2(\mathcal{H}) \mid \text{trace}\, C = 0\}.$$

We have the orthogonal decomposition $S^2(\mathcal{H}) = S_0^2(\mathcal{H}) \oplus \mathbb{R} \cdot I$, where I stands for the identity. In particular, we have dim $S_0^2(\mathcal{H}) = h(h+1)/2 - 1$.

The orthogonal group $O(\mathcal{H})$ acts by linear isometries on $S^2(\mathcal{H})$ via conjugation with fixed point set $\mathbb{R} \cdot I$.

Finally, we define

$$\mathcal{C}_0 = \mathcal{C}_0(\mathcal{H}) = \{C \in S_0^2(\mathcal{H}) \mid C + I \geq 0\},$$

where ≥ 0 means *positive semi-definite*. Positive semi-definiteness is a closed and convex condition so that $\mathcal{C}_0$ is a closed and convex set in $S_0^2(\mathcal{H})$. Moreover, replacing the defining condition of semi-definiteness with positive definiteness gives the interior of $\mathcal{C}_0$, in particular, $0 \in \text{int}\,\mathcal{C}_0$. For $C \in \mathcal{C}_0$, the conditions trace $C = 0$ and $C + I \geq 0$ mean that the sum of eigenvalues of C is zero and the eigenvalues are ≥ -1. Thus, all the eigenvalues are contained in the interval $[-1, h-1]$. The orthogonal group $O(\mathcal{H})$ leaves $\mathcal{C}_0$ invariant and diagonalizes each endomorphism within $\mathcal{C}_0$. In particular, we see that $\mathcal{C}_0$ is bounded, hence compact.

Summarizing, we obtain that $\mathcal{C}_0$ is a *convex body* in $S_0^2(\mathcal{H})$.

Since the only fixed point of $O(\mathcal{H})$ in $\mathcal{C}_0$ is the origin, the latter is the *centroid* of $\mathcal{C}_0$. It also follows that the *critical set* $\mathcal{C}_0^*$ consists of the origin only: $\mathfrak{m}^* = \inf_{O \in \text{int}\,\mathcal{C}_0} \mathfrak{m}(O) = \mathfrak{m}(0)$.

We claim that, for $C \in \partial\mathcal{C}_0$, the distortion ratio $\Lambda(C, 0) = \Lambda_{\mathcal{C}_0}(C, 0)$ (with respect to the origin 0) is the *maximal eigenvalue* of C. Indeed, by definition, $\Lambda(C, 0) = 1/t$, where $t > 0$ is the largest number such that $I - tC \geq 0$ but $I - tC \not> 0$. Replacing the endomorphisms by the respective eigenvalues, the claim follows.

$\mathcal{C}_0 \in \mathfrak{B}$ carries a natural stratification defined by the images $\text{im}(C + I) \subset \mathcal{H}$, $C \in \mathcal{C}_0$. The interior $\text{int}\,\mathcal{C}_0$ corresponds to the stratum with image the entire $\mathcal{H}$. Moreover, for $C_0, C_1 \in \partial\mathcal{C}_0$ and $0 < \lambda < 1$, we have

$$\text{im}((1-\lambda)C_0 + \lambda C_1 + I) = \text{im}((1-\lambda)(C_0 + I) + \lambda(C_1 + I)) = \text{im}(C_0 + I) + \text{im}(C_1 + I).$$

(In the last equality we used the elementary fact that, for positive semi-definite endomorphisms Q, Q', we have $\text{im}\,(Q + Q') = \text{im}\,Q + \text{im}\,Q'$.)

As a byproduct, we see that $C \in \partial\mathcal{C}_0$ is an *extremal* point if the image of $C + I$ is one-dimensional. If this image is given by the line $\mathbb{R} \cdot v \subset \mathcal{H}$, $0 \neq v \in \mathcal{H}$, then the corresponding endomorphism is $C_v = v \odot v - I \in \partial\mathcal{C}_0$, where we normalized $v \in \mathcal{H}$ as $|v|^2 = h$. Note that C_v has maximal eigenvalue $h - 1$ with multiplicity 1 and the only other (minimal) eigenvalue is -1 with multiplicity $h - 1$. We see that maximum distortion appears precisely at C_v, $|v| = h$, $v \in \mathcal{H}$. Thus, we have

$$m^* = m(0) = h - 1.$$

Let $\{e_i\}_{i=1}^h \subset \mathcal{H}$ be an orthonormal basis and consider the set $\{C_{\sqrt{h}e_i}\}_{i=1}^h \subset \partial\mathcal{C}_0$. Since $\sum_{i=1}^h (1/h) C_{\sqrt{h}e_i} = \sum_{i=1}^h (e_i \odot e_i) - I = 0$, this set is an $(h-1)$-configuration (with respect to 0) which is clearly minimal as its elements have maximal distortion. We obtain

$$\sigma_{h-1}(0) = \sum_{i=1}^h \frac{1}{\Lambda(C_{\sqrt{h}e_i}, 0) + 1} = \sum_{i=1}^h \frac{1}{h} = 1.$$

Thus, $d(0) \geq h - 1$ (Theorem 4.1.5). Since $m(0) = h - 1$, by (4.1.11), equality holds: $d(0) = h - 1$. Theorem 4.1.5 thus asserts that the $(h-1)$-simplex $\Delta_0 = [C_{\sqrt{h}e_1}, \ldots, C_{\sqrt{h}e_h}]$ is a *maximal* simplicial slice of $\mathcal{C}_0$ across 0. (Note that this also follows by inspecting the images of the corresponding endomorphisms, but we preferred to point out the connection with the distortion ratio.)

Since $d(0) = m(0)$, the last statement in Theorem 4.1.5 now gives

$$\sigma_k(0) = \frac{k + 1}{h}, \quad k \geq h. \tag{4.1.13}$$

Summarizing, (4.1.13) means that the sequence $\{\sigma_k(0)\}_{k \geq 1}$ has the simplest possible structure: $\sigma_k(0) = \max(1, (k + 1)/h)$, $k \geq 1$; after an initial string of 1's of length $h - 1$, it is arithmetic with difference $1/h$.

The simplex Δ_0 consists of all endomorphisms in $\mathcal{C}_0$ that are *diagonal* with respect to the orthonormal basis $\{e_i\}_{i=1}^h$. Diagonalizability of all the endomorphisms by $O(\mathcal{H})$ means that the $O(\mathcal{H})$-orbit of Δ_0 is the *entire* $\mathcal{C}_0$. In particular, $\sigma_{h-1} = 1$ *identically* on $\text{int}\,\mathcal{C}_0$.

We finally note that the functions σ_k have unique absolute maximum at 0 so that we also have $\sigma_k^* = \max(1, (k + 1)/h)$, $k \geq 1$. Indeed, given $0 \neq O \in \text{int}\,\mathcal{C}_0$, by (4.1.3), for $k \geq h$, we have

$$\sigma_k(O) \leq \sigma_{h-1}(O) + \frac{k - h + 1}{m(O) + 1} = 1 + \frac{k - h + 1}{m(O) + 1} < \frac{k + 1}{h} = \sigma_k(0).$$

Note that sharp inequality holds in $m(O) > \inf_{\text{int}\mathcal{C}_0} m = m^* = h - 1$ since the critical set is a singleton: $\mathcal{C}_0^* = \{0\}$.

4.2   Dual Mean Minkowski Measures

In this section we introduce a "dual" sequence $\{\sigma_k^o\}_{k \geq 1}$ of mean Minkowski measures whose original idea is due to Qi Guo. We summarize some basic properties of these measures, and point out some striking new properties that the sequence $\{\sigma_k\}_{k \geq 1}$ does not have. In particular, we show that Klee's work in Section 2.4 gives a direct and explicit evaluation of σ_n^o on the critical set of the convex body. This leads to a solution of the Grünbaum Conjecture (Remark 1 at the end of Section 2.4) under a much weaker condition than Klee's.

The material in this section is not directly related to our main goal, therefore, in our exposition, we limit ourselves to sketch the main points only. For more details, see the joint work of [Guo–Toth].

Let $C \in \mathfrak{B} = \mathfrak{B}_{\mathcal{X}}$. Recall from Section 1.2 that an affine functional $f : \mathcal{X} \to \mathbb{R}$ is said to be normalized for C if $f(C) = [0, 1]$. The (compact) space of affine functionals normalized for C is denoted by aff_C.

Given $k \geq 1$, a multi-set $\{f_0, \ldots, f_k\} \subset \mathrm{aff}_C$ (repetition allowed) is called a *dual k-configuration* of C if

$$\bigcap_{i=0}^{k} \{X \in \mathcal{X} \,|\, f_i(X) \leq 0\} = \emptyset. \tag{4.2.1}$$

The set of all dual k-configurations of C is denoted by $\mathfrak{C}_k^o = \mathfrak{C}_{C,k}^o$.

We define the *kth dual mean Minkowski measure* $\sigma_k^o = \sigma_{C,k}^o : \mathrm{int}\,C \to \mathbb{R}$ by

$$\sigma_k^o(O) = \inf_{\{f_0, \ldots, f_k\} \in \mathfrak{C}_k^o} \sum_{i=0}^{k} f_i(O), \quad O \in \mathrm{int}\,C. \tag{4.2.2}$$

(As usual, the subscript C will be suppressed when no ambiguity is present.)

For $k \geq 1$, the subspace $\mathfrak{C}_k^o \subset (\mathrm{aff}_C)^k$ is closed and hence compact. Therefore the infimum in (4.2.2) is attained (for every $O \in \mathrm{int}\,C$), and the dual mean Minkowski measures $\sigma_k^o : \mathrm{int}\,C \to \mathbb{R}, k \geq 1$, are continuous functions. In addition, being infima of affine functionals, they are automatically *concave*.

As before, a k-configuration at which $\sigma_k^o(O)$ attains its minimum in (4.2.2) is called *minimizing* or *minimal* (for O).

Remark. A dual 1-configuration is an antipodal pair $\{f, 1 - f\} \subset \mathrm{aff}_C$, so that we have $\sigma_1^o = 1$ identically on $\mathrm{int}\,C$.

We have the following "pointwise duality" between the two mean Minkowski measures:

Theorem 4.2.1. *Let* $C \in \mathfrak{B}$ *and* $O \in \mathrm{int}\,C$. *For* $k \geq 1$, *we have*

$$\sigma_{C,k}^o(O) = \sigma_{C^O,k}(O), \tag{4.2.3}$$

where C^O *is the dual of* C *with respect to* O *(Section 3.1).*

Remark 1. On the right-hand side of (4.2.3), the mean Minkowski measure has *double dependency* on the point O; it is present not only in the argument but also plays the role of the base point in forming the dual C^O. In particular, it does *not* follow (and is actually not true in general) that $\sigma_{C,k} : \text{int}\,C \to \mathbb{R}$ is a concave function. We will treat the problem of concavity of the mean Minkowski measures in Section 4.4 below.

Remark 2. The crux of the proof of Theorem 4.2.1 is the equivalences

$$\{C_0, \ldots, C_k\} \in \mathfrak{C}_{C,k}(O) \quad \Leftrightarrow \quad \{C_0^\flat, \ldots, C_k^\flat\} \in \mathfrak{C}_{C^O,k}^o$$

$$\{f_0, \ldots, f_k\} \in \mathfrak{C}_{C,k}^o \quad \Leftrightarrow \quad \{f_0^\sharp, \ldots, f_k^\sharp\} \in \mathfrak{C}_{C^O,k}(O),$$

where $\flat = \flat_{C,O} : \partial C \to \text{aff}_{C^O}$ and $\sharp = \sharp_{C,O} : \text{aff}_C \to \partial C^O$ are the musical equivalences introduced in Section 3.1. In addition, by (3.1.6), under these equivalences, minimal configurations correspond to each other.

It is possible to derive arithmetic properties of the dual mean Minkowski measures in much the same way as it was done for the original sequence $\{\sigma_k\}_{k \geq 1}$ in the previous section. But the pointwise duality in (4.2.3) allows these properties of the mean Minkowski measures to carry directly over to the duals.

Using (4.1.3) for C^O instead of C, by (4.2.3) and (3.1.9), we obtain *sub-arithmeticity*:

$$\sigma_{k+l}^o(O) \leq \sigma_k^o(O) + \frac{l}{\mathfrak{m}(O)+1}, \qquad O \in \text{int}\,C, \; k,l \geq 1. \tag{4.2.4}$$

In addition, the sequence $\{\sigma_k^o\}_{k \geq 1}$ is arithmetic with difference $1/(\mathfrak{m}+1)$ from the nth term onwards.

Remark. The direct proof of arithmeticity (without the use of duality) beyond $n = \dim \mathcal{X}$ is an application of (the contrapositive of) Helly's theorem (Section 1.4) (instead of Carathéodory's as in Section 4.1): For $k > n$, any dual k-configuration (characterized by (4.2.1)) contains an n-configuration.

To state the dual version of (4.1.4), for $1 \leq k \leq n$, we denote by $\mathfrak{P}_k = \mathfrak{P}_{\mathcal{X},k}$ the space of all orthogonal projections $\Pi : \mathcal{X} \to \mathcal{X}$ onto k-dimensional affine subspaces $\Pi(\mathcal{X}) = \mathcal{E} \subset \mathcal{X}$. We then have

$$\sigma_{C,k}^o(O) = \inf_{\Pi \in \mathfrak{P}_k} \sigma_{\Pi(C),k}^o(\Pi(O)), \qquad O \in \text{int}\,C. \tag{4.2.5}$$

(In the infimum $\Pi(O)$ can be replaced by O if we require $O \in \Pi(\mathcal{X}) = \mathcal{E}$.)

By duality, the bounds in (4.1.5) stay the same for the dual mean Minkowski measures. To characterize the convex bodies for which the lower bound holds is essentially based on (4.2.5).

Theorem 4.2.2. *Let* $C \in \mathfrak{B}$. *For* $k \geq 1$, *we have*

$$1 \leq \sigma_k^o \leq \frac{k+1}{2}. \tag{4.2.6}$$

Assuming $k \geq 2$, *we have* $\sigma_k^o(O) = (k+1)/2$ *for some* $O \in \mathrm{int}\,C$ *if and only if* C *is symmetric with respect to* O. *If, for some* $k \geq 1$, $\sigma_k^o(O) = 1$ *at* $O \in \mathrm{int}\,C$ *then* $k \leq n$, *and* $\sigma_k^o = 1$ *identically on* $\mathrm{int}\,C$, *and* C *has an orthogonal projection to a* k-*simplex.*

The dual mean Minkowski measures are not only continuous in the interior of the convex body but also extend continuously to the boundary (with value 1) via the formula

$$\lim_{d(O,\partial C) \to 0} \sigma_k^o(O) = 1. \tag{4.2.7}$$

To show this, we use the lower bound in (4.2.6) along with sub-arithmeticity ($k = 1$ and $l = k - 1$ in (4.2.4) with $\sigma_1^o = 1$) to obtain

$$1 \leq \sigma_k^o(O) \leq 1 + \frac{k-1}{\mathfrak{m}(O) + 1}.$$

By Lemma 2.1.6, the limit of the right-hand side is 1 as $d(O, \partial C) \to 0$. Thus, (4.2.7) follows.

Finally, we have *super-additivity*

$$\sigma_{k+l}^o - \sigma_{k+1}^o \geq \sigma_l^o - \sigma_1^o, \quad k, l \geq 1,$$

and, as a direct consequence, monotonicity: $\sigma_k^o \leq \sigma_{k+1}^o, k \geq 1$.

As a striking application of Klee's analysis of the critical set, we have the following:

Theorem 4.2.3. *Let* $C \in \mathfrak{B}$ *with critical set* $C^* \subset C$. *For any critical point* $O^* \in C^*$, *we have*

$$\sigma_n^o(O^*) = \frac{n+1}{\mathfrak{m}^* + 1}. \tag{4.2.8}$$

Remark. In the spirit of Lemma 2.4.1 of Klee, let $\mathcal{N}(O^*) = \mathcal{M}(O^*)^o \subset \partial C$ be the antipodal set of $\mathcal{M}(O^*) = \{C \in \partial C \mid \Lambda(C, O^*) = \mathfrak{m}^*\}$ with respect to O^*. Denote by $\mathfrak{G}$ the family of closed half-spaces that intersect $\mathcal{N}(O^*)$ but disjoint from $\mathrm{int}\,C$. Clearly, for each $\mathcal{G} \in \mathfrak{G}$, the boundary $\mathcal{H} = \partial\mathcal{G}$ is a hyperplane supporting C at a point in $\mathcal{N}(O^*)$. Conversely, for any hyperplane $\mathcal{H}$ supporting C at a point in $\mathcal{N}(O^*)$, the closed half-space $\mathcal{G}$ with boundary $\mathcal{H}$ and disjoint from $\mathrm{int}\,C$ belongs to $\mathfrak{G}$. By Lemma 2.4.1, we have

$$\bigcap \mathfrak{G} = \bigcap_{\mathcal{G} \in \mathfrak{G}} \mathcal{G} = \emptyset.$$

Taking interiors, the family

$$\mathfrak{I} = \text{int}\,\mathfrak{G} = \{\text{int}\,\mathcal{G} \mid \mathcal{G} \in \mathfrak{G}\}$$

of open half-spaces is interior-complete (Section 1.6). Recall that this means that, for any sequence $\{\mathcal{I}_k\}_{k \geq 1} \subset \mathfrak{G}$ which is Painlevé–Kuratowski convergent to a limit $\mathcal{I}$, we have $\text{int}\,\mathcal{I} \in \mathfrak{G}$. (Note that, by definition, any Painlevé–Kuratowski limit is a closed set.)

Applying Lemma 1.6.3 (in Klee's extension of Helly's theorem), we see that there are $n + 1$ open half-spaces $\mathcal{I}_0, \ldots, \mathcal{I}_n \in \mathfrak{I}$ such that $\bigcap_{i=0}^{n} \mathcal{I}_k = \emptyset$.

For $i = 0, \ldots, n$, we can select $C_i \in \mathcal{M}(O^*)$ such that $C_i^o \in \bar{\mathcal{I}}_i$ (where the antipodal is with respect to O^*). Then $[C_i, C_i^o]$ is an affine diameter with $\Lambda(C_i, O^*) = \mathfrak{m}^*$. Finally, let $f_i \in \text{aff}_C$ be the (unique) normalized affine functional with zero-set $\partial \mathcal{I}_i$. Now, a simple computation shows that $\{f_0, \ldots, f_n\} \in \mathcal{C}_n^o$ is *minimal* for O^*, and (3.1.2) (with $O = O^*$) gives (4.2.8).

Substituting $k = n-1$ and $l = 1$ into (4.2.4), by Theorem 4.2.3, sub-arithmeticity at a critical point $O^* \in C^*$ reduces to the inequality

$$\frac{n}{\mathfrak{m}^* + 1} \leq \sigma_{n-1}^o(O^*). \qquad (4.2.9)$$

This holds for *any* convex body C. According to one of the main results of [Guo–Toth], if the inequality in (4.2.9) is sharp then the Grünbaum Conjecture holds; that is $n + 1$ affine diameters meet at O^* (Section 2.2). Note that sharp inequality in (4.2.9) is automatic for $\mathfrak{m}^* > n - 1$ (by the lower bound in (4.2.6)), so that Klee's solution of the Grünbaum Conjecture is a special case of this. (For details, see again [Guo–Toth].)

4.3   The Mean Minkowski Measure of Convex Bodies of Constant Width

In this short section we continue our discussion on convex bodies of constant width started in Section 2.5. Our first result gives an explicit formula for the mean Minkowski measure $\sigma = \sigma_n$ of a convex body of constant width *at the critical point* in terms of the Minkowski measure itself. (Recall from Theorem 2.5.3 that the critical set of a convex body of constant width is a singleton.) As a byproduct, this immediately leads to a sharp lower bound for σ in our case. Finally, using this method we will be able to calculate this mean Minkowski measure for a variety of classical examples such as the regular Reuleaux polygons and the Meissner tetrahedra.

Theorem 4.3.1. *Let $C \in \mathfrak{B}$ be a convex body of constant width, and $O^* \in \text{int}\, C$ the (unique) critical point. Then we have*

$$\sigma(O^*) = \frac{n+1}{\mathfrak{m}^* + 1}. \tag{4.3.1}$$

Proof. Recall from Theorem 2.5.4 that the critical point O^* is both the circumcenter and the incenter of C. Moreover, the sum of the circumradius $R = R_C$ and the inradius $r = r_C$ is equal to the (constant) width d; and, according to (2.5.2), the Minkowski measure $\mathfrak{m}^* = R/r$.

Let $\bar{B}_R(O^*)$ be the circumball, and $\mathcal{S}_R(O^*)$ the circumsphere of C. By the critical property of the circumsphere in (1.5.1), we have

$$O^* \in [C \cap \mathcal{S}_R(O^*)].$$

In particular, by Carathéodory's theorem (Section 1.3), there exist $C_0, \ldots, C_n \in \partial C \cap \mathcal{S}_R(O^*)$ such that $O^* \in [C_0, \ldots, C_n]$. By definition, this means that $\{C_0, \ldots, C_n\} \in \mathfrak{C}(O^*)$, an n-configuration of C with respect to O^*.

We now calculate the distortion ratio $\Lambda(C_i, O^*) = d(C_i, O^*)/d(C_i^o, O^*)$, $i = 0, \ldots, n$, where the antipodal is with respect to O^*. Clearly, $d(C_i, O^*) = R$. Moreover, the tangent hyperplane to $\mathcal{S}_R(O^*)$ at C_i supports C, and, since $R + r = d$ is the constant width, the (parallel) tangent hyperplane to the insphere $\mathcal{S}_r(O^*)$ at C_i^o also supports C. We thus have $d(C^o, O^*) = d - R = r$. We conclude that $\Lambda(C_i, O^*) = R/r = \mathfrak{m}^*$. Consequently, we have $\sigma(O^*) \leq (n+1)/(\mathfrak{m}^*+1)$. By (4.1.8), equality holds, and the theorem follows.

Theorem 2.5.5 and the subsequent Remark 2 give the precise range of the Minkowski measure $\mathfrak{m}^*$ of convex bodies of constant width. Since the mean Minkowski measure (at the critical point) can be expressed by the Minkowski measure itself via (4.3.1), we immediately arrive at the following:

Corollary 4.3.2. *Let $C \in \mathfrak{B}$ be a convex body of constant width, and O^* the (unique) critical point. Then we have*

$$n + 1 - \frac{\sqrt{2n(n+1)}}{2} \leq \sigma(O^*) \leq \frac{n+1}{2}. \tag{4.3.2}$$

The upper bound is attained if and only if C is a Euclidean ball. The lower bound is attained if and only if C is the completion of a regular simplex.

Example 4.3.3. As a simple application (of the lower bound in (4.3.2)), we obtain the mean Minkowski measures of the Reuleaux triangle $\mathcal{Q}_3$ and the Meissner tetrahedra $\mathcal{M}$ as

$$\sigma_{\mathcal{Q}_3}(O^*) = 3 - \sqrt{3} \quad \text{and} \quad \sigma_{\mathcal{M}}(O^*) = 4 - \sqrt{6}. \tag{4.3.3}$$

The first formula can immediately be generalized to obtain the mean Minkowski measure of a regular $(2m + 1)$-Reuleaux polygon $\mathcal{Q}_{2m+1}$. A minimizing configuration can be chosen to be comprised by any three vertices whose convex hull contains the center O^*. Using the computations in Example 2.5.1 (cont.), we obtain

$$\sigma_{\mathcal{Q}_{2m+1}}(O^*) = \frac{3}{m^*_{\mathcal{Q}_{2m+1}} + 1} = 3 - \frac{3}{2\sin(\frac{m\pi}{2m+1})}, \quad m \geq 1. \tag{4.3.4}$$

In two dimensions ($n = 2$), the completion of a regular triangle is the *unique* regular Reuleaux triangle. In this case the corollary above reduces to the following:

Corollary 4.3.4. *Let C be a planar convex body of constant width and critical point O^*. Then, we have*

$$3 - \sqrt{3} \leq \sigma(O^*) \leq \frac{3}{2}.$$

The upper bound is attained if and only if C is a (closed) disk, and the lower bound is attained if and only if C is a regular Reuleaux triangle.

In three dimensions ($n = 3$), we see that the mean Minkowski measure of the Meissner tetrahedra (second formula in (4.3.3)) is the lower bound for the respective mean Minkowski measure of any spatial convex body of constant width (at the critical point).

Remark. Corollary 4.3.2 is due to [Jin–Guo 4] with a direct proof.

4.4 Concavity of Mean Minkowski Measures

We now return to the main line and discuss functional properties of the mean Minkowski measures $\{\sigma_k\}_{k \geq 1}$. We first prove continuity of these measures up to the boundary of the convex body.

Despite (pointwise) duality with the *concave* dual mean Minkowski measures (Section 4.2), the main result of the present section asserts that $\sigma = \sigma_n$ is, in general, *not concave* for $n \geq 3$.

This failure of concavity is intimately connected to the possible *singular limiting behavior* of minimizing sequences of configurations in the definition of the mean Minkowski measure σ. Accordingly, the interior of the convex body naturally splits into regular and singular sets. A large part of this entire chapter will be devoted to the study of the subtle structure of this splitting. In this section we make a few initial observations on the regular set.

We begin with the following elementary result:

Theorem 4.4.1. *Let $C \in \mathfrak{B}$. For $k \geq 1$, the function $\sigma_k : \mathrm{int}\, C \to \mathbb{R}$ is continuous and extends continuously to ∂C by setting it equal to 1 on ∂C.*

Before the proof we make some preparations. Given $C \in \mathfrak{B}$ and $O \in \operatorname{int} C$, we define the map $\phi_O : \operatorname{int} C \times \partial C \to \partial C$ as follows. For $O' \in \operatorname{int} C$ and $C \in \partial C$, $\phi_O(O', C)$ is the unique point on the boundary of C such that $C - O = \mu(\phi_O(O', C) - O')$ for some (unique) $\mu > 0$. (See Figure 4.4.1.)

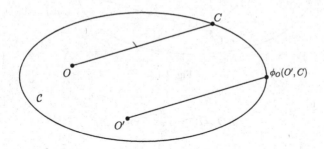

Fig. 4.4.1

The existence (and unicity) of $\phi_O(O', C)$ follows from Corollary 1.1.9. Clearly, $\phi_O(O, C) = C$. Moreover, for $O, O', O'' \in \operatorname{int} C$, we have

$$\phi_{O'}(O'', \phi_O(O', C)) = \phi_O(O'', C). \tag{4.4.1}$$

Consequently, we have

$$\phi_{O'}(O, \phi_O(O', C)) = C.$$

(See Figure 4.4.2.)

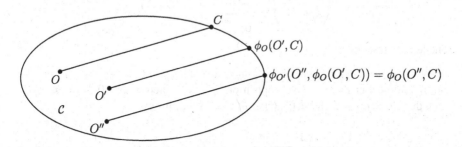

Fig. 4.4.2

Lemma 4.4.2. *For fixed $C \in \partial C$, $\phi_O(., C) : \operatorname{int} C \to \partial C$ is continuous.*

Proof. By (4.4.1), it is enough to prove continuity at O. Let $O' \in \operatorname{int} C$ and $\alpha = \angle CO\phi_O(O', C)$. Then $\alpha = \angle O\phi_O(O', C)O'$. (See Figure 4.4.3.)

Mimicking the proof of Proposition 2.1.1/(b) for the triangle $[O, O', \phi_O(O', C)]$ (and using the notations there), we obtain

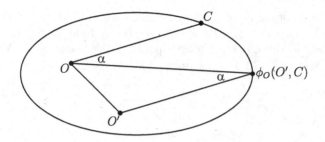

Fig. 4.4.3

$$d(O', O) \geq \sqrt{2}r_O \sin(\alpha/2),$$

provided that $d(O', O) < r < r_O/2$. Thus, $\alpha \to 0$ as $O' \to O$. By Proposition 1.1.10, we then have $\phi_O(O', C) \to C$ as $\alpha \to 0$. The lemma follows.

Lemma 4.4.3. *Let* $C_0, \ldots, C_k \in \partial C$. *Then*

$$O \in [C_0, \ldots, C_k] \Rightarrow O' \in [\phi_O(O', C_0), \ldots, \phi_O(O'C_k)].$$

Proof. We write $O = \sum_{i=0}^k \lambda_i C_i$, $\sum_{i=0}^k \lambda_i = 1$, $\{\lambda_0, \ldots, \lambda_k\} \subset [0, 1]$. Let $\mu_i > 0$ be such that $C_i - O = \mu_i(\phi_O(O', C_i) - O')$, $i = 0, \ldots, k$. Multiplying this equality by λ_i and summing up with respect to $i = 0, \ldots, k$, we obtain

$$\left(\sum_{i=0}^k \lambda_i \mu_i\right) O' = \sum_{i=0}^k \lambda_i \mu_i \phi_O(O', C_i).$$

The lemma follows.

Proof of Theorem 4.4.1. Let $O \in \text{int}\,C$ and $\{C_0, \ldots, C_k\} \in \mathfrak{C}_k(O)$ any k-configuration. Let $\epsilon > 0$. By Proposition 2.1.1/(c), there exists $0 < \delta < r_O/2$ such that $d(O', O) < \delta$ and $d(C', C) < \delta$, $C, C' \in \partial C$, imply

$$\left| \frac{1}{\Lambda(C', O') + 1} - \frac{1}{\Lambda(C, O) + 1} \right| < \frac{\epsilon}{k+1}. \tag{4.4.2}$$

Let $C_i' = \phi_O(O', C_i)$, $i = 0, \ldots, k$. By Lemma 4.4.3, we have $\{C_0', \ldots, C_k'\} \in \mathfrak{C}_k(O')$. Since $\phi_O(., C_i)$ is continuous at O for every $i = 0, \ldots, k$ (Lemma 4.4.2), there exists $0 < \delta' < \delta$ such that $d(O', O) < \delta'$ implies $d(C_i', C_i) < \delta$ for all $i = 0, \ldots, k$. Using (4.4.2), for $d(O', O) < \delta'$, we also have

$$\left| \sum_{i=0}^k \frac{1}{\Lambda(C_i', O') + 1} - \sum_{i=0}^k \frac{1}{\Lambda(C_i, O) + 1} \right| < \epsilon. \tag{4.4.3}$$

We now first choose $\{C_0, \ldots, C_k\} \in \mathfrak{C}_k(O)$ to be minimal (with respect to O). Substituting into (4.4.3), we obtain

$$\left| \sum_{i=0}^{k} \frac{1}{\Lambda(C_i', O') + 1} - \sigma_k(O) \right| < \epsilon.$$

Taking the infimum in the first sum over $\mathfrak{C}_k(O')$, we obtain

$$\sigma_k(O') - \sigma_k(O) < \epsilon.$$

Second, we choose $\{C_0', \ldots, C_k'\} \in \mathfrak{C}_k(O')$ to be minimal (with respect to O'). (This is possible by first choosing C_i' and then defining $C_i = \phi_{O'}(O, C_i')$, since then $\phi_0(O', C_i) = C_i'$.) Again by (4.4.3), we have

$$\sum_{i=0}^{k} \frac{1}{\Lambda(C_i, O) + 1} - \sigma_k(O') < \epsilon.$$

Taking the infimum in the first sum over $\mathfrak{C}_k(O)$, we obtain

$$\sigma_k(O) - \sigma_k(O') < \epsilon.$$

Combining these, for $d(O', O) < \delta'$, we have

$$|\sigma_k(O') - \sigma_k(O)| < \epsilon.$$

Continuity of σ_k follows.

For the last statement we note that

$$\lim_{d(O, \partial C) \to 0} \sigma_k(O) = 1. \tag{4.4.4}$$

The proof is the same as in the dual case (Section 4.2). By (4.1.3) and (4.1.5), we have

$$1 \leq \sigma_k(O) \leq 1 + \frac{k-1}{\mathfrak{m}(O) + 1}, \qquad O \in \operatorname{int} C.$$

Using Lemma 2.1.6, we see that the right-hand side converges to 1 as $d(O, \partial C) \to 0$. This gives (4.4.4). The theorem follows.

We have seen in Section 4.2 that the dual mean Minkowski measures $\sigma_k^o, k \geq 1$, are concave functions on the interior of the convex body. In addition, Theorem 2.1.2 asserts that the function $1/(\mathfrak{m} + 1) : \operatorname{int} C \to \mathbb{R}$ is also concave. It is therefore natural to study concavity properties of the original mean Minkowski measures σ_k, $k \geq 1$. We begin with the (initially) surprising counter-example:

Example 4.4.4. For the n-dimensional cube $\mathcal{C}_n$, $n \geq 3$, the function $\sigma = \sigma_{\mathcal{C}_n}$: $\text{int}\,\mathcal{C}_n \to \mathbb{R}$ is *not* concave.

We realize the cube as $\mathcal{C}_n = [-1, 1]^n \subset \mathbb{R}^n$. The set of n vertices $\{C_1, \ldots, C_n\}$ adjacent to the vertex $V = (1, \ldots, 1)$ forms a simplicial configuration whose centroid is $O = (1 - 2/n)V$. The convex hull $[C_1, \ldots, C_n]$ is a simplicial intersection of $\mathcal{C}_n$ across O. (Up to scaling, this is called the *vertex-figure* of V. The vertex-figures for all the vertices define the cross-polytope dual of $\mathcal{C}_n$.) By Theorem 4.1.1, we have $\sigma_{n-1}(O) = 1$.

Since the antipodal of V (with respect to O) is $V^o = -V$, the distortion ratio at V^o is

$$\Lambda(V^o, O) = \frac{|V + (1 - 2/n)V|}{|V - (1 - 2/n)V|} = \frac{2 - 2/n}{2/n} = n - 1 \leq \mathfrak{m}(O).$$

Using sub-arithmeticity in (4.1.3), we obtain

$$\sigma(O) \leq \sigma_{n-1}(O) + \frac{1}{\mathfrak{m}(O) + 1} \leq 1 + \frac{1}{n}.$$

On the other hand, $\mathcal{C}_n$ is symmetric with respect to the origin 0, so that we have $\sigma(0) = (n + 1)/2$. Finally, using the continuous extension of σ to $\partial \mathcal{C}_n$, we have $\sigma(V) = 1$ (Theorem 4.4.1). Thus, for $n \geq 3$, on the line segment $[0, V]$, we have

$$\sigma(O) \leq 1 + \frac{1}{n} < \frac{2n + 1}{n\ 2} + \left(1 - \frac{2}{n}\right) = \frac{2}{n}\sigma(0) + \left(1 - \frac{2}{n}\right)\sigma(V).$$

This shows that $\sigma\,|[0, V]$ is not concave.

Remark. For a generalization of this example, see Corollary 4.6.8.

The previous example does not work for $n = 2$. In fact, concavity holds in 2 dimensions:

Theorem 4.4.5. *For any planar convex body $\mathcal{C}$ ($n = 2$), the function $\sigma_{\mathcal{C}}$: $\text{int}\,\mathcal{C} \to \mathbb{R}$ is concave.*

The proof of this theorem will be obtained at the end of this section. In view of future applications, we first begin with some observations in the general setting $n \geq 2$ on partial concavity:

Proposition 4.4.6. *Let $\mathcal{C} \in \mathfrak{B}$ and $O \in \text{int}\,\mathcal{C}$. Given $k \geq 1$, let $\{C_0, \ldots, C_k\} \in \mathfrak{C}_k(O)$ be a minimal configuration. Then, for any $O_0, O_1 \in [C_0, \ldots, C_k]$ and $0 < \lambda < 1$ such that $O = (1 - \lambda)O_0 + \lambda O_1$, we have*

$$\sigma_k(O) \geq (1 - \lambda)\sigma_k(O_0) + \lambda\sigma_k(O_1).$$

Proof. This is a simple consequence of Theorem 2.1.2. Indeed, we have

$$\sigma_k(O) = \sum_{i=0}^{k} \frac{1}{\Lambda(C_i, O) + 1} \geq \sum_{i=0}^{k} \frac{1 - \lambda}{\Lambda(C_i, O_0) + 1} + \sum_{i=0}^{k} \frac{\lambda}{\Lambda(C_i, O_1) + 1}$$

since each term in the sum of the left-hand side is concave. By assumption, we have $\{C_0, \ldots, C_k\} \in \mathfrak{C}_k(O_0) \cap \mathfrak{C}_k(O_1)$. Taking infima on the right-hand side, the proposition follows.

In the rest of the section we set $k = n$, and study the function $\sigma = \sigma_n : \operatorname{int} \mathcal{C} \to \mathbb{R}$. (Recall that, unless necessary, the dimension n will be suppressed.)

Given $O \in \operatorname{int} \mathcal{C}$, a configuration $\{C_0, \ldots, C_n\} \in \mathfrak{C}(O)$ is called *simplicial* if $[C_0, \ldots, C_n]$ is an n-simplex with O in its interior. The set of all simplicial configurations (with respect to O) is denoted by $\Delta(O)$. Since $\Delta(O)$ is dense in $\mathfrak{C}(O)$, in the definition (4.1.1) of σ, the set of all configurations $\mathfrak{C}(O)$ can be replaced by $\Delta(O)$, but a minimizing sequence may not subconverge.

We call an interior point $O \in \operatorname{int} \mathcal{C}$ *regular* if the infimum is attained only in $\Delta(O)$. Equivalently, $O \in \operatorname{int} \mathcal{C}$ is a regular point if, for *any* minimal configuration $\{C_0, \ldots, C_n\} \in \mathfrak{C}(O)$, the convex hull $[C_0, \ldots, C_n]$ is an n-simplex with O in its interior. The set of regular points, the *regular set*, is denoted by $\mathcal{R} = \mathcal{R}_\mathcal{C} \subset \operatorname{int} \mathcal{C}$.

An interior point of $\mathcal{C}$ which is not regular is called *singular*, and we have the *singular set* $\mathcal{S} = \mathcal{S}_\mathcal{C} = \operatorname{int} \mathcal{C} \setminus \mathcal{R}$.

The regular set $\mathcal{R}$ is open. Indeed, this is because compactness of $\partial \mathcal{C}$ implies that a sequence of *non-simplicial* configurations may only subconverge to a non-simplicial configuration. Thus, given a sequence $\{O_j\}_{j \geq 1} \subset \mathcal{S}$ of singular points, for each $j \geq 1$, we can select a *minimal* non-simplicial configuration $\{C_0^j, \ldots, C_n^j\} \in \mathfrak{C}(O_j)$. Selecting a convergent subsequence if necessary, we may assume that, for $i = 0, \ldots, n$, $C_i^j \to C_i$ as $j \to \infty$. Clearly, $\{C_0, \ldots, C_n\} \in \mathfrak{C}(O)$ is non-simplicial. It is also minimal by continuity of σ. Thus, O is a singular point.

Remark. Let $O \in \operatorname{int} \mathcal{C}$ be a regular point and $\{C_0, \ldots, C_n\} \in \Delta(O)$ a minimal (simplicial) configuration. Since O is in the interior of the simplex $[C_0, \ldots, C_n]$, for each $i = 0, \ldots, n$, C_i is local *maximum* of the distortion $\Lambda(., O)$. By Corollary 2.2.2, $[C_i, C_i^o]$ is an affine diameter. We obtain that $(n + 1)$ affine diameters pass through any regular point. The question of existence of this kind of points has been raised in Grünbaum's Conjecture. (See the end of Section 2.2.) We see that if the regular set is non-empty then the Grünbaum Conjecture holds.

The existence and non-existence of regular points will be discussed in the next two sections.

Theorem 4.4.7. *On the regular set $\mathcal{R}$, the function σ is concave, that is, for $[O_0, O_1] \subset \mathcal{R}$, we have*

$$\sigma((1 - \lambda)O_0 + \lambda O_1) \geq (1 - \lambda)\sigma(O_0) + \lambda \sigma(O_1).$$

Remark. In the next section we show, by way of an example, that $\mathcal{R}$ may not be convex.

Before the proof of Theorem 4.4.7, for $O \in \mathcal{R}$, we define $\rho(O)$ as the infimum of the distance $d(O, \partial[C_0, \ldots, C_n])$ over all minimal (simplicial) configurations $\{C_0, \ldots, C_n\} \in \mathfrak{C}(O)$. Clearly, $\rho > 0$ pointwise on $\mathcal{R}$.

We can state somewhat more:

Lemma 4.4.8. *For $\mathcal{K} \subset \mathcal{R}$ compact, we have $\inf_{O \in \mathcal{K}} \rho(O) > 0$.*

Proof. Assuming the contrary, let $\{O_j\}_{j \geq 1} \subset \mathcal{K}$ be such that $\rho(O_j) \to 0$ as $j \to \infty$. Since $\mathcal{K}$ is compact, we may assume $O_j \to O \in \mathcal{K} \subset \mathcal{R}$ as $j \to \infty$. In addition, since $\partial \mathcal{C}$ is compact, we can select minimal configurations $\{C_0^j, \ldots, C_n^j\} \in \Delta(O_j)$ such that $d(O_j, \partial[C_0^j, \ldots, C_n^j]) \to 0$ and $C_i^j \to C_i$, $i = 0, \ldots, n$, as $j \to \infty$. The limiting configuration $\{C_0, \ldots, C_n\} \in \mathfrak{C}(O)$ is minimal since σ is continuous. We also have $d(O, \partial[C_0, \ldots, C_n]) = 0$ so that $O \in \partial[C_0, \ldots, C_n]$. This means that $\{C_0, \ldots, C_n\}$ is not simplicial, a contradiction to $O \in \mathcal{R}$.

Proof of Theorem 4.4.7. Let $O_0, O_1 \subset \mathcal{R}$ such that $[O_0, O_1] \subset \mathcal{R}$, and $O = (1 - \lambda) O_0 + \lambda O_1$, $0 < \lambda < 1$. Proposition 4.4.6 asserts the concavity property for σ if O_0 and O_1 are sufficiently close to be contained in the convex hull of a minimal simplicial configuration with respect to O. Lemma 4.4.8 states that O_0 and O_1 can be made sufficiently close universally in this sense provided that O varies within a compact subset of $\mathcal{R}$. Restricting σ to the maximal line segment in $\mathcal{R}$ that passes through O_0 and O_1, concavity is the consequence of the following:

Lemma 4.4.9. *Let $f : (a_0, b_0) \to \mathbb{R}$, $a_0 < b_0$, be a continuous function satisfying the following property:*

For each $c \in (a_0, b_0)$ there exists $\epsilon_c > 0$ such that $(c - \epsilon_c, c + \epsilon_c) \subset (a_0, b_0)$, and for any $[c_0, c_1] \subset (c - \epsilon_c, c + \epsilon_c)$ with $c \in [c_0, c_1]$ we have

$$f(c) \geq (1 - \lambda)f(c_0) + \lambda f(c_1), \qquad (4.4.5)$$

where $c = (1 - \lambda)c_0 + \lambda c_1$. Moreover, given $[a, b] \subset (a_0, b_0)$, there is a universal choice of $\epsilon > 0$ for all $c \in [a, b]$. Then f is concave on (a_0, b_0).

Proof. Let $[a, b] \subset (a_0, b_0)$ and $\epsilon > 0$ a universal choice for $[a, b]$. Let $0 < \lambda < 1$. We need to show that

$$f((1 - \lambda)a + \lambda b) \geq (1 - \lambda)f(a) + \lambda f(b). \qquad (4.4.6)$$

Let $c = (1 - \lambda)c_0 + \lambda c_1$. Let $N \geq 1$ and subdivide $[a, c]$ and $[c, b]$ into $(N + 1)$ equal parts:

$$a = a_0 < a_1 < \ldots < a_N < a_{N+1} = c = b_{N+1} < b_N < \ldots < b_1 < b_0 = b.$$

Note that $c = (1 - \lambda)a_i + \lambda b_i$, $i = 0, \ldots, N$. Choose N large enough so that the ϵ-neighborhood of any subdivision point inside (a, b) contains at least three subdivision points. Then (4.4.5), applied to each triplet of consecutive subdivision points, gives

$$f(a_i) \geq \frac{1}{2}f(a_{i-1}) + \frac{1}{2}f(a_{i+1}), \quad i = 1, \ldots, N,$$

$$f(c) \geq (1 - \lambda)f(a_N) + \lambda f(b_N),$$

$$f(b_i) \geq \frac{1}{2}f(b_{i-1}) + \frac{1}{2}f(b_{i+1}), \quad i = 1, \ldots, N.$$

We refer to the ith inequality in the first and last set of N inequalities as $(A)_i$ and $(B)_i$, $i = 1, \ldots, N$, and the inequality for $f(c)$ as the middle inequality.

We now claim that

$$f(c) \geq (1 - \lambda)f(a_{N-i}) + \lambda f(b_{N-i}), \quad i = 0, \ldots, N. \tag{4.4.7}$$

Observe that, for $i = N$ this is our statement in (4.4.6). Proceeding by induction to prove (4.4.7), we first note that, for $i = 0$, (4.4.7) is the middle inequality. For the general induction step $i \Rightarrow i + 1$, we consider the inequality given algebraically as

$$(1 - \lambda)((A)_{N-i} + \ldots (A)_N) + \lambda((B)_{N-i} + \ldots (B)_N).$$

After simplifications, this becomes

$$(1 - \lambda)(f(a_{N-i}) + f(a_N)) + \lambda(f(b_{N-i}) - f(b_N))$$
$$\geq (1 - \lambda)f(a_{N-i-1}) + \lambda f(b_{N-i-1}) + f(c).$$

The left-hand side is $\leq 2f(c)$ by (4.4.7) and the middle inequality. Equation (4.4.7) now follows for $i + 1$. The lemma and hence Theorem 4.4.7 follows.

Summarizing, the interior of a convex body $C \in \mathfrak{B}$ splits into the disjoint union of the open regular set $\mathcal{R}$ and the (relatively) closed singular set $\mathcal{S}$ and, by Theorem 4.4.7 just proved, $\sigma | \mathcal{R}$ is concave. Example 4.4.4 indicates, however, that, in general, we cannot expect $\sigma | \mathcal{S}$ to be the restriction of a concave function (on $\operatorname{int} C$).

The definition of regularity can be paraphrased as follows: $O \in \mathcal{R}$ if no *minimal* (n-) configuration contains a *proper* subconfiguration. Thus, an interior point O of C is singular if there exists a minimal configuration $\{C_0, \ldots, C_n\} \in \mathfrak{C}(O)$ which contains a (necessarily minimal) k-subconfiguration, say $\{C_0, \ldots, C_k\} \in \mathfrak{C}_k(O)$, for some $1 \leq k < n$. Since at the complementary points C_{k+j}, $j = 1, \ldots, n - k$, the distortion ratio $\Lambda(., O)$ must take its maximum $\mathfrak{m}(O)$, we see that $O \in \mathcal{S}$ if and only if, for some $1 \leq k \leq n - 1$, we have

$$\sigma(O) = \sigma_k(O) + \frac{n-k}{\mathfrak{m}(O) + 1}. \tag{4.4.8}$$

By sub-arithmeticity in (4.1.3), in general, we have

$$\sigma \le \sigma_{n-1} + \frac{1}{\mathfrak{m} + 1} \le \sigma_k + \frac{n-k}{\mathfrak{m} + 1}, \quad 1 \le k \le n - 1. \tag{4.4.9}$$

The following characterization of the singular set emerges:

$$\mathcal{S} = \left\{ O \in \mathrm{int}\,\mathcal{C} \,\middle|\, \sigma(O) = \sigma_{n-1}(O) + \frac{1}{\mathfrak{m}(O) + 1} \right\}. \tag{4.4.10}$$

We see that, on the singular set, the restriction $\sigma \mid \mathcal{S}$ is equal to $\sigma_{n-1} + 1/(\mathfrak{m} + 1)$. In this sum the second term is concave (Theorem 2.1.2), but the first, in general, is not.

In fact, the reason that, for the n-dimensional cube $\mathcal{C}_n$, the function σ is not concave (Example 4.4.4) is that σ_{n-1} is identically 1 on the complement of the inscribed dual cross-polytope (as $\mathcal{C}_n$ has codimension 1 simplicial slices across these points), whereas at the interior points of the cross-polytope $\sigma_{n-1} > 1$.

The following definition is therefore natural. A convex body $\mathcal{C} \in \mathfrak{B}$ is called *simplicial in codimension* k, $0 \le k \le n - 1$, if $\mathcal{C}$ possesses an $(n - k)$-dimensional simplicial slice across any point $O \in \mathrm{int}\,\mathcal{C}$.

Clearly, any convex body is automatically simplicial in codimension $n - 1$, and a convex body is simplicial in codimension 0 if and only if it is a simplex.

By Theorem 4.1.1, $\mathcal{C}$ is simplicial in codimension k if and only if $\sigma_{n-k} = 1$ *identically* on $\mathrm{int}\,\mathcal{C}$.

As an application, in Example 4.1.6, the convex body $\mathcal{C}_0(\mathcal{H})$ of endomorphisms of a Euclidean vector space $\mathcal{H}$ is simplicial in codimension $h(h - 1)/2$. Indeed, as noted there, we have $\dim \mathcal{C}_0(\mathcal{H}) = h(h + 1)/2 - 1$ and $\sigma_{h-1} = 1$. The set $\Delta_0 \subset \mathcal{C}_0$ of diagonal endomorphisms (with respect to a fixed orthonormal basis in $\mathcal{H}$), and its $O(\mathcal{H})$-images, comprise simplicial slices of dimension $h - 1$.

Theorem 4.4.10. *Let $\mathcal{C} \in \mathfrak{B}$ be a codimension 1 simplicial convex body. Then $\sigma : \mathrm{int}\,\mathcal{C} \to \mathbb{R}$ is a concave function.*

Proof. By assumption, we have $\sigma_{n-1} = 1$. By sub-arithmeticity in (4.1.3) gives

$$\sigma \le 1 + \frac{1}{\mathfrak{m} + 1},$$

where the function on the right-hand side is concave (Theorem 2.1.2). Equality holds on the singular set $\mathcal{S}$, and strict inequality holds on the regular set $\mathcal{R}$ on which σ is also concave (Theorem 4.4.7).

To show that σ is concave on the entire interior of $\mathcal{C}$, let $O_0, O_1 \in \mathrm{int}\,\mathcal{C}$, and consider the line segment $\langle O_0, O_1 \rangle \cap \mathcal{C}$. Let f denote the restriction of $1 + 1/(\mathfrak{m} + 1)$ to $\mathcal{S} \cap \langle C_0, C_1 \rangle$, and g the restriction of σ to $\mathcal{R} \cap \langle C_0, C_1 \rangle$. On $\partial \mathcal{C}$ both functions assume their limit equal to 1 (Lemma 2.1.6 and Theorem 4.4.1). We have $g \leq f$ and, since $\mathcal{R}$ is open, g differs from f on a countable union of (disjoint) open intervals. The theorem is now the consequence of the following:

Lemma 4.4.11. *Let $f : [a,b] \to \mathbb{R}$ be a continuous and concave function with $f(a) = f(b) = 1$. Let $U \subset [a,b]$ be an open set. Write U as a countable union of (disjoint) open intervals (a_i, b_i), $i \geq 1$. For each $i \geq 1$, let $g_i : [a_i, b_i] \to \mathbb{R}$ be continuous and concave, $g_i < f$ on (a_i, b_i), $g(a_i) = f(a_i)$ and $g(b_i) = f(b_i)$. Define $g : [a,b] \to \mathbb{R}$ as g_i on $[a_i, b_i]$, $i \geq 1$, and f otherwise. Then g is (continuous and) concave on $[a,b]$.*

Proof. Let $f_0 = f$. Proceeding inductively, we define f_i as f_{i-1}, and replaced by g_i on $[a_i, b_i]$. Since the subintervals are mutually disjoint, the decreasing sequence $\{f_i\}_{i \geq 1}$ is pointwise convergent, and it converges to g.

It remains to show that f_i is concave for each $i \geq 1$. Proceeding inductively again, we need to show that concavity of f_{i-1} implies concavity of f_i. Let $[c_0, c_1] \subset [a, b]$. We need to show

$$f_i((1 - \lambda)c_0 + \lambda c_1) \geq (1 - \lambda)f(c_0) + \lambda f(c_1), \quad 0 \leq \lambda \leq 1.$$

This is an easy case-by-case verification depending on the mutual position of $[c_0, c_1]$ and $[a_i, b_i]$ on which f_{i-1} is modified to f_i. The lemma follows.

Note that Theorem 4.4.5 follows as a special case since any planar convex body is simplicial in codimension 1. In addition, we also have the following:

Corollary 4.4.12. *For three-dimensional cones, σ is a concave function.*

Example 4.4.13. There exists a four-dimensional cone C on which σ is not concave. The construction of this cone is very technical so that we just make a short note here. Full details are in [Toth 7]. The base of the cone is the convex body $C_0 \subset \mathbb{R}^3$ obtained as the intersection of the closed unit ball $\bar{B} \subset \mathbb{R}^3$ and the vertical cylinder $\Delta \times [-1, 1]$ with base $\Delta \subset \mathbb{R}^2$, an equilateral triangle inscribed in the unit circle $S \subset \mathbb{R}^2$. Note that σ_2 is (highly) non-concave as Δ is the only triangular slice of C across points in its interior (therefore $\sigma_2 = 1$ on int Δ) but in the two components of the complement int $C \setminus \Delta$ we have $\sigma_2 > 1$ (Theorem 4.1.1).

4.5 Singular Points

In our study of concavity of the function σ we introduced the concepts of regular and singular points. In the present section we study the existence and structure of the regular and singular sets. It is almost immediate from the definitions that the interior of a simplex consists of regular points only. At the other extreme, the main result of this section asserts that, in a *symmetric* convex body, all interior points are *singular*.

Another purpose of this section is to give a variety of explicitly worked out examples; in particular, to show that the converse of this result is false, that is, there exist non-symmetric convex bodies all of whose interior points are singular.

We begin with a refinement of our discussion on singular points of the previous section.

Let $\mathcal{C} \in \mathfrak{B}$. First note that, by sub-arithmeticity in (4.1.3) and (4.4.10), the regular set is given by

$$\mathcal{R} = \left\{ O \in \operatorname{int} \mathcal{C} \,\middle|\, \sigma(O) < \sigma_{n-1}(O) + \frac{1}{\mathfrak{m}(O) + 1} \right\}. \tag{4.5.1}$$

In particular, due to continuity of the functions in the defining inequality, openness of $\mathcal{R}$ (established in the previous section) follows directly.

For $1 \le m < n$, we define

$$\mathcal{S}_m = \left\{ O \in \operatorname{int} \mathcal{C} \,\middle|\, \sigma(O) = \sigma_{n-m}(O) + \frac{m}{\mathfrak{m}(O) + 1} \right\}.$$

By (4.4.10) again, $\mathcal{S} = \mathcal{S}_1 = \operatorname{int} \mathcal{C} \setminus \mathcal{R}$ is the singular set, and, by sub-arithmeticity in (4.1.3), we have

$$\mathcal{S}_{m+1} \subset \mathcal{S}_m, \quad m = 1, \ldots, n - 2.$$

Let $O \in \mathcal{S}$ be a singular point. We define the *degree of singularity* of O as the largest m such that $O \in \mathcal{S}_m$. Thus, the degree of singularity of O is $n - k$ if and only if the sequence $\{\sigma_j(O)\}_{j \ge 1}$ is arithmetic (with difference $1/(\mathfrak{m}(O) + 1)$) exactly from the kth term onwards.

The next proposition follows from the definitions:

Proposition 4.5.1. *Let $O \in \mathcal{S}$ be a singular point. Then, $O \in \mathcal{S}_{n-k}$, $1 \le k < n$, if and only if there exists a minimal n-configuration which contains a k-configuration. In this case, the k-configuration is also minimal, and, at each n-configuration point complementary to the k-configuration, $\Lambda(., O)$ attains its absolute maximum $\mathfrak{m}(O)$. In addition, if the degree of singularity is $n - k$ (equivalently, $O \in \mathcal{S}_{n-k+1}$), then the k-configuration is simplicial in the sense that its convex hull is a k-simplex with O in its relative interior, and $\Lambda(., O)$, restricted to the affine span of this simplex, attains local maxima at each k-configuration point.*

Remark. The degree of singularity of a singular point $O \in \mathcal{S}$ cannot exceed the codimension of a simplicial slice across O. Indeed, if $O \in \mathcal{S}_m$, $1 \le m < n$, and there is an m'-dimensional simplicial slice across O, then $\sigma_{m'}(O) = 1$. Using sub-arithmeticity in (4.1.3), we have

$$1 + \frac{m}{\mathfrak{m}(O) + 1} \le \sigma_{n-m}(O) + \frac{m}{\mathfrak{m}(O) + 1} = \sigma(O) \le \sigma_{m'}(O) + \frac{n - m'}{\mathfrak{m}(O) + 1} = 1 + \frac{n - m'}{\mathfrak{m}(O) + 1}.$$

Thus, $m \le n - m'$ follows.

Returning to the general setting, recall from Theorem 4.1.1 that, for $O \in \text{int}\,C$, we have

$$1 \le \sigma(O) \le \frac{n+1}{2},$$

with the lower bound attained for a *simplex*, and the upper bound attained for a *symmetric body* (with center of symmetry at O).

According to our next theorem, these two bounds also correspond to the two extreme cases in which the entire interior of C is the *regular* or *singular* set.

Theorem 4.5.2. *Let $C \in \mathfrak{B}$ and $O \in \text{int}\,C$. If $\sigma(O) = 1$ then C is a simplex and the interior of C consists of regular points only. If $\sigma(O) = (n+1)/2$ then C is symmetric with respect to O and the interior of C consists of singular points only.*

Proof. It is clear that the interior of a simplex consists of regular points only with unique minimal configuration comprised of the vertices of the simplex. Thus, we need to prove the last statement only.

Let C be a symmetric convex body with center of symmetry at $O_0 \in \text{int}\,C$. Then the distortion ratio $\Lambda(.,O_0) = 1$ identically on ∂C. Clearly, any configuration for O_0 is minimal so that O_0 is a singular point.

Let $O \in \text{int}\,C$, $O \ne O_0$, be another point. Assume that O is regular. In what follows, the antipodal of a point $C \in \partial C$ will be denoted by C^o. Let $\{C_0, \ldots, C_n\} \in \mathfrak{C}(O)$ be a minimal configuration. By assumption, $[C_0, \ldots, C_n]$ is an n-simplex with O in its interior. At each vertex C_i, $i = 0, \ldots, n$, $\Lambda(.,O)$ attains a local maximum and therefore $[C_i, C_i^o]$ is an affine diameter (Corollary 2.2.2). Let $\mathcal{H}_i$ and $\mathcal{H}_i^o$ be parallel hyperplanes supporting C at C_i and C_i^o, $i = 0, \ldots, n$.

Let A be the intersection of the ray emanating from O and passing through O_0 with the boundary. We claim that $[A, C_i] \subset \mathcal{H}_i \cap \partial C$ provided that $C_i \notin \langle O, O_0 \rangle$.

For simplicity, we suppress the subscript, assume that $\Lambda(.,O)$ assumes a local maximum at the boundary point $C \notin \langle O, O_0 \rangle$ of a minimal configuration, and the affine diameter $[C, C^o]$ has parallel hyperplanes $\mathcal{H}$ and $\mathcal{H}^o$ at its endpoints. Let $C_0^o \in \partial C$ be the antipodal of C *with respect to* O_0. By symmetry at O_0, the hyperplane parallel to $\mathcal{H}$ passing through C_0^o must support C. Hence, it must coincide with $\mathcal{H}^o$. We obtain that $[C^o, C_0^o] \subset \mathcal{H}^o \cap \partial C$.

We now define a sequence of points $\{A_j\}_{j \ge 1} \subset \partial C$ as follows. Let A_1 be the antipodal of C^o with respect to O_0. With A_j defined, let A_{j+1} be the antipodal of A_j^o with respect to O_0. (See Figure 4.5.1.)

We claim that $[A_j, C] \subset \mathcal{H} \cap \partial C$, $j \ge 1$.

First, since $[C^o, C_0^o] \subset \mathcal{H}^o \cap \partial C$, taking antipodals with respect to O_0, we obtain $[A_1, C] \subset \mathcal{H} \cap \partial C$. For the general induction step, assume that, for some $j \ge 1$, we have $[A_j, C] \subset \mathcal{H} \cap \partial C$. Consider a point C' moving continuously from C to A_j along the line segment $[A_j, C]$. Since $\mathcal{H}$ and $\mathcal{H}^o$ are parallel, we have

$$\Lambda(C', O) \ge \Lambda(C, O). \tag{4.5.2}$$

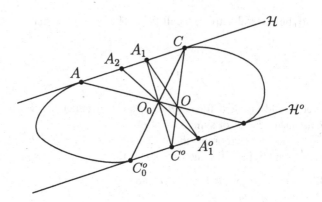

Fig. 4.5.1

Hence, for every specific C' we can replace $C(= C_i)$ by C' in the configuration $\{C_0, \ldots, C_n\}$ as long as the configuration condition $O \in [C_0, \ldots, C_n]$ stays intact. In this case, by minimality, we also have $\Lambda(C', O) = \Lambda(C, O)$. We now observe that this replacement does keep the configuration condition intact for any C' since otherwise for some specific C' we would get a modified minimal configuration with O on the boundary of its convex hull, contradicting to the regularity of O.

We obtain that the distortion ratio $\Lambda(., O)$ is *constant* along $[A_j, C]$, or equivalently, $[A_j^o, C^o]$ is parallel to $[A_j, C]$. This means that

$$[A_j^o, C^o] \subset \mathcal{H}^o \cap \partial \mathcal{C}.$$

Reflecting $[A_j^o, C^o]$ to the center of symmetry O_0, we see that $[A_{j+1}, A_1] \subset \mathcal{H} \cap \partial \mathcal{C}$. Since the entire construction takes place in the plane $\langle O, O_0, C \rangle$, we see that the points C, A_1, A_{j+1} are collinear and $[A_{j+1}, C] \subset \mathcal{H} \cap \partial \mathcal{C}$. The claim follows.

The sequence $\{A_j\}_{j \geq 1}$ converges to A. (In fact, an elementary computation shows that the sequence $\{d(A_j, A)\}_{j \geq 1}$ of distances is geometric.)

Taking the limit, we obtain $[A, C] \subset \mathcal{H} \cap \partial \mathcal{C}$. As in the proof, we can replace C by A in the minimal configuration as long as $C \notin \langle O, O_0 \rangle$.

Finally, since the configuration $\{C_0, \ldots, C_n\}$ is simplicial, it must have at least two configuration points away from the line $\langle O, O_0 \rangle$. Replacing these two points with A we obtain a minimal configuration whose convex hull is *not* a simplex. This contradicts to the regularity of O. The theorem follows.

Remark. Theorem 4.5.2, along with its proof, is taken from [Toth 9]. A different proof is outlined in Problem 8.

As the continuation of Example 4.4.4, we now use Theorem 4.5.2 just proved to determine the degrees of singularity of points in the interior of the cube.

Example 4.5.3. Let $C = C_n = [-1, 1]^n \subset \mathbb{R}^n$, $n \geq 3$, be the n-dimensional cube in standard position. By Theorem 4.5.2, the entire interior of C consists of singular points: $S = \text{int}\,C$. We are interested in determining the degrees of singularity.

First, for the origin 0, the centroid of C_n, we have $\sigma(0) = (n + 1)/2$ with $\mathfrak{m}(0) = 1$ so that $0 \in S_{n-1}$. Thus, 0 is a singular point of degree $n - 1$.

Let $C^0 \subset C$ denote the dual of C viewed as the inscribed cross-polytope with vertices being the centroids of the $(n - 1)$-cells on ∂C. Clearly, through any point $O \in \text{int}\,C \setminus \text{int}\,C^0$ in the complement of the interior of C^0 there passes a codimension one simplicial slice (say, parallel to one of the vertex figures). By Theorem 4.1.1, we have $\sigma_{n-1}(O) = 1$. Since O is a singular point, we then have $\sigma(O) = \sigma_n(O) = 1 + 1/(\mathfrak{m}(O) + 1)$. Since $\sigma_{n-1}(O) = \sigma_{n-2}(O) = \ldots = \sigma_1(O) = 1$, the degree of singularity of O is 1.

It remains to study the degree of singularity for points in the interior of the cross-polytope C^0. For simplicity, we will do this only for $n = 3$.

We claim that the set of degree 2 singular points $\mathcal{A}$ is the intersection of the three coordinate axes with the interior of C. The rest of the interior is comprised by degree 1 singular points.

By symmetry, we may restrict O to the fundamental tetrahedron $\mathcal{T} = [0, V, E, F]$, where $V = (1, 1, 1)$ is a vertex, $E = (0, 1, 1)$ is the midpoint of an edge, and $F = (0, 1, 0)$ is the midpoint of a face. We then let $O = O_{a,b,c} = aV + bE + cF$, where $0 \leq a, b, c \leq 1$ and $a + b + c < 1$. An easy calculation gives

$$\mathfrak{m}(O_{a,b,c}) = \frac{1 + a + b + c}{1 - a - b - c}. \tag{4.5.3}$$

By the above, we need to consider points in the intersection $\mathcal{T}_0 = \mathcal{T} \cap C^o$ only. This is another tetrahedron cut out from $\mathcal{T}$ by the plane extension of the Vertex figure that has normal vector V and passes through the point $V/3$. (See Figure 4.5.2.)

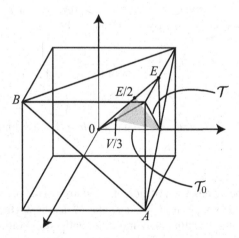

Fig. 4.5.2

Since this plane can also be written as the affine span $\langle V/3, E/2, F \rangle$, the restriction $O_{a,b,c} \in \mathcal{T}_0$ amounts to $0 \leq 3a + 2b + c \leq 1$.

We now consider a specific plane intersection $\mathcal{C}_{a,b,c}$ of $\mathcal{C}$ by a plane that passes through $O_{a,b,c}$ and the vertices $A = (1, 1, -1)$ and $B = (1, -1, 1)$.

For $3a + 2b + c = 1$, $\mathcal{C}_{a,b,c}$ is the vertex figure considered above. For $3a + 2b + c < 1$, $\mathcal{C}_{a,b,c}$ is a symmetric trapezoid (which, for $a = b = c = 0$, becomes a rectangle). As computation shows, we have $\mathcal{C}_{a,b,c} = [A, B, P_{a,b,c}, Q_{a,b,c}]$, where

$$P_{a,b,c} = \left(-1, 1, \frac{5a + 4b + 2c - 1}{1 - a}\right) \quad \text{and} \quad Q_{a,b,c} = \left(-1, \frac{5a + 4b + 2c - 1}{1 - a}, 1\right).$$

Now, a somewhat more tedious computation shows that

$$\sigma_{\mathcal{C}_{a,b,c}}(O_{a,b,c}) = \frac{3 - 3a - 2b - c}{2}. \tag{4.5.4}$$

(In fact, $O_{a,b,c}$ is a regular point of $\mathcal{C}_{a,b,c}$ with a minimizing configuration $\{A, B, P_{a,b,c}\}$.)

We now compute

$$\sigma_{\mathcal{C}}(O_{a,b,c}) \leq \sigma_{\mathcal{C},2}(O_{a,b,c}) + \frac{1}{\mathfrak{m}_{\mathcal{C}}(O_{a,b,c}) + 1}$$

$$\leq \sigma_{\mathcal{C}_{a,b,c}}(O_{a,b,c}) + \frac{1}{\mathfrak{m}_{\mathcal{C}}(O_{a,b,c}) + 1}$$

$$= 2 - \frac{4a + 3b + 2c}{2}.$$

where we used (4.1.3), (4.1.4) and (4.5.3), (4.5.4).

On the other hand, assume that $O_{a,b,c} \in \mathcal{S}_2$. (We already know from Theorem 4.5.2 that $O_{a,b,c} \in \mathcal{S} = \mathcal{S}_1$.) By definition, we have

$$\sigma(O_{a,b,c}) = 1 + \frac{2}{\mathfrak{m}(O_{a,b,c}) + 1}$$

$$= 2 - a - b - c.$$

Comparing these two computations, we get $a = b = 0$, the defining equality for $\mathcal{A}$. We obtain

$$0 \in \mathcal{S}_2 \subset \mathcal{A}.$$

Finally, due to the symmetric position of $\mathcal{A}$ relative to the parallel pairs of faces, it is easy to see that every point in $\mathcal{A}$ is singular of degree 2. (In fact, the points of any configuration can be continuously moved to antipodal position with increasing distortion.) Thus $\mathcal{S}_2 = \mathcal{A}$.

Continuing our study of the singular set, a natural question to ask is whether the converse of the second statement of Theorem 4.5.2 is true: If the interior of a convex body C consists of singular points only, is then C symmetric?

This is true for $n = 2$ and false for $n \geq 3$. First, we will give two simple (albeit computational) three-dimensional counter-examples, both simplicial in codimension 1 (Section 4.4).

In determining the regular set of a given convex body we can restrict ourselves to minimal configurations consisting of *extremal points* only (Section 1.2). This is the content of the following:

Proposition 4.5.4. *Let $C \in \mathfrak{B}$. If O is a regular point of C then there exists a minimal configuration $\{C_0, \dots, C_n\} \in \mathfrak{C}(O)$ consisting of extremal points.*

Proof. Let $\{C_0, \dots, C_n\} \in \mathfrak{C}(O)$ be a minimal configuration. Since O is a regular point, by definition, $[C_0, \dots, C_n]$ is an n-simplex containing O in its interior. Consequently, the distortion ratio $\Lambda(., O)$ attains a relative maximum at each C_i, $i = 0, \dots, n$.

Suppressing the index for simplicity, assume that a configuration point C is not an extremal point. Then C is a k-flat point for some $k > 0$ having a (unique) supporting maximal k-dimensional flat $\mathcal{A}_C$ (with C being in the (relative) interior of the convex body $\partial C \cap \mathcal{A}_C$ in $\mathcal{A}_C$). (See Section 2.2 as well as the proof of Corollary 2.2.3.)

Since $\Lambda(., O)$ attains a relative maximum at C, by Proposition 2.2.1, the antipodal point C^o is l-flat, $l \geq k$, and $\mathcal{A}_C$ is parallel to $\mathcal{A}_{C^o}$.

Choose a point C' on the boundary of the convex body $\partial C \cap \mathcal{A}_C$ in $\mathcal{A}$. Then, C' is a lower dimensional flat point than C. Since $\mathcal{A}_C$ is parallel to $\mathcal{A}_{C^o}$, $\Lambda(., O)$ is constant on $[C, C']$. Moving C toward C' and replacing C with the moved point, the configuration condition $O \in [C_0, \dots, C_n]$ stays intact since O is a regular point. Thus, replacing C by C' in the configuration, we arrive at a minimal configuration with C' being a lower dimensional flat point than C.

Proceeding inductively, we can replace C with and extremal point without altering minimality. Finally, performing this on all configuration points, the proposition follows.

Example 4.5.5. Let $C = C_+ \cup C_-$ be the double of two *regular* tetrahedra $C_{\pm} = [C_0, C_1, C_2, C_{\pm}] \subset \mathcal{X}$, $n = \dim \mathcal{X} = 3$. We claim that $\mathcal{S} = \operatorname{int} C$.

By symmetry, we may assume that $O \in \operatorname{int} C \cap C_+$, and write $O = O_\lambda = (1 - \lambda) O_0 + \lambda C_+$, where $0 \leq \lambda < 1$ and $O_0 \in \operatorname{int} [C_0, C_1, C_2]$. By symmetry again, we may also assume that O_λ is in the fundamental tetrahedron $\mathcal{T} = [C_0, (C_0 + C_1)/2, (C_0 + C_1 + C_2)/3, C_+]$. (See Figure 4.5.3.)

For computational convenience we may assume that $(C_0 + C_1 + C_2)/3$ is the origin. Then the restriction $O = O_\lambda \in \mathcal{T}$ amounts to $O_0 = O_0^{a,b} = aC_0 + b(C_0 + C_1)/2$, where $a, b \geq 0$ and $a + b < 1$. To show this latter dependency, we also write $O = O_\lambda = O_\lambda^{a,b}$. Since $\mathcal{R}$ is open, to prove that $\mathcal{S} = \operatorname{int} C$, it is enough to show that $O_\lambda^{a,b} \in \mathcal{S}$ for $\lambda > 0$ and $a, b > 0$.

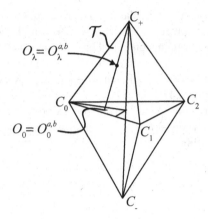

Fig. 4.5.3

Assume, on the contrary, that $O_\lambda^{a,b}$ is a regular point. Consider a minimal configuration. By Proposition 4.5.4, we may assume that the configuration points are extremal points, in our case, vertices of $\mathcal{C}$.

Due to the position of $O_\lambda^{a,b}$, for the minimal configuration, we have only two choices:

$$\{C_0, C_1, C_2, C_+\} \quad \text{or} \quad \{C_0, C_1, C_-, C_+\}.$$

We first rule out the second case. Indeed, parametrize the line segment $[C_2, C_-]$ by $C(t) = (1-t)C_2 + tC_-, 0 \le t \le 1$. Then a brief computation gives

$$\Lambda(C(t), O_\lambda^{a,b}) = \frac{2 - t + (1-\lambda)(a+b) + \lambda}{(1-\lambda)(1-a-b)}. \tag{4.5.5}$$

This shows that the distortion decreases from C_2 to C_-. This rules out the second choice.

We conclude that $\{C_0, C_1, C_2, C_+\}$ is the unique minimal configuration for the regular point $O = O_\lambda^{a,b}$. By regularity, we have

$$\sigma(O) = \sum_{i=0}^{2} \frac{1}{\Lambda(C_i, O) + 1} + \frac{1}{\Lambda(C_+, O) + 1} < 1 + \frac{1}{\mathfrak{m}(O) + 1}. \tag{4.5.6}$$

We calculate these terms as follows. First, $\Lambda(C_i, O) = \Lambda_{\mathcal{C}}(C_i, O) = \Lambda_{\mathcal{C}_+}(C_i, O)$, $i = 0, 1, 2$, and since $\mathcal{C}_+$ is a simplex, we have

$$\sigma_{\mathcal{C}_+}(O) = \sum_{i=0}^{2} \frac{1}{\Lambda(C_i, O) + 1} + \frac{1}{\Lambda_{\mathcal{C}_+}(C_+, O) + 1} = 1.$$

On the other hand, we clearly have $\Lambda_{C_+}(C_+, O) = (1 - \lambda)/\lambda$. Using this, we find that the first sum on the left-hand side of the inequality in (4.5.6) is equal to $1 - \lambda$.

To calculate $\Lambda(C_+, O) = \Lambda_C(C_+, O)$, we write the antipodal of C_+ as $C_+^o = (1 - s)C_+ + sO_0$, with $s > 1$. Since this antipodal is on the face $[C_0, C_1, C_-]$, using regularity of the participating simplices $C_\pm$, a simple computation gives that $s = 2/(a + b + 1)$. Thus, we have

$$\Lambda(C_+, O) = \frac{1 - \lambda}{s - (1 - \lambda)} = \frac{1 - \lambda}{\frac{2}{a+b+1} - (1 - \lambda)}. \tag{4.5.7}$$

With this, the left-hand side of the inequality in (4.5.6) is calculated. To determine the right-hand side in (4.5.6), we first note that maximum distortion of $\Lambda(., O)$, $O = O_\lambda^{a,b}$, occurs at a vertex (Corollary 2.2.3). A simple comparison gives that this vertex is C_2. (See Figure 4.5.3 again.) The distortion ratio at C_2 has already been calculated in (4.5.5) ($t = 0$), so that we obtain

$$\mathfrak{m}(O) = \mathfrak{m}(O_\lambda^{a,b}) = \frac{3}{(1 - \lambda)(1 - a - b)} - 1. \tag{4.5.8}$$

Substituting all the ingredients in (4.5.7), (4.5.8) back to (4.5.6), a simple computation reduces this inequality to $1 < a + b$. This is a contradiction. We conclude that O cannot be a regular point. Thus, $\operatorname{int} C = S$ follows.

Now, since every interior point is singular, in summary, we obtain

$$\sigma(O_\lambda^{a,b}) = 1 + \frac{(1 - \lambda)(1 - a - b)}{3}, \quad O_\lambda^{a,b} = (1 - \lambda)\left(aC_i + \frac{b}{2}(C_i + C_j)\right) + \lambda C_\pm,$$

where $i, j = 0, 1, 2$, $i \neq j$, and the centroid of C is at the origin.

In conclusion, we also see that $\mathfrak{m}^* = \inf_{\operatorname{int} C} \mathfrak{m}(O) = 2$ attained only for $\lambda = a = b = 0$. This point, making up the critical set C^*, is the centroid of C. In particular, as $\dim C^* = 0$, and, in Klee's estimate (2.4.1), sharp inequality holds.

Finally, we also see that σ is a concave function which is linear along line segments emanating from the centroid. The measure of symmetry $\sigma^* = \max_{\operatorname{int} C} \sigma = 4/3$ is attained at the centroid.

Our second example is the continuation of Example 2.4.3 (and uses the notation there):

Example 4.5.6. Let $C = [\Delta^-, \Delta^+]$ be the vertical cylinder of height $h > 0$ on an equilateral triangle $\Delta = [C_1, C_2, C_3] \subset \mathbb{R}^2$, where $\Delta^\pm = [C_1^\pm, C_2^\pm, C_3^\pm]$, $C_i^\pm = (C_i, \pm h/2)$, $i = 1, 2, 3$. As usual, for computational convenience, we assume that the centroid is at the origin.

In Example 2.4.3, we showed that $C^* = [O^-, O^+]$, where $O^\pm = (0, 0, \pm h/6)$, the middle third of the vertical axis.

We claim that all interior points of C are singular.

Assume, on the contrary, that there is a regular point $O \in \operatorname{int} C$. As in the previous example, by Proposition 4.5.4, there exists a minimal configuration consisting of extremal points (vertices) of C. Up to symmetries of C, we then have two possibilities for this configuration:

$$\{C_1^-, C_2^-, C_3^-, C_3^+\} \quad \text{or} \quad \{C_1^-, C_2^-, C_2^+, C_3^+\}.$$

(See Figure 4.5.4.)

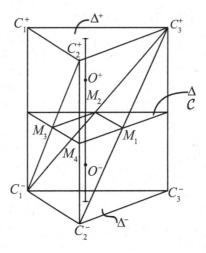

Fig. 4.5.4

Case I. $O \in [C_1^-, C_2^-, C_3^-, C_3^+] \cap \text{int}\, C$. Since O is regular, C_3^+ can be moved freely on Δ^+ with the configuration condition staying intact. The corresponding distortion ratio either stays constant or increases when the ray emanating from this moved point through O intersects one of the sides of C. Since the configuration is minimal, the latter cannot happen. Thus, we must have

$$O \in \bigcap_{i=1}^{3} [C_1^-, C_2^-, C_3^-, C_i^+] = [C_1^-, C_2^-, C_3^-, O^-].$$

Finally, in this case we can raise the configuration points C_1^-, C_2^-, and C_3^- along the vertical edges to the level of O without changing the distortions and obtain a singular configuration. This contradicts to the regularity of O.

Case II. $O \in [C_1^-, C_2^-, C_2^+, C_3^+] \cap \text{int}\, C$. The middle section Δ intersects the tetrahedron $[C_1^-, C_2^-, C_2^+, C_3^+]$ in a square whose vertices are the midpoints M_1, M_2, M_3 of the three sides of C and the midpoint M_4 of the vertical edge $[C_2^-, C_2^+]$. (See again Figure 4.5.4.) By symmetry, without loss of generality, we may assume that O is in the lower polyhedron $[M_1, M_2, M_3, M_4, C_1^-, C_2^-]$.

If $O \in [M_1, M_3, M_4, C_1^-, C_2^-]$ then we can move C_3^+ to C_3^- along their vertical line segment without changing the distortion ratio, and arriving at Case I.

Thus, we may assume that O is in the tetrahedron $[M_1, M_2, M_3, C_1^-]$. We can now move C_2^+ to C_2^- along their vertical line segment, and get a contradiction to regularity.

Thus, the claim follows in both cases.

Since all interior points are singular, by definition, we have $\sigma = \sigma_2 + 1/(m+1)$ identically on $\text{int}\,\mathcal{C}$. In addition, $\mathcal{C}$ is simplicial of codimension 1 so that $\sigma_2 = 1$ holds identically on $\text{int}\,\mathcal{C}$. Combining these, we obtain

$$\sigma = 1 + \frac{1}{m+1}. \tag{4.5.9}$$

Actually, the maximum distortion $m : \text{int}\,\mathcal{C} \to \mathbb{R}$ (and thereby σ) can be determined explicitly as follows. By symmetry, we may restrict ourselves to the fundamental cylinder $\mathcal{T} = \mathcal{T}_0 \times [0, h/2]$, where $\mathcal{T}_0 = [C_1, (C_1 + C_2)/2, 0] \subset \Delta$. (See Figure 4.5.5.)

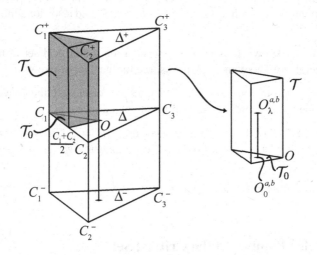

Fig. 4.5.5

Given $O \in \text{int}\,\mathcal{C} \cap \mathcal{T}$, we set $O = O_\lambda^{a,b} \in \mathcal{T}_0 \times \{\lambda(h/2)\}, 0 \le \lambda < 1$, with vertical projection to $\mathcal{T}_0$ as $O_0^{a,b} = aC_1 + b(C_1 + C_2)/2, a, b \ge 0, a + b < 1$.

With this, we have

$$m(O_\lambda^{a,b}) = \max\left(m_\Delta(O_0^{a,b}), \frac{1+\lambda}{1-\lambda}\right) = \max\left(\frac{2+a+b}{1-a-b}, \frac{1+\lambda}{1-\lambda}\right). \tag{4.5.10}$$

To show this we first note that maximum distortion occurs at an extremal point (Corollary 2.2.3), which then must be one of the vertices C_i^-, $i = 1, 2, 3$. If the antipodal of C_i^- with respect to $O_\lambda^{a,b}$ is contained in one of the three sides of $\mathcal{C}$ in $\partial\Delta \times [-h/2, h/2]$ then $\Lambda(C_i^-, O_\lambda^{a,b}) = \Lambda_\Delta(C_i, O_0^{a,b})$. If the antipodal is contained in

Δ^+ then $\Lambda(C_i^-, O_\lambda^{a,b}) = (1 + \lambda)/(1 - \lambda)$, the latter corresponding to the distortion ratio of the vertical chord of C through $O_\lambda^{a,b}$. Thus, the first equality in (4.5.10) follows.

Now, by elementary calculation, we have

$$m_\Delta(O_0^{a,b}) = \frac{2 + a + b}{1 - a - b},$$

and the second equality in (4.5.10) follows.

Substituting (4.5.10) to (4.5.9), we arrive at the formula

$$\sigma(O_\lambda^{a,b}) = 1 + \min\left(\frac{1 - a - b}{3}, \frac{1 - \lambda}{2}\right).$$

The following transparent picture emerges: σ attains its maximum $\sigma^* = 4/3$ along the critical set $C^* = [O^-, O^+]$. In addition, σ is linear along line segments connecting points in C^* with the side $\partial\Delta \times [-h/2, h/2]$ and also along line segments from $O^\pm$ to $\Delta^\pm$.

Remark. It is not known whether σ_n is constant on the critical set of the convex body, in general. By Theorem 4.2.3, this is true for the dual Minkowski measure σ_n^o.

Interestingly and in contrast to the examples just discussed, the converse of the second statement of Theorem 4.5.2 is affirmative in dimension 2:

Proposition 4.5.7. *Let C be a planar convex body ($n = 2$). Then the interior of C consists of singular points if and only if C is symmetric (with respect to its single critical point O^*, $C^* = \{O^*\}$).*

We will prove this in the next section.

4.6 Regular Points and the Critical Set

The examples in the previous section were simple because all the interior points of the convex body were singular. In the present section we consider more complex cases when both regular and singular points may be present in the interior of the convex body.

The first natural question about regular points is the converse of the first statement of Theorem 4.5.2: If the interior of a convex body C consists of regular points only, is then C a simplex?

As opposed to the case of singular points, an affirmative answer can be given under mild conditions as follows:

Theorem 4.6.1. *Let $C \in \mathfrak{B}$ with all its interior points regular. Assume that one of the following conditions hold:*

(I) *There is a flat point on ∂C, that is a point $C \in \partial C$ with a hyperplane $\mathcal{H} \subset \mathcal{X}$*
supporting C such that C is contained in the (non-empty) relative interior of
$\partial C \cap \mathcal{H}$ in $\mathcal{H}$.

(II) *C has (at least) n isolated extremal points (isolated points in the extremal set*
$C^\wedge \subset \partial C$).

Then C is a simplex.

Theorem 4.6.1 can clearly be rephrased as a statement on existence of singular
points for non-simplicial convex bodies. It is an open question whether the theorem
is true without *any* conditions.

The proof of Theorem 4.6.1 is long and technical, and will be deferred to the next
section.

In the present section we explore the connection among the regular set $\mathcal{R}$ (and
thereby the singular set $\mathcal{S}$), the Minkowski measure m^*, and the critical set C^*. We
begin with the question of existence of regular points. The next result relies heavily
on Klee's analysis of the critical set (Section 2.4):

Theorem 4.6.2. *Let $C \in \mathfrak{B}$ and assume that $m^* > n - 1$ holds. Then the critical*
set C^* consists of a single regular point $O^* \in \mathcal{R}$ and

$$\sigma(O^*) = \frac{n+1}{m^* + 1}. \tag{4.6.1}$$

Proof. By Corollary 2.4.13, the assumption implies that $C^* = \{O^*\}$ and $O^* \in$
$\text{int}\,[\mathcal{M}^*]$ with $\langle \mathcal{M}^* \rangle = \mathcal{X}$, where $\mathcal{M}^* = \mathcal{M}(O^*) = \{C \in \partial C \mid \Lambda(C, O^*) = m^*\}$.
Hence there exists a configuration $\{C_0, \dots, C_n\} \in \mathfrak{C}(O^*)$ *contained in* $\mathcal{M}^*$. (The
fact that no more than $n + 1$ points are needed follows from Carathéodory's theorem
in Section 1.3.) Since $C_i \in \mathcal{M}^*$ we have $\Lambda(C_i, O^*) = m^*$, $i = 0, \dots, n$, and the
definition of σ gives

$$\sigma(O^*) \le \sum_{i=0}^{n} \frac{1}{\Lambda(C_i, O^*) + 1} = \frac{n+1}{m^* + 1}.$$

The opposite inequality is the obvious lower estimate for σ in (4.1.8). Thus, (4.6.1)
follows.

It remains to show regularity of O^*. Using (4.6.1) already proved and our
condition $m^* > n - 1$, we estimate

$$\sigma(O^*) = \frac{n}{m^* + 1} + \frac{1}{m^* + 1} < 1 + \frac{1}{m^* + 1} \le \sigma_{n-1}(O^*) + \frac{1}{m^* + 1}.$$

Regularity of O^* follows by (4.5.1).

Remark 1. Since the regular set $\mathcal{R}$ is open, Theorem 4.6.2 implies that, if $m^* > n - 1$
then an *open neighborhood* of O^* belongs to $\mathcal{R}$. In Example 4.6.14 below, we will
construct a sequence $\{C_m\}_{m \ge 2}$ of three-dimensional cones converging to a circular

cone such that $C_m^* = \{O_m^*\}$, $O_m^* \in \mathcal{R}_{C_m}$, $m \geq 2$, and $d(O_m^*, S_{C_m}) \to 0$ and $m_{C_m}^* \to 2$ as $m \to \infty$. This indicates that, for a possible a priori lower bound on this open neighborhood, one should assume $m^* \geq \epsilon + n - 1$, for some $\epsilon > 0$.

Remark 2. Theorem 4.6.2 immediately implies Proposition 4.5.7 stated at the end of the previous section. In fact, if the interior of a convex body C consists of singular points only, then, by what we just proved, $(1 \leq) m^* \leq n - 1$. This, for $n = 2$, gives $m^* = 1$. Symmetry of C follows.

Remark 3. Examples 4.5.5 and 4.5.6 show that the inequality $m^* > n - 1$ in Theorem 4.6.2 is sharp in the sense that there are convex bodies with $m^* = n - 1$ and no regular points.

Example 4.6.3. We briefly revisit the unit half-disk

$$C = \{(x, y) \in \mathbb{R}^2 \mid x^2 + y^2 \leq 1, \, y \geq 0\}$$

of Example 3.4.1. As noted there, $m^* = \sqrt{2}$ with the (unique) critical point at $O^* = (0, \sqrt{2} - 1)$, $\{O^*\} = C^*$, and centroid $g(C) = (0, 4/3\pi)$, different from O^*. By Theorem 4.6.2, O^* is a regular point with $\sigma(O^*) = 3/(\sqrt{2} + 1)$.

We now claim that $\mathcal{R} = \text{int} \, \Delta$, where $\Delta = [C_0, C_-, C_+]$ with $C_0 = (0, 1)$ and $C_\pm = (\pm 1, 0)$.

This can be seen as follows. Given $(a, b) \in \text{int} \, C$, there may be *at most* three affine diameters passing through (a, b), those that also pass through C_0, C_-, and C_+. (See Figure 4.6.1.)

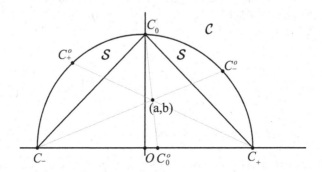

Fig. 4.6.1

Since, at a regular point, there must be at least three, this immediately implies that the complement $\text{int} \, C \setminus \Delta$ must belong to the singular set: $\mathcal{R} \subset \text{int} \, \Delta$.

For the opposite inclusion, we let $(a, b) \in \text{int} \, \Delta$. By (4.5.1), for $(a, b) \in \mathcal{R}$, it is enough to show that

$$\frac{1}{\Lambda(C_0,(a,b))+1} + \frac{1}{\Lambda(C_-,(a,b))+1} + \frac{1}{\Lambda(C_+,(a,b))+1} < 1 + \frac{1}{m(a,b)+1},$$
$$(4.6.2)$$

since the left-hand side is $\geq \sigma(a,b)$. Once we have this, the existence and uniqueness of the three affine diameters through (a,b) imply that $\{C_0, C_-, C_+\} \in \mathfrak{C}(a,b)$ is a *minimal* configuration so that the left-hand side is actually equal to $\sigma(a,b)$.

By symmetry, we may assume that $a \geq 0$. Then, by simple comparison, we have $m(a,b) = \max(\Lambda(C_0,(a,b)), \Lambda(C_-,(a,b)))$. In addition, extending the respective chords to horizontal and vertical lines through C_0 and $C_\pm$, we obtain

$$\frac{1}{\Lambda(C_0,(a,b))+1} + \frac{1}{\Lambda(C_+,(a,b))+1} < 1 \text{ and } \frac{1}{\Lambda(C_-,(a,b))+1} + \frac{1}{\Lambda(C_+,(a,b))+1} < 1.$$

With these (4.6.2), and hence $\mathcal{R} = \text{int}\, C$ follows.

As a byproduct, a simple computation of the left-hand side in (4.6.2) gives the explicit formula

$$\sigma(a,b) = \begin{cases} \frac{1-a^2-b^2}{1-a^2} + b & \text{if } (a,b) \in \text{int } \Delta \\ 1 + \frac{1-a^2-b^2}{2(1+|a|)} & \text{if } (a,b) \notin \text{int } \Delta \end{cases} \qquad (4.6.3)$$

Finally, note that $\sigma^* = \max_{\text{int}\, C} \sigma = 5/4$ is attained at $(0, 1/2)$, yet another point distinct from $O^* = (0, \sqrt{2}-1)$ and $g(C) = (0, 4/3\pi)$.

Returning to the main line, recall from Section 4.4 that $C \in \mathfrak{B}$ is simplicial in codimension k, $0 \leq k < n$ if C has an $(n-k)$-dimensional simplicial slice across any interior point, or equivalently, if $\sigma_{n-k} = 1$ identically on the interior of C.

For a codimension k simplicial convex body there is a simple interplay between the Minkowski measure, the critical set, and the set of degree $\geq k$ singular points:

Proposition 4.6.4. *Let* $C \in \mathfrak{B}$ *be a codimension k simplicial convex body, $1 \leq k < n$. Then* $m^* \geq n - k$ *and equality holds if and only if $C^* \subset S_k$.*

Proof. By sub-arithmeticity in (4.1.3) and (4.1.8), we have

$$\frac{n+1}{m^*+1} \leq \sigma(O^*) \leq \sigma_{n-k}(O^*) + \frac{k}{m^*+1} = 1 + \frac{k}{m^*+1}, \quad O^* \in C^*. \qquad (4.6.4)$$

Comparing the two sides, we obtain $m^* \geq n - k$. We have $m^* = n - k$ if and only if equality holds everywhere in (4.6.4). By definition, this means that $C^* \subset S_k$.

We now briefly revisit our study of the concavity properties of σ:

Proposition 4.6.5. *Let* $C \in \mathfrak{B}$ *be a codimension k simplicial convex body, $1 \leq k < n$. Let $O \in S_k$ and choose $C \in \partial C$ such that $\Lambda(., O)$ assumes its global maximum at C. Then $\sigma|[O, C^o)$ is concave if and only if $[O, C^o) \subset S_k$. In this case σ is linear on $[O, C^o)$.*

Remark. Note that in the two propositions above $\mathcal{S}_{k+1} = \emptyset$. (See the remark after Proposition 4.5.1.)

Before the proof of Proposition 4.6.5 we develop a useful *comparison formula.* Let $\mathcal{C} \in \mathfrak{B}$ and $O \in \text{int}\,\mathcal{C}$. For $C, C' \in \partial\mathcal{C}$, $C' \neq C, C^o$, we let $\mathcal{K}_{C,C'} = \langle O, C, C'\rangle$. By assumption, $\mathcal{K}_{C,C'}$ is an affine plane in $\mathcal{X}$. Within $\mathcal{K}_{C,C'}$ we let $\mathcal{G}_{C,C'}$ be the closed half-plane which contains C' and has boundary line $\langle O, C\rangle$. Taking antipodals, we also have the closed half-plane $\mathcal{G}_{C'^o,C^o}$ which contains C^o and has boundary line $\langle O, C'^o\rangle = \langle O, C'\rangle$. We now define

$$\mathcal{C}_{C,C'} = \mathcal{G}_{C,C'} \cap \mathcal{G}_{C'^o,C^o} \cap \text{int}\,\mathcal{C}.$$

Clearly, $\bar{\mathcal{C}}_{C,C'}$ is a convex body in $\mathcal{G}_{C,C'}$ with boundary consisting of the line segments $[O, C']$, $[O, C^o]$ and a boundary arc of $\mathcal{C} \cap \mathcal{G}_{C,C'}$. (See Figure 4.6.2.)

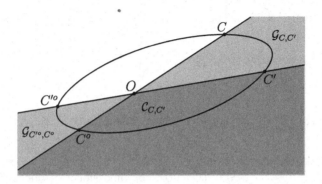

Fig. 4.6.2

Clearly $\mathcal{C}_{C,C'} = \mathcal{C}_{C'^o,C^o}$, and, for fixed $C \in \partial\mathcal{C}$, we have

$$\bigcap_{C' \in \partial\mathcal{C},\, C' \neq C, C^o} \mathcal{C}_{C,C'} = [O, C^o).$$

With these, we now state the following *Comparison Lemma*:

Lemma 4.6.6. *Let* $\mathcal{C} \in \mathfrak{B}$ *and* $O \in \text{int}\,\mathcal{C}$. *Let* $C, C' \in \partial\mathcal{C}$, $C' \neq C, C^o$, *and assume that*

$$\Lambda(C, O) \geq \Lambda(C', O). \tag{4.6.5}$$

Then, for $O' \in \mathcal{C}_{C,C'}$, *we have*

$$\Lambda(C, O') \geq \Lambda(C', O'). \tag{4.6.6}$$

Moreover, sharp inequality in (4.6.5) implies sharp inequality in (4.6.6).

Proof. Consider the ray ρ emanating from C and passing through C', and the ray ρ^o emanating from C'^o and passing through C^o.

First, assume that sharp inequality holds in (4.6.5). This means that ρ and ρ^o intersect at a point P. By convexity of C, we have

$$\mathcal{C}_{C,C'} \subset [C', O, C^o, P].$$

Let $O' \in \mathcal{C}_{C,C'}$, and let B and B' be the antipodals of C and C' with respect to O'. Consider the ray ρ' emanating from B' and passing through B, and the ray ρ'' emanating from B and passing through B'. Along the boundary $\partial C \cap \mathcal{C}_{C,C'}$, the point B follows C^o in the same direction as B' follows C'^o. (See Figure 4.6.3.)

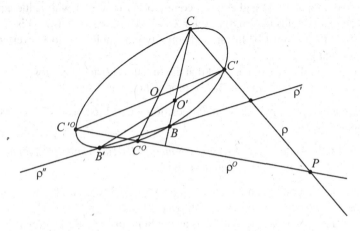

Fig. 4.6.3

Therefore, by convexity of C, ρ'' intersects ρ^o. Since $\rho' \cup \rho'' = \langle B, B' \rangle$, ρ' must intersect ρ. The sharp inequality in (4.6.6) follows.

If equality holds in (4.6.5) then ρ and ρ' are parallel, and the proof can be adjusted accordingly.

An immediate consequence is the following:

Corollary 4.6.7. *Let $C \in \mathfrak{B}$ and $O \in \mathrm{int}\, C$. Assume that $\Lambda(., O)$ attains its local (global) maximum at $C \in \partial C$. Then, for any $O' \in [O, C^o)$, $\Lambda(., O')$ also attains its local (global) maximum at C.*

Proof of Proposition 4.6.5. Parametrize the line segment $[O, C^o]$ as

$$\lambda \mapsto O_\lambda = (1 - \lambda)O + \lambda C^o, \quad 0 \le \lambda \le 1.$$

Consider the function $\lambda \mapsto \sigma(O_\lambda)$, $\lambda \in [0, 1)$. Sub-arithmeticity in (4.1.3) and $\sigma_{n-k} = 1$ give

$$\sigma(O_\lambda) \le 1 + \frac{k}{\mathfrak{m}(O_\lambda) + 1}, \qquad 0 \le \lambda < 1. \tag{4.6.7}$$

By Corollary 4.6.7, if the distortion $\Lambda(., O)$ attains its global maximum $\mathfrak{m}(O)$ at $C \in \partial\mathcal{C}$ then, for each $0 \le \lambda < 1$, $\Lambda(., O_\lambda)$ also attains its global maximum $\mathfrak{m}(O_\lambda)$ at C. Hence (4.6.7) rewrites as

$$\sigma(O_\lambda) \le 1 + \frac{k}{\Lambda(C, O_\lambda) + 1} = 1 + (1-\lambda)\frac{k}{\Lambda(C, O) + 1}, \qquad 0 \le \lambda < 1. \tag{4.6.8}$$

By assumption, $O = O_0$ is a singular point of degree k, so that, at $\lambda = 0$, equality holds in (4.6.8). On the other hand, by Theorem 4.4.1, the left-hand side in (4.6.8) is continuous in $\lambda \in [0, 1)$ and extends continuously to $\lambda = 1$ with value equal to 1. Hence, equality holds in (4.6.8) also for $\lambda = 1$. Thus, if $\lambda \mapsto \sigma(., O_\lambda)$ is concave for $\lambda \in [0, 1]$ then it must be linear, and equality must hold in (4.6.8) everywhere. This means that $[O, C^o) \subset \mathcal{S}_k$.

The converse follows from the fact that the function $1/(\mathfrak{m}+1)$ is always concave in the interior of $\mathcal{C}$ (Theorem 2.1.2). The proposition follows.

We derive several applications of Proposition 4.6.5. First, $k = n - 1$ gives the following:

Corollary 4.6.8. *Assume that $C \in \mathfrak{B}$ is symmetric. If C has a simplicial intersection of dimension ≥ 2 then σ is not concave in $\mathrm{int}\,C$.*

Proof. For the center of symmetry, we have $O \in \mathcal{S}_{n-1}$. Therefore, Proposition 4.6.5 applies for $k = n - 1$ (and for any $C \in \partial\mathcal{C}$ since $\Lambda(., O) = \mathfrak{m}(O) = 1$).

Let $O' \in \mathrm{int}\,C$, and assume that C has a simplicial intersection of dimension ≥ 2 across O'. Let $\langle O, O' \rangle \cap C = [C, C^o]$ such that $O \in [O', C]$. Since $O' \in [O, C^o)$, if σ were concave we would have $O' \in \mathcal{S}_{n-1}$. By the remark after Proposition 4.5.1, the degree of singularity $(n-1)$ of O' cannot exceed the codimension of a simplicial slice across O'. This is a contradiction, and the corollary follows.

Remark. Corollary 4.6.8 is a substantial generalization of the observation that for the n-dimensional cube, $n \ge 3$, the function σ is not concave (Example 4.4.4).

Since σ is concave for codimension 1 simplicial convex bodies (Theorem 4.4.10), for $k = 1$, Proposition 4.6.5 takes a particularly simple form as well as reveals information about the topology of the singular set:

Corollary 4.6.9. *Let $C \in \mathfrak{B}$ be a codimension 1 simplicial convex body. Given $O \in \mathrm{int}\,C$, for any $C \in \mathcal{M}(O) = \{C \in \partial\mathcal{C} \mid \Lambda(C, O) = \mathfrak{m}(O)\}$, the line segment $[O, C^o)$ intersects $\mathcal{R}$ and $\mathcal{S}$ in single intervals (one of which may be empty). In particular, $\mathcal{S} \cup \partial\mathcal{C}$ is path-connected.*

Remark. As Example 4.6.3 shows, the singular set itself may be disconnected. The first statement of Corollary 4.6.9 also implies that, along any topological $(n-1)$-sphere within $\mathcal{R}$, its interior is also contained in $\mathcal{R}$.

We summarize our findings in dimension 2 as follows:

Corollary 4.6.10. *Let C be a planar convex body. Then $1 \leq m^* \leq 2$. If $m^* = 1$ then C is symmetric with all interior points singular. If $1 < m^* < 2$ then the regular set $\mathcal{R}$ is non-empty and simply connected, and $\mathcal{S} \cup \partial C$ is connected. If $m^* = 2$ then C is a triangle with all interior points regular.*

Next, we have an important example revealing a few subtleties of the structure of the regular set $\mathcal{R}$. We need a preparatory step:

Lemma 4.6.11. *Let C be a planar convex body with a supporting line $\mathcal{H}$ which has a non-trivial maximal line segment $\mathcal{F} = [C_1, C_2] = \mathcal{H} \cap \partial C$ common with the boundary of C. Assume that the second supporting line $\mathcal{H}^o$ parallel to $\mathcal{H}$ contains exactly one boundary point C_0. Then the set*

$$\mathcal{T} = \{O \in \mathrm{int}\,[C_0, C_1, C_2] \mid \Lambda(C_0, O) \leq \max(\Lambda(C_1, O), \Lambda(C_2, O))\}$$

is contained in the regular set $\mathcal{R}$.

Remark. Recall that a boundary point C_0 with the property in the lemma is called *exposed* (Section 2.1).

The set $\mathcal{T}$ has non-empty interior. Indeed, if $M_1 \in [C_0, C_1]$ and $M_2 \in [C_0, C_2]$ are the midpoints of the respective line segments then

$$[C_0, M_1, M_2] \cap \mathrm{int}\,[C_0, C_1, C_2] \subset \mathcal{T}.$$

Proof of Lemma 4.6.11. Let $O \in \mathcal{T}$. We first claim that

$$\max_{\mathcal{F}} \Lambda(., O) = \max(\Lambda(C_1, O), \Lambda(C_2, O)). \tag{4.6.9}$$

Let $C_1' \in [C_0^o, C_1]$. Then, with obvious notations, we have

$$\Lambda(C_1', O) \leq \Lambda_{[C_0, C_1, C_0^o, C_1^o]}(C_1', O) \leq \Lambda(C_1, O),$$

where the second inequality is because $\mathcal{H}^o$ is parallel to $\mathcal{H}$. (See Figure 4.6.4.)

Similarly, for $C_2' \in [C_0^o, C_2]$, we have $\Lambda(C_2', O) \leq \Lambda(C_2, O)$. The claim follows.

Let $\mathcal{F}^o$ be the antipodal of $\mathcal{F}$ with respect to $O \in \mathcal{T}$. Then $\mathcal{F}^o$ is a (connected) continuous curve on the boundary of C with endpoints C_1^o and C_2^o.

Once again the existence of the supporting line $\mathcal{H}^o$ to C at C_0 parallel to $\mathcal{H}$ implies

$$\max_{\mathcal{F}^o} \Lambda(., O) = \Lambda(C_0, O).$$

By the definition of $\mathcal{T}$, we thus have

$$\max_{\mathcal{F}^o} \Lambda(., O) \leq \max(\Lambda(C_1, O), \Lambda(C_2, O)). \tag{4.6.10}$$

Equations (4.6.9)–(4.6.10) imply that the maximum $m(O)$ of $\Lambda(., O)$ is attained on one of the two connected components of $\partial C \setminus ([C_1, C_2] \cup \mathcal{F}^o)$.

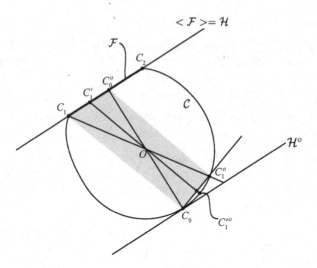

Fig. 4.6.4

Without loss of generality, we may assume that the maximum is attained at a point $\bar{C}$ which belongs to the component with endpoints C_1 and C_2^o. Since $O \in [C_0, C_1, C_2]$ and (automatically) $O \in [C_0, C_2, C_2^o]$, by convexity, we also have $O \in [C_0, \bar{C}, C_2]$. (See Figure 4.6.5.)

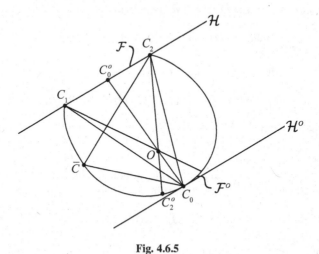

Fig. 4.6.5

Thus, $\{C_0, \bar{C}, C_2\}$ is a configuration for O. We have

$$\sigma(\mathcal{C}, O) \leq \frac{1}{\Lambda(C_0, O) + 1} + \frac{1}{\Lambda(C_2, O) + 1} + \frac{1}{\mathfrak{m}(O) + 1}.$$

To prove that O is a regular point, by (4.5.1), it is enough to show that

$$\frac{1}{\Lambda(C_0, O) + 1} + \frac{1}{\Lambda(C_2, O) + 1} < 1, \tag{4.6.11}$$

or equivalently

$$\frac{1}{\Lambda(C_2, O) + 1} < \frac{1}{\Lambda(C_0^o, O) + 1}.$$

Again by the existence of the supporting line $\mathcal{H}^o$ at C_0, non-strict inequality certainly holds here. Finally, if $\Lambda(C_2, O) = \Lambda(C_0^o)$ then $\langle C_0, C_2^o \rangle$ is parallel to $\mathcal{H}$. This means that C_0 is not unique in $\partial \mathcal{C} \cap \mathcal{H}^o$, contradicting to our assumption. The lemma follows.

Example 4.6.12. Let $\mathcal{P}_p = \{p\} \subset \mathbb{R}^2$, $p \geq 3$, be a regular p-sided polygon which, for simplicity, we assume to be inscribed in the unit circle. By the first statement of Theorem 4.5.2, for $p = 3$, the interior of the triangle $\mathcal{P}_3$ consists of regular points only. By the second statement of Theorem 4.5.2, for $p = 2m$ even, the interior of $\mathcal{P}_{2m} = \{2m\}$ consists of singular points only.

In contrast, for $p = 2m + 1$, $m \geq 2$, odd, we now claim that the regular set of $\mathcal{P}_{2m+1} = \{2m + 1\}$ is the interior of the *star-polygon* $\{\frac{2m+1}{m}\}$.

The frontispiece depicts the case $m = 6$. The union of the 13 shaded triangles is the singular set.

It is interesting to note that the ratio of the areas of $\{\frac{2m+1}{m}\}$ and $\{2m + 1\}$ tends to $2/3$ as $m \to \infty$ whereas the limiting polygon is the unit disk $\mathcal{B}$ with no regular points (Theorem 4.5.2 again). More specifically, in each polygon $\mathcal{P}_{2m+1}$ the open central disk $\mathcal{B}_{1/3}$ of radius $1/3$ is contained in the regular set $\mathcal{R}$, whereas in the limiting unit disk every point becomes singular.

As in Example 2.5.1, let

$$V_k = \left(\cos\left(\frac{2\pi k}{2m + 1} \right), \sin\left(\frac{2\pi k}{2m + 1} \right) \right), \quad k = 0, \ldots, 2m,$$

be the vertices of $\mathcal{P}_{2m+1}$. Let O be any interior point of $\mathcal{P}_{2m+1}$ and assume that, for some $j = 0, \ldots, 2m$, $[V_j, V_j^o]$ is an affine diameter (passing through O). Comparing supporting lines, we see that V_j^o must be in the opposite side to V_j, that is, we have $V_j^o \in [V_{j+m}, V_{j+m+1}]$. Thus, O must be contained in the triangle $[V_j, V_{j+m}, V_{j+m+1}]$.

The intersection of any *three* of these triangles (corresponding to three different indices $j = 0, \ldots, 2m$) is contained in the star-polygon $\{\frac{2m+1}{m}\}$. Thus, any interior point O of $\mathcal{P}_{2m+1}$ *complementary* to this star-polygon must be singular.

It remains to show that the interior points of the star-polygon are regular. Let

$$X_k = [V_k, V_{m+k}] \cap [V_{k+1}, V_{m+k+1}],$$

where the indices are counted modulo $2m + 1$. We now let $C_0 = V_k$, $C_1 = V_{m+k}$, and $C_2 = V_{m+k+1}$ and apply Lemma 4.6.11. We obtain

$$\mathcal{T} = [V_k, X_k, 0] \cup [V_k, X_{m+k}, 0].$$

The union of these for $k = 0, \ldots, 2m$ is the star-polygon $\{\frac{2m+1}{m}\}$ and we are done.
Note that the critical set $\mathcal{P}^*_{2m+1} = \{0\}$, and we have

$$\mathfrak{m}^* = \mathfrak{m}(0) = \Lambda(V_k, 0) = \sec\left(\frac{\pi}{2m+1}\right), \quad \text{for any } k = 0, \ldots, 2m.$$

In a perfect match with Theorem 4.5.2, we obtain

$$\sigma(0) = \frac{3}{\sec\left(\frac{\pi}{2m+1}\right) + 1}.$$

To calculate the area $\mathcal{A}_{2m+1}$ of the singular set, the complement of the star-polygon $\{\frac{2m+1}{m}\}$ in $\mathcal{P}_{2m+1}$, is elementary. A simple computation gives

$$\mathcal{A}_{2m+1} = (2m+1)\sin\left(\frac{2m}{2m+1}\pi\right)\left(\frac{1}{2\cos\left(\frac{2m}{2m+1}\pi\right) - 1} - \cos\left(\frac{2m}{2m+1}\pi\right)\right).$$

The limit as $m \to \infty$ is clearly $2\pi/3$.
   The minimum distance of the singular set from the center is

$$\frac{1}{1 - 2\cos\left(\frac{2m}{2m+1}\pi\right)}.$$

As $m \to \infty$, this decreases to $1/3$. Thus, the regular set of $\mathcal{P}_{2m+1}$ contains the open disk with center at the origin and radius $1/3$. In the limit, each point of this disk will turn singular.

   The example we just discussed will be the base of a cone constructed next. Before, we need the following general fact:

Lemma 4.6.13. *Let $1 \le k < n$ and assume that $O \notin \mathcal{S}_{n-k}$. If $\{C_0, \ldots, C_n\} \in \mathfrak{C}(O)$ is a minimal configuration then, among the configuration points C_i, $i = 0, \ldots, n$, there are at most k such that*

$$\Lambda(C_i, O) \le \frac{k+1}{\sigma_k(O)} - 1.$$

Remark. For $k = 1$, Lemma 4.6.13 asserts that, if $O \notin \mathcal{S}_{n-1}$ then, with the possible exception of *one* configuration point, we have $\Lambda(C_i, O) > 1$.

The exceptional point cannot be avoided as can be seen by taking C a simplex and O close to one of the vertices.

The condition $O \notin S_{n-1}$ is also essential: If C is symmetric with center of symmetry O then $O \in S_{n-1}$ and $\Lambda(., O)$ is identically 1 on ∂C.

Proof of Lemma 4.6.13. The proof is by contradiction. Renumbering if necessary, we may assume that, for $i = 0, \ldots, k$, we have $\Lambda(C_i, O) \leq (k+1)/\sigma_k(O) - 1$. Then

$$\sum_{i=0}^{k} \frac{1}{\Lambda(C_i, O) + 1} \geq \sum_{i=0}^{k} \frac{\sigma_k(O)}{k+1} = \sigma_k(O).$$

Let $\{B_0, \ldots, B_k\} \in \mathfrak{C}_k(O)$ be a minimal configuration, so that we have

$$\sigma_k(O) = \sum_{i=0}^{k} \frac{1}{\Lambda(B_i, O) + 1}.$$

We extend this to the n-configuration $\{B_0, \ldots, B_k, C_{k+1}, \ldots, C_n\}$. Using this and the previous inequality, we have

$$\sum_{i=0}^{k} \frac{1}{\Lambda(B_i, O) + 1} + \sum_{j=k+1}^{n} \frac{1}{\Lambda(C_j, O) + 1} = \sigma_k(O) + \sum_{j=k+1}^{n} \frac{1}{\Lambda(C_j, O) + 1}$$

$$\leq \sum_{i=0}^{n} \frac{1}{\Lambda(C_i, O) + 1} = \sigma(O).$$

Since the extension is a configuration, equality holds here. Thus, $O \in S_{n-k}$. This contradicts to our assumption.

Example 4.6.14. Let $\mathcal{P}_{2m+1}, m \geq 2$, be as in the previous Example 4.6.12. Consider a cone $C_m = [\mathcal{P}_{2m+1}, V] \subset \mathbb{R}^3$, where $V \in \mathbb{R}^3 \setminus \mathbb{R}^2$. Let $O_\lambda = \lambda V, 0 \leq \lambda \leq 1$. (Note that $O_0 = 0$ is the center of $\mathcal{P}_{2m+1}$.) An easy application of Lemma 2.1.5 shows that

$$\Lambda(V_k, O_\lambda) = \frac{\sec\left(\frac{\pi}{2m+1}\right) + \lambda}{1 - \lambda}, \quad k = 0, \ldots, 2m. \tag{4.6.12}$$

We claim that for

$$\lambda \leq \frac{2 - \sec\left(\frac{\pi}{2m+1}\right)}{3} \tag{4.6.13}$$

the point O_λ must be singular. Indeed, if O_λ were regular then *three* vertices of $\mathcal{P}_{2m+1}$ (and V) must contribute to a minimal configuration. By (4.6.13), the distortion ratio in (4.6.12) on these vertices is ≤ 2. This contradicts to Lemma 4.6.13 for $k = 2$ since $O \notin S = S_1$, and $\sigma_2 = 1$ identically on $\operatorname{int} C_m$.

We denote the right-hand side of (4.6.13) by λ_m so that $O_m = O_{\lambda_m} \in \mathcal{S}$.

It is geometrically plausible that the point O_λ belongs to the critical set $\mathcal{C}_m^*$ if $\Lambda(V_k, O_\lambda) = \Lambda(V, O_\lambda) = (1 - \lambda)/\lambda$, for all $k = 0, \ldots, 2m$. (See Problem 6 at the end of Chapter 2.) By (4.6.12), this happens for $\lambda = \lambda_m^* = 1/(\sec(\pi/(2m+1))+2)$. Letting $O_m^* = O_{\lambda_m^*}$, we then have

$$\mathrm{m}_{\mathcal{C}_m}^* = \mathrm{m}(O_m^*) = \frac{1 - \lambda_m^*}{\lambda_m^*} = \sec\left(\frac{\pi}{2m+1}\right) + 1.$$

Since $\mathrm{m}^* > 2$, by Theorem 4.6.2, the critical set $\mathcal{C}_m^*$ consists of the single regular point $O_m^* \in \mathcal{R}$ and we have

$$\sigma(O_m^*) = \frac{4}{\sec\left(\frac{\pi}{2m+1}\right) + 2}.$$

Now, the crux is that the sequences $\{\lambda_m\}_{m \geq 2}$ and $\{\lambda_m^*\}_{m \geq 2}$ *equiconverge* to $1/3$, so that $d(O_m, O_m^*) \to 0$ as $m \to \infty$. We finally obtain $d(\mathcal{C}_m^*, \mathcal{S}_{\mathcal{C}_m}) \to 0$ and $\mathrm{m}_{\mathcal{C}_m}^* \to 2$ as $m \to \infty$.

4.7   A Characterization of the Simplex

In this section we will prove Theorem 4.6.1. The assumptions (I) and (II) are very different and so are the respective proofs. Accordingly, we will split this section into two parts.

Part I. We assume that $\mathcal{C} \in \mathfrak{B}$ has a flat point on its boundary: $O_0 \in \partial\mathcal{C}$.

We first study the existence of regular points near O_0.

Lemma 4.7.1. *Let $\mathcal{C} \in \mathfrak{B}$ with a flat point O_0 on its boundary. Let $\{O_k\}_{k \geq 1} \subset \mathcal{R}$ such that $\lim_{k \to \infty} O_k = O_0$. Denote by $\mathcal{H}_0$ the unique hyperplane supporting $\mathcal{C}$ at O_0. Then $\mathcal{C}_0 = \mathcal{H}_0 \cap \partial\mathcal{C}$ is an $(n-1)$-simplex. In addition, if $\mathcal{C}_0 = [C_1, \ldots, C_n]$ then, for each $1 \leq i \leq n$, there exist parallel hyperplanes $\mathcal{H}_i$ and $\mathcal{H}_i'$ supporting $\mathcal{C}$ such that $[C_1, \ldots, \widehat{C_i}, \ldots, C_n] \subset \mathcal{H}_i$ and $C_i \in \mathcal{H}_i'$.*

Remark. According to Example 4.6.12, the regular polygon $\mathcal{P}_{2m+1}$, $m \geq 2$, has no sequence of regular points converging to a boundary flat point and, clearly, there are no parallel supporting lines at the endpoints of any side.

As another example, the diagonals of a proper trapezoid $\mathcal{C}$ split the trapezoid into four triangles, and the regular set is the interior of the triangle with one side being the *longer* parallel side. Hence, a flat point on this longer side can be approximated by regular points, while the flat points on the other sides cannot be. This is also confirmed by the lemma above as parallel supporting hyperplanes exist at the endpoints of the longer parallel side but not at the endpoints of the shorter side.

Proof of Lemma 4.7.1. For each $k \geq 1$, the point O_k, is regular so that there exists a minimal *simplicial* configuration $\{C_{0,k}, \ldots, C_{n,k}\} \in \mathfrak{C}(O_k)$. Using Proposition 4.5.4, without loss of generality, we may assume that the configuration points $C_{i,k}$, $0 \leq i \leq n$, $k \geq 1$, are all extremal points of $\mathcal{C}$.

Moreover, by compactness, selecting a subsequence if necessary we may also assume that, for each $0 \leq i \leq n$, we have $C_{i,k} \to C_i \in \partial \mathcal{C}$ as $k \to \infty$.

Our present aim is to find the possible location of each limit point $C_i \in \partial \mathcal{C}$, $0 \leq i \leq n$. Clearly, C_i cannot be in the relative interior of $\mathcal{C}_0$ since all configuration points are extremal points.

First, we consider the case when $C_i \notin \mathcal{C}_0$.

Since O_0 is in the relative interior of $\mathcal{C}_0$, for k large, the antipodal $C_{i,k}^{O_k}$ of $C_{i,k}$ with respect to O_k must be in the relative interior of $\mathcal{C}_0$. For any of these, $\mathcal{H}_0$ is the unique supporting hyperplane of $\mathcal{C}$ at $C_{i,k}^{O_k}$. Since O_k is a regular point, the chord $[C_{i,k}, C_{i,k}^{O_k}]$ is an *affine diameter*. Hence, by definition, there exists a hyperplane $\mathcal{H}_0'$ supporting $\mathcal{C}$ at $C_{i,k}$ and parallel to $\mathcal{H}_0$. Clearly, $\mathcal{H}_0'$ depends only on $\mathcal{H}_0$ and $\mathcal{C}$.

There cannot be any additional point $C_j \notin \mathcal{C}_0$, $0 \leq j \neq i \leq n$.

In fact, if C_i, $C_j \notin \mathcal{C}_0$ then, by what was said above, for k large, the points $C_{i,k}$ and $C_{j,k}$ are both in $\mathcal{H}_0'$. We then slide $C_{j,k}$ to $C_{i,k}$ along the line segment connecting them and obtain contradiction to regularity of O_k.

More precisely, we consider the 1-parameter family of multi-sets

$$t \mapsto \{C_{0,k}, \ldots, C_{i,k}, \ldots, \widehat{C_{j,k}}, (1-t)C_{j,k} + tC_{i,k}, \ldots, C_{n,k}\}, \quad 0 \leq t \leq 1.$$

Since $\mathcal{H}_0$ and $\mathcal{H}_0'$ are parallel, $\Lambda(., O_k)$ evaluated on this 1-parameter family does not depend on t. The configuration condition that O_k is in the respective convex hull is valid at $t = 0$. Let $0 < t_0 \leq 1$ be the last parameter for which the configuration condition holds. If $t_0 < 1$ then the configuration at t_0 is minimal but not simplicial (as the latter is an open condition), a contradiction to regularity of O_k. Therefore $t_0 = 1$. But then the once again minimal configuration has the point $C_{i,k}$ listed twice, also a contradiction to regularity.

Thus, we obtain that there may be *at most one* C_i, $0 \leq i \leq n$, with $C_i \notin \mathcal{C}_0$. If there is one, renumbering if necessary, we may assume this to be C_0, and let $I = \{1, \ldots, n\}$; otherwise, we let $I = \{0, \ldots, n\}$. We will show below that this latter case cannot happen.

With this we have $C_i \in \partial \mathcal{C}_0$, $i \in I$. By continuity of the distortion ratio, we have $\Lambda_{\mathcal{C}}(C_{i,k}, O_k) \to \Lambda_{\mathcal{C}_0}(C_i, O_0)$ as $k \to \infty$. (In the exceptional case of C_0, we have $\Lambda(C_{0,k}, O_k) \to \infty$, as $k \to \infty$.)

Since the configurations are minimal, we obtain

$$\sigma(O_k) = \sum_{i=0}^{n} \frac{1}{1 + \Lambda_{\mathcal{C}}(C_{i,k}, O_k)} \to \sum_{i \in I} \frac{1}{1 + \Lambda_{\mathcal{C}_0}(C_i, O_0)} = 1, \quad \text{as} \quad k \to \infty,$$

(4.7.1)

where the last equality is because of Theorem 4.4.1. (See the limit in (4.4.4).) From the study of the possible exceptional point, it is clear that $\{C_i\}_{i \in I}$ is a configuration for O_0 in C_0. Since C_0 is $(n-1)$-dimensional, by Theorem 4.1.1, the only way the last equality in (4.7.1) can hold is that $I = \{1, \ldots, n\}$ and C_0 is an $(n-1)$-simplex with vertices $C_1, \ldots, C_n$. The first statement of the lemma follows.

We now define $\mathcal{V}$ as the set of those boundary points $C \in \partial C \setminus C_0$ at which there is a supporting hyperplane parallel to $\mathcal{H}_0$. By the proof above, it is clear that $C_0 \in \mathcal{V}$ and the supporting hyperplane for any point in $\mathcal{V}$ must be $\mathcal{H}'_0$. Thus, we have

$$\mathcal{V} = C \cap \mathcal{H}'_0 = \partial C \cap \mathcal{H}'_0.$$

(See Figure 4.7.1.)

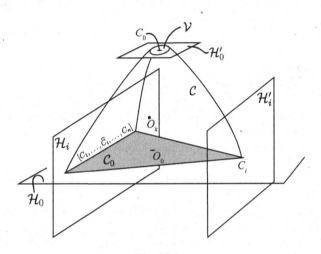

Fig. 4.7.1

For the second part of the proof of Lemma 4.7.1 as well as for the sequel, we need the following:

Lemma 4.7.2. *Given $O \in \text{int} C$, assume that $\Lambda(., O)$ assumes a local maximum at $C \in \partial C$. Then, we have the implication*

$$C^o \in C_0 \Rightarrow C \in \mathcal{V}. \tag{4.7.2}$$

Proof. If $C^o \in C_0$ then $C \notin C_0$. Let $\mathcal{H}'$ be the hyperplane passing through C and parallel to $\mathcal{H}_0$. Since $\mathcal{H}_0$ supports C at C^o and $\Lambda(., O)$ attains a local maximum at C, Corollary 2.2.2 applies. We obtain that $\mathcal{H}'$ supports C. Thus $\mathcal{H}' = \mathcal{H}'_0$ and $C \in \mathcal{V}$.

We now return to the proof of Lemma 4.7.1. Recall that, for k large, $C^o_{0,k}$ is in the (relative) interior of C_0, and therefore, by Lemma 4.7.2, we have $C_{0,k} \in \mathcal{V}$. Since O_k is regular, $C_{0,k}$ can be any point in $\mathcal{V}$, in particular, we can choose $C_{0,k} = C_0$ constant.

Recall now from the first part of the proof that, for $1 \le i \le n$, $C_{i,k} \to C_i \in C_0$, as $k \to \infty$, and $C_0 = [C_1, \ldots, C_n]$. Clearly, for k large, $C_{i,k} \notin \mathcal{V}$. Hence, by

Lemma 4.7.2, $C_{i,k}^{o_k} \notin C_0$. Since $[C_{i,k}, C_{i,k}^{o_k}]$ is an affine diameter there exist parallel hyperplanes $\mathcal{H}_{i,k} \ni C_{i,k}^{o}$ and $\mathcal{H}'_{i,k} \ni C_{i,k}$ both supporting C.

Denote by $\delta_{i,k}$ the *dihedral angle* of the angular sector given by the (transversal) hyperplanes $\mathcal{H}_0$ and $\mathcal{H}_{i,k}$ containing C. Define $\delta'_{i,k}$ similarly (with $\mathcal{H}'_{i,k}$ in place of $\mathcal{H}_{i,k}$). Clearly, $0 < \delta_{i,k}, \delta'_{i,k} < \pi$, and since $\mathcal{H}_{i,k}$ and $\mathcal{H}'_{i,k}$ are parallel, we also have $\delta_{i,k} + \delta'_{i,k} = \pi$. Selecting subsequences, we may assume that $\delta_{i,k} \to \delta_i$ and $\delta'_{i,k} \to \delta'_i$ as $k \to \infty$. Taking the respective limits, we obtain $\delta_i + \delta'_i = \pi$. By convexity, we also have $0 < \delta_i, \delta'_i < \pi$.

Let $\mathcal{H}_i$ be the hyperplane containing $[C_1, \ldots, \widehat{C}_i, \ldots, C_n]$ and having dihedral angle δ_i with $\mathcal{H}_0$. By construction, $\mathcal{H}_i$ is the limit of the supporting hyperplanes $\mathcal{H}_{i,k}$, and so it must also support C. Denote by $\mathcal{H}'_i$ the hyperplane containing C_i and parallel to $\mathcal{H}_i$. Again by construction, $\mathcal{H}'_i$ supports C at C_i. The second statement of Lemma 4.7.1 follows.

Remark. Since C is between the parallel supporting hyperplanes $\mathcal{H}_i$ and $\mathcal{H}'_i$, a simple comparison of the distortion ratios shows that, for k large, $\Lambda(C_i, O_k) \geq \Lambda(C_{i,k}, O_k)$. Hence, for large k, $\{C_0, \ldots, C_n\} \in \mathfrak{C}(O_k)$ (with $C_0 \in \mathcal{V}$ arbitrary) is a *minimizing* configuration.

From now on, (without loss of generality) we will assume that each $\mathcal{H}_i$, $1 \leq i \leq n$, is *closest to* C in the sense that there is no supporting hyperplane between $\mathcal{H}_i$ and $\partial C \setminus C_0$.

Lemma 4.7.3. *Any affine diameter of C disjoint from C_0 has endpoints on a pair $\mathcal{H}_i$ and $\mathcal{H}'_i$, for some $i = 1, \ldots, n$, or on a pair $\mathcal{H}_i$ and $\mathcal{H}_j$, for some distinct $1 \leq i, j \leq n$.*

Proof. Let $[B, B'] \subset C$ be an affine diameter disjoint from C_0. Let $\mathcal{H}$ and $\mathcal{H}'$ be parallel hyperplanes supporting C with $B \in \mathcal{H}$ and $B' \in \mathcal{H}'$.

Assume first that $B \notin \mathcal{H}_i$ for $1 \leq i \leq n$.

We fix $1 \leq i \leq n$. The hyperplane $\mathcal{K}_i = \langle B, C_1, \ldots, C_{i-1}, \widehat{C}_i, C_{i+1}, \ldots, C_n \rangle$ intersects $\mathcal{H}_i$ in $A = \langle C_1, \ldots, C_{i-1}, \widehat{C}_i, C_{i+1}, \ldots, C_n \rangle$. This hyperplane $\mathcal{K}_i$ is transversal to $\mathcal{H}$. (Otherwise, having B as a common point, they would be equal, $\mathcal{K}_i = \mathcal{H}$, and, due to the minimal choice of $\mathcal{H}_i$ above, we would also have $\mathcal{H} = \mathcal{H}_i$, contradicting to $B \notin \mathcal{H}_i$.) We now rotate $\mathcal{K}_i$ about A to $\mathcal{H}_i$ staying on one side of C_0. We consider whether during this rotation the rotated hyperplanes stay transversal to $\mathcal{H}$. Assume not. Then, at one stage of the rotation, a rotated hyperplane is parallel to $\mathcal{H}$. Since this rotated hyperplane along with $\mathcal{H}$, $\mathcal{H}_0$, $\mathcal{H}_i$, and $\mathcal{H}'_i$ all contain a (translated) copy of A, the entire configuration can be understood via its intersection with the two-dimensional $A^{\perp}$. (See Figure 4.7.2.)

Projecting C_i and B to $A^{\perp}$ (along A) we see that we must have $\mathcal{H} = \mathcal{H}'_i$, and so $B \in \mathcal{H}'_i$. In this case, we also have $\mathcal{H}' = \mathcal{H}_i$ (since they are parallel and both supporting) and so $B' \in \mathcal{H}_i$. We arrive at the first stated scenarios: the affine diameter $[B', B]$ connects $\mathcal{H}_i$ and $\mathcal{H}'_i$.

Hence, during the rotation, the rotated hyperplanes stay transversal to $\mathcal{H}$, in particular, $\mathcal{H}_i$ and $\mathcal{H}$ intersect transversally.

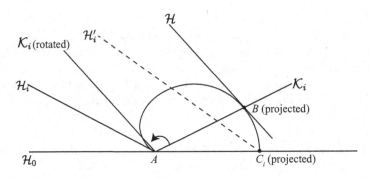

Fig. 4.7.2

Based on our initial assumption, this holds *for all* $1 \le i \le n$; that is, $\mathcal{H}_i$ and $\mathcal{H}$ intersect transversally for all $1 \le i \le n$.

The intersections $\mathcal{H}_i \cap \mathcal{H}$, $1 \le i \le n$, bound an $(n-1)$-simplex Δ in $\mathcal{H}$ with B in its interior since $B \notin \mathcal{H}_i$, $1 \le i \le n$.

Switching the roles of B and B', if $B' \notin \mathcal{H}_j$ for all $1 \le j \le n$, then, repeating the argument above and discarding the stated scenarios, we obtain that $\mathcal{H}_j \cap \mathcal{H}'$, $1 \le j \le n$, bound a simplex Δ' in $\mathcal{H}'$ with B' in its interior.

Since all the participating hyperplanes support $\mathcal{C}$, it follows that the convex hull $[\Delta, \Delta']$ contains $\mathcal{C}$. This convex hull is a polytope which, in addition to its parallel simplicial faces Δ and Δ', has n other side faces supported by $\mathcal{H}_i$, $1 \le i \le n$. On the other hand, the configuration of the side faces intersected with the hyperplane $\mathcal{H}_0$ cuts out the $(n-1)$-simplex $\mathcal{C}_0$ (Lemma 4.7.1).

By assumption, $\mathcal{C}_0$ is disjoint from $[B, B']$. Since $\mathcal{H}_0$ is supporting $\mathcal{C}$, this implies that $\mathcal{H}_0 = \mathcal{H}$ or $\mathcal{H}_0 = \mathcal{H}'$. Thus, B or B' is in $\mathcal{C}_0$, a contradiction. The lemma follows.

To finish the proof of part I of Theorem 4.6.1, from now on we assume that *all* interior points of $\mathcal{C}$ are regular (and $O_0 \in \partial\mathcal{C}$ is a flat point).

Lemma 4.7.4. *Let $B \in \partial\mathcal{C} \setminus \mathcal{C}_0$. Assume that $B \in \mathcal{H}_i$ for some $1 \le i \le n$. Then the intersection $\mathcal{C}_i = \mathcal{H}_i \cap \mathcal{C}$ is an $(n-1)$-simplex $[C, C_1, \ldots, \widehat{C_i}, \ldots, C_n]$ with $C \in \mathcal{V}$.*

Proof. Due to its minimal choice made above, $\mathcal{H}_i$ is a supporting hyperplane of $\mathcal{C}$ at B. The n-simplex $[B, C_1, \ldots, \widehat{C_i}, \ldots, C_n] \subset \mathcal{H}_i$ must then be contained in the boundary of $\mathcal{C}$. Since *any* point in the relative interior of this simplex is a flat point and, by assumption, all interior points are regular, we can now apply Lemma 4.7.1. We obtain that $\mathcal{C}_i = \mathcal{H}_i \cap \mathcal{C}$ is an $(n-1)$-simplex. The intersection $\mathcal{C}_i \cap \mathcal{C}_0$ is the $(n-2)$-simplex $[C_1, \ldots, \widehat{C_i}, \ldots, C_n]$. We denote by C the missing vertex of $\mathcal{C}_i$. Applying the last statement of Lemma 4.7.1 to this situation (with $\mathcal{C}_i$ in place of $\mathcal{C}_0$), we see that $\mathcal{C}$ has a supporting hyperplane at C, parallel to $\mathcal{H}_0$. By convexity, this can only be $\mathcal{H}'_0$, so that $C \in \mathcal{V}$. The lemma follows.

Lemma 4.7.5. *For any $O \in \text{int}\,\mathcal{C}$, the vertices $C_1, \ldots, C_n$ along with a point $C_0 \in \mathcal{V}$ form a minimal configuration with respect to O. In particular, we have $\mathcal{C} = [\mathcal{C}_0, \mathcal{V}]$.*

Proof. Given $O \in \text{int}\, C$, by regularity, we may choose a minimizing configuration $\{B_0, \ldots, B_n\} \in \mathfrak{C}(O)$ consisting of extremal points (Proposition 4.5.4). Fix $0 \le i \le n$.

If $B_i \in C_0$ then, since B_i is extremal, it must be one of the vertices $\{C_1, \ldots, C_n\}$ of C_0.

If $B_i \notin C_0$ but $B_i^o \in C_0$ then, by Lemma 4.7.2, $B_i \in \mathcal{V}$. (Since O is regular, $\Lambda(., O)$ attains local maximum at B_i.)

In the remaining case $[B_i, B_i^o]$ (with B_i^o with respect to O) is an affine diameter away from C_0. We are in the position to apply Lemma 4.7.3. If $B_i \in \mathcal{H}_j$ for some $j = 1, \ldots, n$, then, by Lemma 4.7.4, B_i must be in the $(n-1)$-simplex $[C, C_1, \ldots, \widehat{C}_j, \ldots, C_n]$ with $C \in \mathcal{V}$. Since B_i is extremal, it must be one of the vertices of this simplex. Once again, we obtain that $B_i = C_k$, for some $k = 1, \ldots, n$, $k \ne j$, or $B_i \in \mathcal{V}$. Finally, if $B_i \in \mathcal{H}_j'$ and $B_i^o \in \mathcal{H}_j$ then B_i can be moved to C_j along the line segment $[B_i, C_j] \subset \mathcal{H}_j'$. This line segment is part of the boundary of C since $\mathcal{H}_j'$ is supporting C. During this move the distortion ratio $\Lambda(., O)$ does not decrease since $\mathcal{H}_j$ is parallel to $\mathcal{H}_j'$ and supports C. In addition, the configuration condition stays intact since O is a regular point. We obtain that B_i can be moved to C_j retaining minimality. (See Figure 4.7.3.)

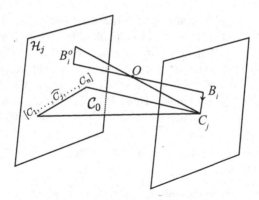

Fig. 4.7.3

Since a minimizing configuration for a regular point cannot contain multiple points, renumbering and making some moves if needed, we conclude that our minimizing configuration may be assumed to have the form

$$\{B_0, \ldots, B_k, C_{i_1}, \ldots, C_{i_l}\}, \quad 1 \le i_1 < \ldots < i_k \le n, \ k + l = n,$$

where $B_0, \ldots, B_k \in \mathcal{V}$. It remains to show that $k = 0$.

Since $O \in [B_0, \ldots, B_k, C_{i_1}, \ldots, C_{i_l}]$, we have the convex combination

$$O = \sum_{i=0}^{k} \lambda_i B_i + \sum_{j=1}^{l} \lambda_{ij} C_{ij}, \quad \sum_{i=0}^{k} \lambda_i + \sum_{j=1}^{l} \lambda_{ij} = 1, \ 0 \le \lambda_i, \lambda_{ij} \le 1.$$

We now compress the first sum in the usual way letting $\mu_0 = \sum_{i=0}^{k} \lambda_i > 0$ and $C_0 = \frac{1}{\mu_0} \sum_{i=1}^{k} \lambda_i B_i \in \mathcal{V}$. We obtain

$$O = \mu_0 C_0 + \sum_{j=1}^{l} \lambda_{i_j} C_{i_j}. \qquad (4.7.3)$$

This implies that the antipodal of C_0 is in $\mathcal{C}_0$, in particular, $\Lambda(C_0, O) \geq \Lambda(B_i, O)$, $0 \leq i \leq k$. Thus, we have

$$\sigma(O) = \sum_{i=0}^{k} \frac{1}{1 + \Lambda(B_i, O)} + \sum_{j=1}^{l} \frac{1}{1 + \Lambda(C_{i_j}, O)} \geq \frac{k+1}{1 + \Lambda(C_0, O)} + \sum_{j=1}^{l} \frac{1}{1 + \Lambda(C_{i_j}, O)}.$$

Again by (4.7.3), $\{C_0, \ldots, C_0, C_{i_1}, \ldots, C_{i_l}\}$ (with C_0 repeated k times) is a configuration with respect to O, so that the opposite inequality also holds. Since O is regular, $k = 0$ must hold. The lemma follows.

We are now ready for the final step as follows:

Lemma 4.7.6. $\mathcal{V}$ *consists of a single point.*

Proof. Let $C_0 \in \mathcal{V}$. If $\mathcal{V}$ consists of more than one point then the simplex $[C_0, \ldots, C_n]$ cannot be the whole $\mathcal{C}$. In particular, there is a point $O \in \text{int}\,\mathcal{C}$ on the boundary of $[C_0, \ldots, C_n]$. Applying Lemma 4.7.5, there is $C_0' \in \mathcal{V}$ such that $\{C_0', C_1, \ldots, C_n\}$ is a minimal simplicial configuration with respect to O. Since the antipodal of C_0 with respect to O is on $\mathcal{C}_0$, we have $\Lambda(C_0, O) \geq \Lambda(C_0', O)$. By minimality, equality must hold. Thus, $\{C_0, C_1, \ldots, C_n\}$ is minimizing with respect to O. This is a contradiction to the regularity of O since it is on the boundary of the simplex $[C_0, C_1, \ldots, C_n]$.

Combining Lemmas 4.7.5 and 4.7.6, we obtain that $\mathcal{C} = [C_0, C_1, \ldots, C_n]$. Part I of Theorem 4.6.1 follows.

Part II. We now change the setting, and start the proof of Theorem 4.6.1 anew under the assumption in (II). We also assume that all interior points of $\mathcal{C}$ are regular. As in Section 1.2, we let $\mathcal{C}^\wedge \subset \partial \mathcal{C}$ denote the set of extremal points of $\mathcal{C}$. We call an extremal point $C \in \mathcal{C}^\wedge$ *isolated* if C has an open neighborhood disjoint from $\mathcal{C}^\wedge \setminus \{C\}$.

Proposition 4.7.7. *Let $C \in \mathfrak{B}$ be a convex body with all its interior points regular. Assume that C has (at least) two isolated extremal points $C_1, C_2 \in \mathcal{C}^\wedge$. Then, for any plane $\mathcal{X}_0 \subset \mathcal{X}$ containing $[C_1, C_2]$ and an interior point of $\mathcal{C}$, the intersection $\mathcal{C}_0 = \mathcal{C} \cap \mathcal{X}_0$ is a triangle with $[C_1, C_2]$ as one side.*

Before the proof of Proposition 4.7.7 we derive a sequence of lemmas. The first lemma states that the local structure of $\mathcal{C}$ near an isolated extremal point is "conical."

Lemma 4.7.8. *Let C be an isolated extremal point of $\mathcal{C} \in \mathfrak{B}$. Then*

$$U_C \doteq \mathcal{C} \setminus \overline{[\mathcal{C}^\wedge \setminus \{C\}]} \qquad (4.7.4)$$

is a relatively open set in $\mathcal{C}$ that contains C. For any $C' \in U_C \cap \partial\mathcal{C}$, $C' \neq C$, the line segment $[C, C']$ is on the boundary of $\mathcal{C}$, and it extends to a boundary line segment $[C, C'']$ with $C'' \in [\mathcal{C}^\wedge \setminus \{C\}]$.

Proof. The first statement can be rephrased as

$$C \notin \overline{[\mathcal{C}^\wedge \setminus \{C\}]}. \qquad (4.7.5)$$

Assume the contrary. We can then select a sequence $\{C_k\}_{k\geq 1} \subset [\mathcal{C}^\wedge \setminus \{C\}]$ converging to C. For each $k \geq 1$, we write C_k as a convex combination $\sum_{i=0}^n \lambda_{ik} C_{i,k}$, $\sum_{i=0}^n \lambda_{ik} = 1, 0 \leq \lambda_{ik} \leq 1, 0 \leq i \leq n$, with $C_{i,k} \in \mathcal{C}^\wedge$, $C_{i,k} \neq C$. By compactness, we may assume that, for each $0 \leq i \leq n$, $C_{i,k} \to C_i$ and $\lambda_{ik} \to \lambda_i$ as $k \to \infty$. Taking the limit, we obtain $C = \sum_{i=0}^n \lambda_i C_i$. Since C is an extremal point, the only way this is possible is that this sum reduces to a single term. We obtain that $C_i = C$ for a specific $0 \leq i \leq n$. With this we have $C_{i,k} \to C_i = C$ as $k \to \infty$. Hence C is not isolated. (4.7.5) follows.

For the second statement, let $C' \notin \overline{[\mathcal{C}^\wedge \setminus \{C\}]}$ be a boundary point of $\mathcal{C}$. Since $[\mathcal{C}^\wedge] = \mathcal{C}$ (Minkowski–Krein–Milman theorem in Section 1.2), we can certainly write C' as a convex combination of C and (finitely many) points in $\mathcal{C}^\wedge \setminus \{C\}$. The point C must participate in this combination with positive coefficient. Hence C' is in the interior of a segment $[C, C'']$, where $C'' \in [\mathcal{C}^\wedge \setminus \{C\}]$. Finally, since C and C' are both boundary points of $\mathcal{C}$, the entire line segment $[C, C'']$ is on the boundary of $\mathcal{C}$. The lemma follows.

Remark. It is instructive to revisit Example 1.2.5 in the specific case when $\mathcal{A} \subset \mathcal{S}$ is a singleton, and identify the isolated extremal points and their conical neighborhoods as in (4.7.4).

Lemma 4.7.9. *Let C be an isolated extremal point of $\mathcal{C}$ with associated open set U_C as in (4.7.4). Then, for every regular point $O \in U_C$, there is a minimal configuration which contains C. In particular, in this case $\Lambda(., O)$ takes a local maximum at C.*

Proof. Let $O \in U_C$ and choose a minimal configuration consisting of extremal points (Proposition 4.5.4). If C does not participate in the configuration then O must be contained in $[\mathcal{C}^\wedge \setminus \{C\}]$. This contradicts to the assumption. Thus, C is a point in the configuration. The second statement is clear.

For the next step, we introduce some notation. Let C be an isolated extremal point of $\mathcal{C}$. Let $\mathcal{X}_0 \subset \mathcal{X}$ be a plane containing C and an interior point $O \in \text{int}\,\mathcal{C}$ which we may assume to be contained in U_C. We consider the planar convex body $\mathcal{C}_0 = \mathcal{C} \cap \mathcal{X}_0$ with isolated extremal point C. As Lemma 4.7.8 asserts, $\mathcal{C}_0$ contains an *angular domain* with vertex at C. We let $[C, P], [C, Q] \subset \partial\mathcal{C}_0$ denote the maximal side segments of this domain. We orient $\mathcal{X}_0$ from O such that the positive orientation corresponds to the sequence P, C, Q.

Since C_0 is convex, with respect to the orientation just fixed, through any boundary point there passes a unique pair of left and right tangent lines. Let α be the angle at C between the line segment $[C, O]$ and the right tangent line at C. Due to the conical structure of C_0 at C asserted by Lemma 4.7.8, this angle is $\angle O C Q$. In a similar vein, we let α^o be the angle at C^o between the line segment $[C^o, O]$ and the right tangent at C^o to the boundary of C_0.

From now on we assume that U_C consists of regular points only. By Lemma 4.7.9, $\Lambda(., O)$ attains a local maximum at C. As a simple consequence of Proposition 2.2.1, we have

$$\alpha \leq \alpha^o. \tag{4.7.6}$$

For any boundary point $B \in \partial C_0$, let $0 \leq \phi(B) \leq \pi$ denote the angle between the left and right tangent lines at B to C_0. Then we have $\phi(C) = \angle PCQ$. For O close to a fixed interior point of $[C, P]$ (within $\text{int}\, C_0$), the right tangent to C_0 at C^o intersects the extension of the line segment $[C, P]$ beyond P. We let R denote this intersection point. (See Figure 4.7.4.)

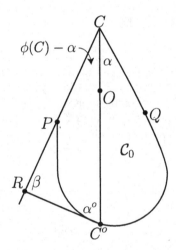

Fig. 4.7.4

From the triangle $[C, C^o, R]$, we obtain

$$\phi(C) - \alpha + \alpha^o + \beta = \pi,$$

where $\beta = \angle C^o R C$. Combining this with (4.7.6), we get

$$\phi(C) + \beta \leq \pi.$$

We now let O converge a fixed interior point of $[C, P]$. We claim that β approaches to $\phi(P)$. In fact, during this convergence, the antipodal C^o approaches to P along the boundary of C_0, and the *right* tangent line at C^o approaches to the *left* tangent line at P. (See Problem 20/(c) at the end of Chapter 1.)

We obtain the following:

Lemma 4.7.10. *Let $C \in \mathfrak{B}$, C an isolated extremal point, and assume that U_C consists of regular points. Then, for any plane $\mathcal{X}_0 \subset \mathcal{X}$ passing through C and an interior point of C, we have*

$$\phi(C) + \phi(P) \le \pi, \tag{4.7.7}$$

where $\phi(C)$ and $\phi(P)$ are the tangential angles of $C \cap \mathcal{X}_0$ at C and P, and $[C, P]$ is a maximal line segment on the boundary of $C \cap \mathcal{X}_0$.

In the lemma above, we call P an *adjacent point* to the isolated extremal point C. Equivalently, P is adjacent to C if $[C, P]$ is a maximal line segment on the boundary of C.

Proof of Proposition 4.7.7. Clearly $C_0 = C \cap \mathcal{X}_0$ is a convex body in $\mathcal{X}_0$ with isolated extremal points C_1 and C_2. Let $P_1, Q_1 \in \partial C_0$ and $P_2, Q_2 \in \partial C_0$ be adjacent to C_1 and C_2, respectively. Orient $\mathcal{X}_0$ and choose the labels such that (with respect to an(y) interior point of C_0) P_1, C_1, Q_1 and P_2, C_2, Q_2 are positively oriented. Assume first that the adjacent points are all distinct, the right tangent at Q_1 and the left tangent at P_2 intersect at a point X, and the left tangent at P_1 and the right tangent at Q_2 intersect at a point Y. (See Figure 4.7.5.)

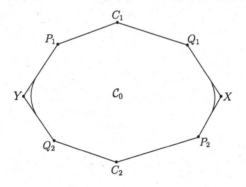

Fig. 4.7.5

For the angle sum of the (convex) octagon $[P_1, C_1, Q_1, X, P_2, C_2, Q_2, Y]$ we have

$$\phi(P_1) + \phi(C_1) + \phi(Q_1) + \beta + \phi(P_2) + \phi(C_2) + \phi(Q_2) + \gamma = 6\pi, \tag{4.7.8}$$

where β and γ are the angles at X and Y, respectively. On the other hand, by (4.7.7), we have

$$\phi(C_1) + \phi(P_1), \ \phi(C_1) + \phi(Q_1), \ \phi(C_2) + \phi(P_2), \ \phi(C_2) + \phi(Q_2) \le \pi.$$

Adding these, we obtain

$$2\phi(C_1) + 2\phi(C_2) + \phi(P_1) + \phi(P_2) + \phi(Q_1) + \phi(Q_2) \le 4\pi.$$

Comparing this with (4.7.8), we get

$$\phi(C_1) + \phi(C_2) + 2\pi - \beta - \gamma \leq 0.$$

This is a contradiction. Notice that we get a contradiction even when $\beta = \pi$ or $\gamma = \pi$ (the cases when the corresponding tangents coincide), and even when $P_1 = Q_2$ but $P_2 \neq Q_1$, or when $P_2 = Q_1$ but $P_1 \neq Q_2$.

If X or Y do not exist, we can add additional supporting lines to boundary points of C_0 and get contradiction again.

Summarizing, we obtain $P_1 = Q_2$ and $P_2 = Q_1$, so that we have

$$C_0 = [P_1, C_1, P_2, C_2].$$

If P_1, C_1, P_2, C_2 are all distinct then, by (4.7.7), $[P_1, C_1, P_2, C_2]$ is a *parallelogram* with $[C_1, C_2]$ as a diagonal.

Finally, if these points are not distinct then C_0 is a triangle with $[C_1, C_2]$ as a side (and P_1 or P_2 is the other vertex), and we are done.

To finish the proof of Proposition 4.7.7 it remains to show that the parallelogram intersection is impossible. Using Lemmas 4.7.8–4.7.10, we let $O \in U_{C_1}$ and consider a minimal configuration $\{C_1, B_1, \ldots, B_n\} \in \mathfrak{C}(O)$ consisting of extremal points only. By the last statement of Lemma 4.7.9, O is contained in the interior of the triangle $[P_1, C_1, P_2]$. Thus, the antipodal P_1^o is contained in $[C_1, P_2]$. (See Figure 4.7.6.)

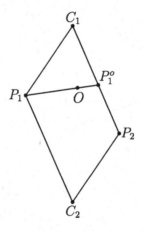

Fig. 4.7.6

Any point on the line segment $[C_1, P_1^o]$ has the same distortion ratio as C_1 since C_0 is a parallelogram. Since O and $\overline{[C^\wedge \setminus C_1]}$ are disjoint, there must be a point $C_1' \in [C_1, P_2]$ for which O is on the boundary of $\overline{[(C^\wedge \setminus \{C_1\}) \cup \{C_1'\}]}$. Thus, O is on the boundary of $[C_1', C_2, \ldots, C_n]$. Hence $\{C_1', C_2, \ldots, C_n\} \in \mathfrak{C}(O)$. It must be minimal with $C_1' \in [C_1, P_1^o]$ since the distortion along $[P_1^o, P_2]$ increases. This, however, contradicts to the regularity of O. Proposition 4.7.7 follows.

With these preparations we now turn to the proof of part II of Theorem 4.6.1. We assume that $\mathcal{C} \in \mathfrak{B}$ with all its interior points regular. We let $C_1, \ldots, C_n$ be a fixed sequence of isolated extremal points.

For $2 \le m \le n$, we let $\mathcal{P}_m$ denote the following statement:

"For any $1 \le i_1, \ldots, i_m \le n$ mutually distinct, and any $O_0 \in \operatorname{int} \mathcal{C} \setminus \langle C_{i_1}, \ldots, C_{i_m} \rangle$, the set $\{C_{i_1}, \ldots, C_{i_m}\}$ is affinely independent, and the intersection $\mathcal{C} \cap \langle C_{i_1}, \ldots, C_{i_m}, O_0 \rangle$ is an m-simplex with $[C_{i_1}, \ldots, C_{i_m}]$ as a side."

Note that $\mathcal{P}_2$ is Proposition 4.7.7. Moreover, for reasons of dimension, $\mathcal{P}_n$ says that $\mathcal{C}$ is an n-simplex; the statement in part II of Theorem 4.6.1. Therefore we can use induction with respect to $m \ge 2$, with the initial step already accomplished.

For the general induction step $m - 1 \Rightarrow m$, $3 \le m \le n$, we assume that $\mathcal{P}_{m-1}$ holds. Rearranging if necessary, we consider $C_1, \ldots, C_m$, and let $\mathcal{X}_0 = \langle C_1, \ldots, C_m, O_0 \rangle$ for some $O_0 = \operatorname{int} \mathcal{C} \setminus \langle C_1, \ldots, C_m \rangle$. For $\mathcal{P}_m$, we need to show that $\{C_1, \ldots, C_m\}$ is affinely independent and $\mathcal{C}_0 = \mathcal{C} \cap \mathcal{X}_0$ is an m-simplex.

First, by the induction hypothesis, $\{C_1, \ldots, C_{m-1}\}$ is affinely independent and $\mathcal{C} \cap \langle C_1, \ldots, C_{m-1}, O_0 \rangle$ is an $(m - 1)$-simplex with $[C_1, \ldots, C_{m-1}]$ as a side. In particular, we have $\mathcal{C} \cap \langle C_1, \ldots, C_{m-1} \rangle = [C_1, \ldots, C_{m-1}] \subset \partial \mathcal{C}$.

If $\{C_1, \ldots, C_m\}$ were affinely *dependent* then we would have $C_m \in \langle C_1, \ldots, C_{m-1} \rangle$ so that $C_m \in [C_1, \ldots, C_{m-1}] \subset \partial \mathcal{C}$. Since $C_1, \ldots, C_m$ are distinct, this would contradict to the assumption that C_m is an extremal point. We obtain that $\{C_1, \ldots, C_m\}$ is an affinely independent set.

It follows that $\dim \mathcal{X}_0 = m$, the set $\Delta = [C_1, \ldots, C_m] \subset \mathcal{C}_0$ is an m-simplex, and $\mathcal{H} = \langle \Delta \rangle = \langle C_1, \ldots, C_m \rangle$ a hyperplane in $\mathcal{X}_0$. (For the most part of the proof below we will work within $\mathcal{X}_0$ so that all the concepts are understood within this affine subspace.) We denote by $\mathcal{G} \subset \mathcal{X}_0$ the closed half-space with $\partial \mathcal{G} = \mathcal{H}$ and $O_0 \in \operatorname{int} \mathcal{G}$.

For $1 \le i \le m$, we let $\Delta_i = [C_1, \ldots, \widehat{C_i}, \ldots, C_m]$, the ith face of Δ opposite to the vertex C_i. For $1 \le i \ne j \le m$, we let $\Delta_{ij} = \Delta_i \cap \Delta_j$.

We will repeatedly use the induction hypothesis in the following setting:

"For $O \in \operatorname{int} \mathcal{C}_0 \cap \operatorname{int} \mathcal{G}$, we have $\mathcal{C}_0 \cap \langle \Delta_i, O \rangle = [\Delta_i, B_i]$ for some $B_i \in \partial \mathcal{C}_0 \cap \operatorname{int} \mathcal{G}$. Taking the respective boundaries, we have $\Delta_i \subset \partial \mathcal{C}_0$, $1 \le i \le m$; in particular $\mathcal{C}_0 \cap \mathcal{H} = \Delta$; and $[\Delta_{ij}, B_i] \subset \partial \mathcal{C}_0$, $1 \le i \ne j \le m$."

We now turn to the proof of the second statement of $\mathcal{P}_m$ above: $\mathcal{C}_0$ is an m-simplex.

Let $\mathcal{H}' \subset \operatorname{int} \mathcal{G}$ be a hyperplane parallel to $\mathcal{H}$ and supporting $\mathcal{C}_0$ at some point $C_0 \in \partial \mathcal{C}_0$. Choose a sequence $\{O_k\}_{k \ge 1} \subset \operatorname{int} \mathcal{C}_0 \cap \operatorname{int} \mathcal{G}$ such that $\lim_{k \to \infty} O_k = C_0$. By the induction hypothesis, for each $1 \le i \le m$, we have

$$\mathcal{C}_0 \cap \langle \Delta_i, O_k \rangle = [\Delta_i, B_{i,k}], \qquad (4.7.9)$$

for some $B_{i,k} \in \partial \mathcal{C}_0 \cap \operatorname{int} \mathcal{G}$, $k \ge 1$. Since $\mathcal{H}'$ supports $\mathcal{C}_0$ at C_0, for each $1 \le i \le m$, we clearly have $\lim_{k \to \infty} B_{i,k} = C_0$, $1 \le i \le m$. (Otherwise, by compactness, $\{B_{i,k}\}_{k \ge 1}$ would subconverge to a point $C_0' \in \partial \mathcal{C} \cap \mathcal{H}'$, $C_0' \ne C_0$, contradicting to $O_k \in [\Delta_i, B_{i,k}]$ and $\lim_{k \to \infty} O_k = C_0$. See Figure 4.7.7.)

Letting $k \to \infty$ in (4.7.9), we obtain

$$\mathcal{C}_0 \cap \langle \Delta_i, C_0 \rangle = [\Delta_i, C_0], \qquad 1 \le i \le m. \qquad (4.7.10)$$

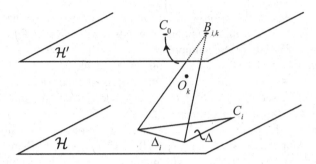

Fig. 4.7.7

Since

$$\partial[\Delta_i, C_0] = \Delta_i \cup \bigcup_{1 \le j \neq i \le m} [\Delta_{ij}, C_0], \tag{4.7.11}$$

as a byproduct (induction hypothesis), we have

$$[\Delta_{ij}, C_0] \subset \partial C_0, \quad 1 \le i \neq j \le m. \tag{4.7.12}$$

We now claim that

$$[\Delta_i, C_0] \subset \partial C_0, \quad 1 \le i \le m. \tag{4.7.13}$$

Assume *on the contrary* that $[\Delta_i, C_0] \not\subset \partial C_0$ for a specific $1 \le i \le m$. This means that the closed half-space $\mathcal{G}_i \subset \mathcal{X}_0$ with boundary hyperplane $\mathcal{H}_i = \langle \Delta_i, C_0 \rangle \subset \mathcal{X}_0$ and $C_i \notin \mathcal{G}_i$ intersects the interior of C_0.

Let $\mathcal{H}'_i \subset \text{int}\,\mathcal{G}_i$ be a hyperplane parallel to $\mathcal{H}_i$ and supporting C_0 at some point $V_i \in \partial C_0 \cap \text{int}\,\mathcal{G}_i$. (See Figure 4.7.8.)

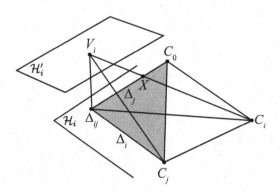

Fig. 4.7.8

Repeating the previous argument (in the use of a sequence $\{O_k\}_{k\geq 1} \subset \operatorname{int} C_0 \cap$ $\operatorname{int} \mathcal{G}_i$ converging to V_i), we obtain $C_0 \cap \langle \Delta_j, V_i \rangle = [\Delta_j, V_i]$, $1 \leq j \leq m$, and $[\Delta_{jk}, V_i] \subset \partial C_0$, $1 \leq j \neq k \leq m$.

Now, C_i and V_i are on different sides of $\mathcal{H}_i$, therefore $[C_i, V_i]$ and $\mathcal{H}_i$ intersect in a point $X \in C_0$. By (4.7.10), $C_0 \cap \mathcal{H}_i = [\Delta_i, C_0]$ so that $X \in [\Delta_i, C_0]$. In addition, since $m \geq 3$, and $C_i \in \Delta_{jk}$ for some (actually any) $1 \leq j \neq k \leq m$ distinct from i, we are in the position to apply (4.7.12) to get $[C_i, V_i] \subset \partial C_0$. In particular, $X \in \partial C_0$. Combining the last two inclusions for X, we have $X \in \partial[\Delta_i, C_0]$. Thus, by (4.7.11), we finally have $X \in [\Delta_{ij}, C_0]$, for some $1 \leq j \neq i \leq m$.

Summarizing, we obtain that $[\Delta_{ij}, C_0]$ and $[C_i, V_i]$ are both contained in the boundary of C_0 and intersect *transversally* at X. By convexity, the convex hull $[\Delta_j, C_0, V_i]$, $\Delta_j = [\Delta_{ij}, C_i]$, is also contained in the boundary of C_0, and, for reasons of dimension, $\langle \Delta_j, C_0, V_i \rangle$ is a supporting hyperplane of C_0.

Once again, let $\{O_k\}_{k \geq 1} \subset \operatorname{int} C_0 \cap \operatorname{int} \mathcal{G}$ be a sequence converging to X. By the induction hypothesis, $C_0 \cap \langle \Delta_j, O_k \rangle$ is an $(m-1)$-simplex with Δ_j as a side. Taking the limit as $k \to \infty$, we obtain that the limiting intersection is an $(m-1)$-simplex with Δ_j as a side and an extra vertex W. On the other hand, the limit of the hyperplanes $\langle \Delta_j, O_k \rangle$ as $k \to \infty$ is the hyperplane $[\Delta_j, C_0, V_i]$ supporting C_0. Thus, the limiting simplex $[\Delta_j, W]$ must contain C_0 and V_i. Due to the extremal choices of the latter two points, we must have $W = C_0$ and $W = V_i$ simultaneously. This is a contradiction, so that we finally arrive at (4.7.13).

Since (4.7.13) holds for all $1 \leq i \leq m$, we see that $C_0 \cap \mathcal{G}$ is the m-simplex $[\Delta, C_0]$. Let $\mathcal{G}'$ be the closed half-space complementary to $\operatorname{int} \mathcal{G}$ in $\mathcal{X}_0$. If $\mathcal{G}'$ is disjoint from the interior of C_0 then C_0 is the m-simplex $[\Delta, C_0]$, and $\mathcal{P}_m$ follows. Otherwise, applying the argument above to $\mathcal{G}'$ instead of $\mathcal{G}$, we obtain that $C_0 \cap \mathcal{G}'$ is another m-simplex $[\Delta, C_0']$. In this case C_0 is then a double m-simplex with base Δ (that is, two m-simplices with disjoint interiors joined at their common side Δ.) It remains to show that this cannot occur.

First, assume that $m < n$. Then $C_{m+1} \in \partial C$ exists. Let $O_0 \in \operatorname{int} \Delta$, and apply the construction above to $\Delta' = [\Delta_1, C_{m+1}] = [C_2, \ldots, C_m, C_{m+1}]$ and $\mathcal{X}_0' = \langle \Delta', O_0 \rangle$. (See Figure 4.7.9.)

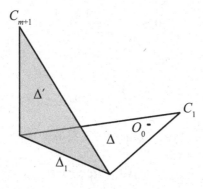

Fig. 4.7.9

We obtain that $C_0' = C \cap \mathcal{X}_0'$ is an m-simplex or a double m-simplex with base Δ'. On the other hand, we have $C \cap \langle \Delta_1, O_0 \rangle = [\Delta_1, C_1] = \Delta$ with O_0 an interior point of C_0' away from Δ'. This contradicts to the extremality of C_1.

Finally, let $m = n$. In this case $C_0 = C = [\Delta, C_0, C_0']$ is a double n-cone in $\mathcal{X}$. This clearly cannot happen as double cones cannot have all their interior points regular. (This is a consequence of Proposition 4.5.4. If $[C_0, C_0']$ intersects the interior of Δ then this intersection point must be singular as it does not have a simplicial minimal configuration consisting of extremal points only. If $[C_0, C_0']$ meets the boundary of Δ then all interior points of Δ are singular for the same reason. Compare this with Example 4.5.5.) Part II of Theorem 4.6.1 follows.

4.8 Stability of the Mean Minkowski Measure

In this section we derive stability estimates for the inequalities:

$$1 \le \sigma_k \le \frac{k+1}{2}, \quad k \ge 2 \tag{4.8.1}$$

(Theorem 4.1.1). We begin with the upper bound as it is much simpler. As usual, we let

$$\sigma_k^* = \max_{O \in \mathrm{int}\, C} \sigma_k(O), \quad k \ge 2.$$

(Note that the maximum is well defined by continuity of σ_k; see Theorem 4.4.1.)

Recall from Theorem 4.1.1 that the upper bound in (4.8.1) is attained, $\sigma_k^* = (k+1)/2$, for some $k \ge 2$, if and only if C is *symmetric* (with respect to the unique point at which σ_k assumes its maximum).

Our first result patterns the stability estimate for the upper bound of the Minkowski measure m_C^* (Theorem 3.2.1) as follows:

Theorem 4.8.1. *Let* $2 \le k \le n$ *and*

$$0 \le \epsilon \le \frac{n-1}{n+1} \frac{k-1}{2}.$$

If a convex body $C \in \mathfrak{B}$ *satisfies*

$$\frac{k+1}{2} - \epsilon \le \sigma_{C,k}^* \tag{4.8.2}$$

then there exists a symmetric convex body $\tilde{C} \in \mathfrak{B}$ *such that*

$$d_H(C, \tilde{C}) \le D_C \frac{n}{k-1} \epsilon, \tag{4.8.3}$$

where D_C *is the diameter of* C.

Remark. Replacing $\sigma^*_{C,k}$ in the assumption (4.8.2) by $\sigma_k(g(C))$, where $g(C)$ is the centroid of C, Theorem 4.8.1 still holds.

Proof of Theorem 4.8.1. Suppressing C, sub-arithmeticity in (4.1.3) (with $k = 1$ and $l = k - 1$ there) gives

$$\sigma_k(O) \le 1 + \frac{k - 1}{\mathfrak{m}(O) + 1}, \quad O \in \mathrm{int}\, C.$$

Taking the maxima on both sides (over the interior of C), we obtain

$$\sigma^*_k \le 1 + \frac{k - 1}{\mathfrak{m}^* + 1}.$$

Combining this with the imposed lower bound (4.8.2), we obtain

$$\mathfrak{m}^* \le \frac{2}{1 - \delta} - 1 = 1 + 2\frac{\delta}{1 - \delta}, \tag{4.8.4}$$

where $\delta = 2\epsilon/(k - 1)$. The imposed restriction on ϵ translates into

$$0 \le \delta \le \frac{n - 1}{n + 1} < 1.$$

Thus, in (4.8.4), we have $2\delta/(1 - \delta) \le n - 1$. Theorem 3.2.1 applies (with ϵ there replaced by $2\delta/(1 - \delta)$) yielding

$$d_H(C, \tilde{C}) \le D_C \frac{n}{n + 1} \frac{\delta}{1 - \delta} \le D_C \frac{n}{n + 1} \frac{\frac{2\epsilon}{k - 1}}{1 - \frac{n - 1}{n + 1}} \le D_C \frac{n}{k - 1} \epsilon.$$

Theorem 4.8.1 follows.

Turning to the lower bound in (4.8.1), recall first that $\sigma_k(O) = 1$, for some $O \in \mathrm{int}\, C$, if and only if C has a k-dimensional simplicial slice across O (Theorem 4.1.1). Hence a stability estimate can only be expected for $\sigma = \sigma_n$. A simple application of Schneider's stability (Theorem 3.2.4) gives the following:

Theorem 4.8.2. *Let $0 \le \epsilon < 1/n(n + 1)$ and $C \in \mathfrak{B}$. Assume that at a critical point $O^* \in C^*$ we have*

$$\sigma(O^*) \le 1 + \epsilon. \tag{4.8.5}$$

Then there exists an n-simplex $\Delta (\subset C)$ such that

$$d_{BM}(C, \Delta) < 1 + \frac{(n + 1)^2 \epsilon}{1 - n(n + 1)\epsilon}.$$

Proof. Using the trivial lower bound in (4.1.8) for σ, we obtain

$$\frac{n+1}{m^*+1} = \frac{n+1}{m(O^*)+1} \le \sigma(O^*) \le 1 + \epsilon.$$

Hence, we have

$$(n+1)(1-\epsilon) \le \frac{n+1}{1+\epsilon} \le m^* + 1.$$

Rearranging, we find

$$n - (n+1)\epsilon \le m^*.$$

Now Theorem 3.2.4 applies with ϵ replaced by $(n+1)\epsilon$. The theorem follows.

To obtain a stronger stability estimate we need to relax the inequality in (4.8.5) as it is clearly too restrictive; for example, Theorem 4.8.2 does not apply to $\sigma(g(\mathcal{C}))$ with $g(\mathcal{C})$, the centroid of $\mathcal{C}$. In the rest of this section we derive a stability estimate with weaker assumptions which thereby has wider range of applications.

Recall from Section 4.4 that an interior point $O \in \operatorname{int} \mathcal{C}, \mathcal{C} \in \mathfrak{B}$, is *regular* if and only if

$$\sigma(O) < \sigma_{n-1}(O) + \frac{1}{m(O)+1}.$$

If $O \in \mathcal{R}$ is regular then the convex hull Δ of any minimal configuration $\{C_0, \ldots, C_n\} \in \mathfrak{C}(O)$ is an n-simplex containing O in its interior: $O \in \operatorname{int} \Delta \subset \mathcal{C}$.

The main result of this section is the following:

Theorem 4.8.3. *Let $\mathcal{C} \in \mathfrak{B}$ and $O \in \operatorname{int} \mathcal{C}$ an interior point satisfying*

$$m(O) \le n. \tag{4.8.6}$$

Assume that, for $0 \le \epsilon < 1/(n+1)$, we have

$$1 \le \sigma(O) \le 1 + \epsilon. \tag{4.8.7}$$

Then $O \in \mathcal{R}$ is a regular point, and we have

$$d_{BM}(\mathcal{C}, \Delta) \le \frac{1}{1 - (n+1)\epsilon}, \tag{4.8.8}$$

where Δ can be chosen as the convex hull of any minimal configuration $\{C_0, \ldots, C_n\} \in \mathfrak{C}(O)$.

Before the proof, we first note that we can lower the value of ϵ (to $\sigma(O) - 1$), and instead of (4.8.7), impose

$$1 < \sigma(O) < 1 + \frac{1}{n+1}. \tag{4.8.9}$$

(For simplicity, we excluded the trivial case $\sigma(O) = 1$.) Assuming now (4.8.6) and (4.8.9), we will prove a much more precise statement than the Banach–Mazur estimate in (4.8.8), namely that

$$\Delta \subset \mathcal{C} \subset \tilde{\Delta} = S_{\tilde{r}, \tilde{c}}(\Delta), \tag{4.8.10}$$

where

$$\tilde{c} = \frac{1}{\sigma(O) - 1} \sum_{i=0}^{n} \left(\frac{1}{\Lambda(C_i, O) + 1} - \frac{1}{\Lambda_\Delta(C_i, O) + 1} \right) C_i \in \Delta \tag{4.8.11}$$

and

$$\tilde{r} = \frac{1}{1 - (n+1)(\sigma(O) - 1)}. \tag{4.8.12}$$

Remark 1. Since $\mathfrak{m}^* \leq n$ the upper bound in (4.8.6) holds on the critical set C^*. The upper bound in (4.8.9) imposed on C^* then implies that the critical set is a singleton. Indeed, by the trivial lower estimate in (4.1.8) and (4.8.9), we have $(n + 1)/(\mathfrak{m}^* + 1) \leq \sigma(O^*) < 1 + 1/(n+1)$, $O^* \in C^*$. This gives $\mathfrak{m}^* + 1 > n$, and Corollary 2.4.13 implies unicity of the critical point.

Remark 2. Even though Theorem 4.8.3 is a stability result for the mean Minkowski measure σ, as a special case ($O = O^*$), it gives a stability result for the Minkowski measure $\mathfrak{m}^*$ itself very close to Schneider's (Theorem 3.2.4). In fact, given $\mathcal{C} \in \mathfrak{B}$, for $0 \leq \epsilon < 1/(n+1)$, we claim

$$\mathfrak{m}^* \geq \frac{n - \epsilon}{1 + \epsilon} \quad \Rightarrow \quad d_{BM}(\mathcal{C}, \Delta) \leq 1 + \frac{(n+1)\epsilon}{1 - (n+1)\epsilon}.$$

Indeed, since $0 \leq \epsilon < 1/(n+1) < 1/n$, we have $n - 1 < (n - \epsilon)/(1 + \epsilon) \leq \mathfrak{m}^*$, and Theorem 4.6.2 implies that, for the unique critical point O^* (which is regular), we have $\sigma(O^*) = (n + 1)/(\mathfrak{m}^* + 1)$. Our assumption on the lower estimate of $\mathfrak{m}^*$ above can then be written as $\sigma(O^*) \leq 1 + \epsilon$. Now, Theorem 4.8.3 applies with $O = O^*$ (Remark 1 above), and the Banach–Mazur estimate above follows.

Remark 3. As in Problem 15 at the end of Chapter 3, the Banach–Mazur distance d_{BM} can be replaced by the dilatation invariant pseudo-metric d constructed there.

Remark 4. The inequality in (4.8.6) holds for the centroid: $\mathfrak{m}(g(\mathcal{C})) \leq n$. (Theorem 3.4.2). Hence (4.8.10) holds provided that $1 < \sigma(g(\mathcal{C})) < 1 + 1/(n+1)$.

Remark 5. For the double regular tetrahedron C in Example 4.5.5 the conditions
(4.8.6) and (4.8.9) are mutually exclusive and complementary ($m(O) \leq 3$ and
$m(O) > 3$, $O \in \mathrm{int}\,C$, since all interior points are singular and C is simplicial in
codimension 1). This also shows that the upper bound in (4.8.9) is sharp in the sense
that if we have equality there then the conclusion of Theorem 4.8.3 does not hold.

Remark 6. For the unit half-disk in Example 4.6.3, a simple computation shows
that (4.8.6) and (4.8.9) hold on a non-empty open subset of $\Delta = [C_0, C_-, C_+]$,
where $\{C_0, C_-, C_+\}$ is the universal minimal configuration for its interior points. In
particular, choosing O on the symmetry axis $O = (0, b)$, we get $1/3 < b < 1/\sqrt{3}$.
For these points, the center of symmetry $\tilde{C}$ is the origin and we have $\tilde{r} = 1/(1 - 3b(1 - b))$. We see that the center $\tilde{C}$ of similarity can be on the boundary of Δ.

We now return to the main line. If (4.8.6) and (4.8.9) hold then, by Theorem 4.1.1,
we have

$$\sigma(O) < 1 + \frac{1}{n+1} \leq \sigma_{n-1}(O) + \frac{1}{m(O) + 1}.$$

This means that O is a regular point. Thus, by definition, any minimal
n-configuration is simplicial, so that Δ in Theorem 4.8.3 is an n-simplex with
O in its interior. In addition, since

$$\sigma(O) = \sum_{i=0}^{n} \frac{1}{\Lambda(C_i, O) + 1} \quad \text{and} \quad \sigma_\Delta(O) = \sum_{i=0}^{n} \frac{1}{\Lambda_\Delta(C_i, O) + 1} = 1,$$

the sum in (4.8.11) is a convex combination, and we have $\tilde{C} \in \Delta$.

Thus, it remains to prove that $C \subset \tilde{\Delta}$, where $\tilde{\Delta}$ is given by (4.8.11)–(4.8.12). The
proof is long and technical and will be carried out in the rest of this section.

The overall plan is as follows. For simplicity, we may assume that O is
the origin 0. Given a minimal simplicial configuration $\{C_0, \ldots, C_n\} \in \mathfrak{C}(0)$
with $\Delta = [C_0, \ldots, C_n]$, the crux is to write C as the union of the *antipodal*
simplex $[C_0^o, \ldots, C_n^o]$ and $(n + 1)$ 'bulges' $\mathcal{B}_i$, $i = 0, \ldots, n$. The bulge $\mathcal{B}_i$ is the
part of C contained in the positive cone $\mathbb{R}_+ \cdot [C_0^o, \ldots, \widehat{C_i^o}, \ldots, C_n^o]$ spanned by
$\{C_0^o, \ldots, \widehat{C_i^o}, \ldots, C_n^o\}$ and truncated by $[C_0^o, \ldots, \widehat{C_i^o}, \ldots, C_n^o]$, the ith face.

In the first step of the proof we embed each bulge $\mathcal{B}_i$ into a polytope $\mathcal{P}_i$
(Theorem 4.8.10). (See Figures 4.8.1 and 4.8.2 for $n = 2, 3$.)

Second, if V_i denotes the outermost vertex of $\mathcal{P}_i$ we will then show that C is
contained in the n-simplex $[V_0, \ldots, V_n]$ (Theorem 4.8.11).

Finally, another estimate will yield $[V_0, \ldots, V_n] \subset \tilde{\Delta} = [\tilde{C}_0, \ldots, \tilde{C}_n]$, with
$\tilde{C}_0, \ldots, \tilde{C}_n$, the vertices of the simplex $\tilde{\Delta} = S_{\tilde{r}, \tilde{C}}(\Delta)$ in (4.8.10)–(4.8.12).
Note that the polytope $\mathcal{P}_i$, $i = 0, \ldots, n$, has a transparent combinatorial structure:
its vertices can be parametrized by subsets of $\{0, \ldots, \hat{i}, \ldots, n\}$, and the larger the
cardinality of the subset is the further the corresponding vertex is from the origin.

We start with a set $\{C_i \mid i \in \mathcal{I}\} \subset \partial C$ of boundary points indexed by a finite
subset $\mathcal{I} \subset \mathbf{Z}$. (During the course of the proof this set will be various *subsets*

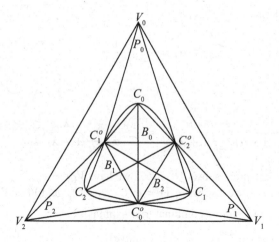

Fig. 4.8.1

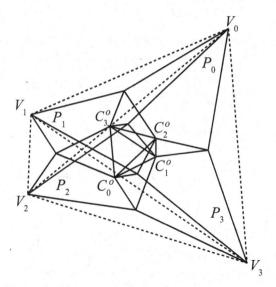

Fig. 4.8.2

indexing a minimal configuration.) Recall that the interior point O is the origin 0. For simplicity, we write $\lambda_i = \Lambda(C_i, 0)$, $i \in \mathcal{I}$, and first assume that

$$\sum_{i \in \mathcal{I}} \frac{1}{1 + \lambda_i} < 1. \tag{4.8.13}$$

For $I \subset \mathcal{I}$, we define

$$\sigma_I = \sum_{i \in I} \frac{1}{1 + \lambda_i} \tag{4.8.14}$$

and

$$V_I = -\frac{1}{1 - \sigma_I} \sum_{i \in I} \frac{1}{1 + \lambda_i} C_i = \frac{1}{1 - \sigma_I} \sum_{i \in I} \frac{\lambda_i}{1 + \lambda_i} C_i^o. \tag{4.8.15}$$

In particular, for a 1-point subset $\{i\}$, $i \in \mathcal{I}$, we have $V_{\{i\}} = C_i^o$.

If $I = \{i_1, \ldots, i_k\}$, we also write $\sigma_I = \sigma(C_{i_1}, \ldots, C_{i_k})$ and $V_I = V(C_{i_1}, \ldots, C_{i_k})$.

We begin with two simple combinatorial lemmas.

Lemma 4.8.4. *For $i \in I \subset \mathcal{I}$, $|I| \geq 2$, we have*

$$V_{I \setminus \{i\}} \in [C_i, V_I].$$

Proof. This is a simple computation. Subtracting i from I in the sum in (4.8.15), we have

$$V_{I \setminus \{i\}} = \frac{1}{1 - \sigma_{I \setminus \{i\}}} \frac{1}{1 + \lambda_i} C_i + \frac{1 - \sigma_I}{1 - \sigma_{I \setminus \{i\}}} V_I.$$

Since the coefficients are positive, by (4.8.14), we obtain

$$1 - \sigma_I + \frac{1}{1 + \lambda_i} = 1 - \sigma_{I \setminus \{i\}}.$$

The lemma follows.

Lemma 4.8.5. *For $I, J \subset \mathcal{I}$ we have the following:*

(i) If $I \cap J \neq \emptyset$ then there exist $0 < t \leq s < 1$ such that

$$(1 - t)V_I + tV_J = (1 - s)V_{I \cup J} + sV_{I \cap J}.$$

(ii) If I and J are disjoint then there exist $r > 1$ and $0 < t < 1$ such that

$$r((1 - t)V_I + tV_J) = V_{I \cup J}.$$

Proof. Given $0 < t < 1$, we first consider the convex combination

$$(1 - t)V_I + tV_J = \frac{1 - t}{1 - \sigma_I} \sum_{i \in I} \frac{\lambda_i}{1 + \lambda_i} C_i^o + \frac{t}{1 - \sigma_J} \sum_{j \in J} \frac{\lambda_j}{1 + \lambda_j} C_j^o.$$

The coefficients in front of the two sums are equal if

$$t = \frac{1 - \sigma_J}{2 - \sigma_I - \sigma_J}.$$

With this, the two sums can be joined. Taking account of the overlap $I \cap J$, we obtain

$$(1 - t)V_I + tV_J = \frac{1}{2 - \sigma_I - \sigma_J} \left(\sum_{k \in I \cup J} \frac{\lambda_k}{1 + \lambda_k} C_k^o + \sum_{l \in I \cap J} \frac{\lambda_l}{1 + \lambda_l} C_l^o \right).$$

On the level of the coefficients, we also have $\sigma_I + \sigma_J = \sigma_{I \cup J} + \sigma_{I \cap J}$. Thus, we obtain

$$(1 - t)V_I + tV_J = (1 - s)V_{I \cup J} + sV_{I \cap J},$$

where

$$s = \frac{1 - \sigma_{I \cap J}}{2 - \sigma_{I \cup J} - \sigma_{I \cap J}}.$$

If $I \cap J = \emptyset$ then the second term is absent. The lemma now follows.

Remark. Note that the cases (i) and (ii) can be united if we define $V_\emptyset = 0$. No significant advantage is gained with this, however.

We now begin the first step constructing the polytopes $\mathcal{P}_i$, $i = 0, \ldots, n$. In addition to the assumptions and the notations above, we will consider only subsets $I \subset \mathcal{I}$ for which $\{C_i \mid i \in I\}$ are *linearly independent*, in particular, $0 \notin \langle \{C_i\}_{i \in I} \rangle$.

For a subset $\mathcal{A} \subset \mathcal{X}$, we denote the *linear* span of $\mathcal{A}$ by $\mathcal{X}(\mathcal{A}) = \langle \mathcal{A}, 0 \rangle$, the smallest linear subspace of $\mathcal{X}$ that contains $\mathcal{A}$. For $I \subset \mathcal{I}$ satisfying the above, we define

$$\mathcal{X}_I = \mathcal{X}(\{C_i\}_{i \in I}) = \mathcal{X}(\{C_i^o\}_{i \in I}).$$

We now let

$$\mathcal{T}_I = \left\{ \sum_{i \in I} \mu_i C_i^o \;\middle|\; \sum_{i \in I} \mu_i \geq 1, \; \mu_i \geq 0, \; i \in I \right\}. \tag{4.8.16}$$

The defining inequalities show that $\mathcal{T}_I \subset \mathcal{X}_I$ is a *truncated (convex) cone*. In addition, by (4.8.14)–(4.8.15), for $|I| \geq 2$, we have $V_I \in \text{int}\, \mathcal{T}_I$ since

$$\frac{1}{1 - \sigma_I} \sum_{i \in I} \frac{\lambda_i}{1 + \lambda_i} = \frac{1}{1 - \sigma_I} \sum_{i \in I} \left(1 - \frac{1}{1 + \lambda_i} \right) = \frac{|I| - \sigma_I}{1 - \sigma_I} > 1. \tag{4.8.17}$$

Moreover, for $J \subset I$, we have $\mathcal{T}_J = \mathcal{T}_I \cap \mathcal{X}_J$, in particular, $\{V_J\}_{J \subset I} \subset \mathcal{T}_I$.

Finally, we define $\mathcal{P}_I \subset \mathcal{X}_I$ inductively (with respect to $|I|$) as follows. For $I = \{i\}$, we set $\mathcal{P}_{\{i\}} = \{V_{\{i\}}\} = \{C_i^o\}$, and, for $|I| \geq 2$, we define $\mathcal{P}_I$ as the convex hull

$$\mathcal{P}_I = \left[\cup_{i \in I} \mathcal{P}_{I \setminus \{i\}}, V_I\right]$$

Clearly, $\mathcal{P}_I \subset \mathcal{T}_I$. In addition, for $J \subset I$ we also have

$$\mathcal{P}_J = \mathcal{P}_I \cap \mathcal{X}_J.$$

As above, for $I = \{i_1, \ldots, i_k\}$, we will also use the notations $\mathcal{X}_I = \mathcal{X}$ $(C_{i_1}, \ldots, C_{i_k})$, $\mathcal{T}_I = \mathcal{T}(C_{i_1}, \ldots, C_{i_k})$, and $\mathcal{P}_I = \mathcal{P}(C_{i_1}, \ldots, C_{i_k})$.

Example 4.8.6. For $I = \{i, j\}$, we have

$$\mathcal{P}(C_i, C_j) = [V(C_i, C_j), V(C_i), V(C_j)],$$

where $V(C_i) = C_i^o$, $V(C_j) = C_j^o$, and

$$V(C_i, C_j) = \frac{1}{1 - \frac{1}{1+\lambda_i} - \frac{1}{1+\lambda_j}} \left(\frac{\lambda_i}{1+\lambda_i} C_i^o + \frac{\lambda_j}{1+\lambda_j} C_j^o\right).$$

Thus, $\mathcal{P}(C_i, C_j)$ is a triangle. (See Figure 4.8.3.)

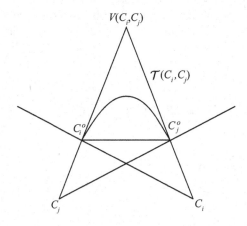

Fig. 4.8.3

Example 4.8.7. For $I = \{i, j, k\}$, $\mathcal{P}(C_i, C_j, C_k)$ is a polyhedron depicted in Figure 4.8.4. The seven vertices of $\mathcal{P}(C_i, C_j, C_k)$ are V_I, and $V_{\{i,j\}}, V_{\{j,k\}}, V_{\{k,i\}}$, and $V_{\{i\}}, V_{\{j\}}, V_{\{k\}}$. In Proposition 4.8.9 below we will prove that there are seven faces:

$F_0 = [V_{\{i\}}, V_{\{j\}}, V_{\{k\}}],$

$F_i = [V_{\{j,k\}}, V_{\{j\}}, V_{\{k\}}], \quad F_j = [V_{\{k,i\}}, V_{\{k\}}, V_{\{i\}}], \quad F_k = [V_{\{i,j\}}, V_{\{i\}}, V_{\{j\}}],$

$F^i = [V_I, V_{\{i,j\}}, V_{\{k,i\}}, V_{\{i\}}], \quad F^j = [V_I, V_{\{j,k\}}, V_{\{i,j\}}, V_{\{j\}}], \quad F^k = [V_I, V_{\{j,k\}}, V_{\{k,i\}}, V_{\{k\}}].$

In particular, $\mathcal{P}_I$ is *not* a simplex for $|I| \geq 3$.

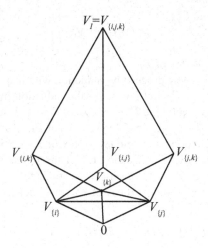

Fig. 4.8.4

In the next two propositions we give a detailed description of the geometry of $\mathcal{P}_I$.

Proposition 4.8.8.   *$\mathcal{P}_I \subset \mathcal{X}_I$ is a convex polytope with vertices $V_J, J \subset I$.*

Proof. It follows directly from the definition that $\mathcal{P}_I$ is the convex hull of the points $V_J, J \subset I$. It remains to show that each $V_J, J \subset I$, is an extremal point (vertex) of $\mathcal{P}_I$. We will do this by induction with respect to $|I|$. A quick look at Examples 4.8.6 and 4.8.7 shows that this holds for $|I| = 2, 3$, so that only the general induction step needs to be discussed.

We first show that V_I is a vertex. To do this consider the hyperplane (in $\mathcal{X}_I$):

$$\mathcal{H}_I = \frac{|I| - \sigma_I}{1 - \sigma_I}(\{C_i^o\}_{i \in I}).$$

By (4.8.15) and (4.8.17), we have $V_I \in \mathcal{H}_I$.
We now claim that $V_J, J \subset I, J \neq I$, are contained in the same *open* half-space with boundary $\mathcal{H}_I$. Comparing

$$V_J = \frac{1}{1 - \sigma_J} \sum_{j \in J} \frac{\lambda_j}{1 + \lambda_j} C_j^o \qquad (4.8.18)$$

with (4.8.15) and (4.8.17), we need to show that

$$\frac{|J| - \sigma_J}{1 - \sigma_J} < \frac{|I| - \sigma_I}{1 - \sigma_I}.$$

This, however, is clear since $|I| > |J|$ and $\sigma_I > \sigma_J$. The claim follows.

Since $\mathcal{P}_I$ is the convex hull of V_J, $J \subset I$, we obtain that $\mathcal{H}_I$ is a hyperplane supporting $\mathcal{P}_I$ at V_I, and

$$\mathcal{P}_I \cap \mathcal{H}_I = \{V_I\}.$$

This means that V_I is a vertex of $\mathcal{P}_I$.

Assuming $|I| \geq 3$, we now proceed by induction with respect to $|I|$ to show that V_J, $J \subset I$, are vertices of $\mathcal{P}_I$. Let $i \in I$. By the induction hypothesis, $\{V_J\}_{J \subset I \setminus \{i\}}$ are vertices of $\mathcal{P}_{I \setminus \{i\}}$. The polytope $\mathcal{P}_{I \setminus \{i\}}$ is a face of $\mathcal{P}_I$ with supporting hyperplane extension $\mathcal{X}_{I \setminus \{i\}}$. Hence $\{V_J\}_{J \subset I \setminus \{i\}}$ are also vertices of $\mathcal{P}_I$. (If $\mathcal{P}$ is a convex polytope and $\mathcal{H}$ is a supporting hyperplane then any vertex of $\mathcal{P} \cap \mathcal{H}$ is also a vertex of $\mathcal{P}$: $(\mathcal{P} \cap \mathcal{H})^{\wedge} = \mathcal{P}^{\wedge} \cap \mathcal{H}$; see (1.2.3) in Section 1.2.) Thus, for each proper subset $J \subset I$, V_J is a vertex of $\mathcal{P}_I$. Since V_I is also a vertex, we are done.

Finally, V_J, $J \subset I$, are all the vertices of $\mathcal{P}_I$ since their convex hull is $\mathcal{P}_I$. The proposition follows.

Proposition 4.8.9. *Let $|I| \geq 3$. Then $\mathcal{P}_I$ has $2|I| + 1$ faces as follows:*

$$F_0 = [\{C_i^o\}_{i \in I}],$$
$$F_i = \mathcal{P}_{I \setminus \{i\}} = [\{V_J \mid i \notin J\}], \ i \in I,$$
$$F^i = [\{V_J \mid i \in J\}], \ i \in I.$$

Proof. First, observe that $\mathcal{P}_I$ is contained in the truncated cone $\mathcal{T}_I$ whose faces are F_0 and $\mathcal{T}_{I \setminus \{i\}}$, $i \in I$. Clearly, F_0 is also a face of $\mathcal{P}_I$.

Since $F_i = \mathcal{P}_{I \setminus \{i\}} = \mathcal{P}_I \cap \mathcal{T}_{I \setminus \{i\}}$, it is clear that F_i, $i \in I$, are also faces of $\mathcal{P}_I$.

Second, we show that F^i, $i \in I$, is a face of $\mathcal{P}_I$. By definition, $V_{\{i\}} \in F^i$, and $V_{\{i,j\}} \in F^i$, $j \in I \setminus \{i\}$. It is easy to see that $V_{\{i,j\}} - V_{\{i\}}$, $j \in I \setminus \{i\}$, are linearly independent.

We claim that, for any $J \subset I$ with $i \in J$, we have

$$V_J \in \langle V_{\{i\}}, \{V_{\{i,j\}}\}_{j \in J \setminus \{i\}} \rangle,$$

or equivalently

$$V_J = \alpha_i V_{\{i\}} + \sum_{j \in J \setminus \{i\}} \alpha_j V_{\{i,j\}} \tag{4.8.19}$$

with

$$\alpha_i + \sum_{j \in J \setminus \{i\}} \alpha_j = 1. \tag{4.8.20}$$

We expand the right-hand side in (4.8.19) and use $V_{\{i\}} = C_i^o$ to obtain

$$V_J = \left(\alpha_i + \sum_{j \in J \setminus \{i\}} \frac{\alpha_j}{1 - \sigma_{\{i,j\}}} \frac{\lambda_i}{1 + \lambda_i} \right) C_i^o + \sum_{j \in J \setminus \{i\}} \frac{\alpha_j}{1 - \sigma_{\{i,j\}}} \frac{\lambda_j}{1 + \lambda_j} C_j^o. \tag{4.8.21}$$

Equating the coefficients with those of V_J in (4.8.18), a simple computation gives

$$\alpha_j = \frac{1 - \sigma_{\{i,j\}}}{1 - \sigma_J}, \ j \in J \setminus \{i\} \quad \text{and} \quad \alpha_i = \frac{2 - |J|}{1 - \sigma_J} \frac{\lambda_i}{1 + \lambda_i}.$$

It remains to check that (4.8.20) is satisfied. Using the values of the coefficients α_j, $j \in J$, just obtained, (4.8.20) can be written as

$$(2 - |J|) \frac{\lambda_i}{1 + \lambda_i} + \sum_{j \in J \setminus \{i\}} (1 - \sigma_{\{i,j\}}) = 1 - \sigma_J.$$

This holds, however, since

$$\sum_{j \in J \setminus \{i\}} \sigma_{\{i,j\}} = \sum_{j \in J \setminus \{i\}} \left(\frac{1}{1 + \lambda_i} + \frac{1}{1 + \lambda_j} \right) = \frac{|J| - 2}{1 + \lambda_i} + \sigma_J.$$

The claim follows.

In particular, we have

$$\dim \langle F^i \rangle = \dim \langle V_{\{i\}}, \{V_{\{i,j\}}\}_{j \in I \setminus \{i\}} \rangle = |I| - 1.$$

Finally, to conclude that F^i is a face of $\mathcal{P}_I$, it remains to show that $\mathcal{P}_I$ is on one side of the hyperplane $\langle F^i \rangle$.

Let V_K be a vertex of $\mathcal{P}_I$ not listed in F^i, that is, $i \notin K \subset I$. Let $J = K \cup \{i\}$. We are now in the position to apply Lemma 4.8.5 (ii) to the disjoint subsets K and $\{i\}$ to obtain

$$r((1 - t)V_K + tV_{\{i\}}) = V_J,$$

for some $r > 1$ and $0 < t < 1$. Since $V_{\{i\}}, V_J \in F^i$, this means that V_K and the origin 0 are on the same side of $\langle F^i \rangle$. Thus, F^i is a face of $\mathcal{P}_I$.

To finish the proof of the proposition, it remains to prove that F_0, and F_i, F^i, $i \in I$, are *all* the faces of $\mathcal{P}_I$.

To do this, we consider the hyperplane extensions of these faces:

$$\mathcal{H}_0 = \langle F_0 \rangle = \langle \{ C_i^o \}_{i \in I} \rangle,$$

$$\mathcal{H}_i = \langle F_i \rangle = \langle \mathcal{P}_{I \setminus \{i\}} \rangle = \mathcal{X}_{I \setminus \{i\}} = \mathcal{X}(\{ C_j^o \}_{j \in I \setminus \{i\}}), \ i \in I,$$

$$\mathcal{H}^i = \langle F^i \rangle = \langle \{ V_J \, | \, i \in J \} \rangle, \ i \in I.$$

(Note that $\mathcal{H}_i$, $i \in I$, are linear.) Each of these hyperplanes is the boundary of a half-space that contains $\mathcal{P}_I$. Let $\mathcal{P}'_I$ denote the intersection of these half-spaces. Clearly, $\mathcal{P}_I \subset \mathcal{P}'_I \subset \mathcal{T}_I$. It remains to show that $\mathcal{P}_I = \mathcal{P}'_I$, or equivalently, that the vertices of $\mathcal{P}'_I$ are the same as the vertices of $\mathcal{P}_I$ (given in Proposition 4.8.8).

To do this, we consider the vertices of $\mathcal{P}'_I$ as (non-redundant) intersections of the hyperplanes above. To obtain a vertex, we need to take at least $|I|$ hyperplanes as $\dim \mathcal{X}_I = |I|$. We split the discussion into two cases according to whether $\mathcal{H}_0$ is participating in the intersection or not.

Case (i): Assume that $\mathcal{H}_0$ is participating in the intersection. We first show that, for each $i \in I$, $\mathcal{H}_0 \cap \mathcal{H}^i$ intersects F_0 at the single point C_i^o so that the remaining part of the intersection is redundant (that is, disjoint from $\bar{\mathcal{P}}_I$).

Let $X \in \mathcal{H}^i$. As above, we have

$$X = \alpha_i V_{\{i\}} + \sum_{j \in I \setminus \{i\}} \alpha_j V_{\{i,j\}} \tag{4.8.22}$$

and

$$\alpha_i + \sum_{j \in I \setminus \{i\}} \alpha_j = 1. \tag{4.8.23}$$

As in (4.8.21), we expand the right-hand side of (4.8.22) to obtain

$$X = \left(\alpha_i + \sum_{j \in I \setminus \{i\}} \frac{\alpha_j}{1 - \sigma_{\{i,j\}}} \frac{\lambda_i}{1 + \lambda_i} \right) C_i^o + \sum_{j \in I \setminus \{i\}} \frac{\alpha_j}{1 - \sigma_{\{i,j\}}} \frac{\lambda_j}{1 + \lambda_j} C_j^o. \tag{4.8.24}$$

Now, $X \in F_0$ if and only if $\alpha_j \geq 0, j \in I \setminus \{i\}$, and

$$\alpha_i + \sum_{j \in I \setminus \{i\}} \frac{\alpha_j}{1 - \sigma_{\{i,j\}}} \frac{\lambda_i}{1 + \lambda_i} \geq 0$$

and

$$\alpha_i + \sum_{j \in I \setminus \{i\}} \frac{\alpha_j}{1 - \sigma_{\{i,j\}}} \frac{\lambda_i}{1 + \lambda_i} + \sum_{j \in I \setminus \{i\}} \frac{\alpha_j}{1 - \sigma_{\{i,j\}}} \frac{\lambda_j}{1 + \lambda_j} = 1.$$

By the definition of $\sigma_{\{i,j\}}$, this last equality reduces to

$$\alpha_i + \sum_{j \in I \setminus \{i\}} \alpha_j \frac{2 - \sigma_{\{i,j\}}}{1 - \sigma_{\{i,j\}}} = 1.$$

Combining this with (4.8.23), we obtain

$$\sum_{j \in I \setminus \{i\}} \alpha_j \frac{1}{1 - \sigma_{\{i,j\}}} = 0.$$

Since the coefficients are non-negative, this is possible only if $\alpha_j = 0, j \in I \setminus \{i\}$. By (4.8.23), $\alpha_i = 1$ and $X = V_{\{i\}} = C_i^o$ follow.

Thus, besides $\mathcal{H}_0$, the only participating hyperplanes we need to consider are $\mathcal{H}_i$, $i \in I$. There must be at least $|I| - 1$ of these. On the other hand, there cannot be $|I|$ of these as their intersection is the redundant origin $0 \notin \mathcal{T}_I$. Hence, there exists $i \in I$, such that the participating hyperplanes are $\mathcal{H}_0$ and $\mathcal{H}_j, j \in I \setminus \{i\}$. The intersection of these is clearly C_i^o. Case (i) follows.

Case (ii): We first show that the hyperplanes $\mathcal{H}_i$ and $\mathcal{H}^i$ (with the same $i \in I$) cannot participate together in the intersection; in particular, that there are exactly $|I|$ participating hyperplanes, one for each index in I.

Let $X \in \mathcal{H}^i \cap \mathcal{H}_i$. Write X as in (4.8.22) with (4.8.23). Expanding as in (4.8.24), $X \in \mathcal{H}_i$ forces the coefficient of C_i^o to vanish:

$$\alpha_i + \sum_{j \in I \setminus \{i\}} \frac{\alpha_j}{1 - \sigma_{\{i,j\}}} \frac{\lambda_i}{1 + \lambda_i} = 0.$$

This, combined with (4.8.23) gives

$$\sum_{j \in I \setminus \{i\}} \alpha_j \left(1 - \frac{1}{1 - \sigma_{\{i,j\}}} \frac{\lambda_i}{1 + \lambda_i} \right) = 1.$$

The coefficient of α_j is negative since $(1 - \sigma_{\{i\}})/(1 - \sigma_{\{i,j\}}) > 1$. For non-redundancy, X must be in $\mathcal{T}_I$, in particular, $\alpha_j \geq 0, j \in I \setminus \{i\}$. This is a contradiction.

Let $J \subset I$ index the participating hyperplanes $\mathcal{H}^j, j \in J$, and let its complement, $K = I \setminus J$, index the participating hyperplanes $\mathcal{H}_k, k \in K$. As above we may assume that J is non-empty. By definition, V_J is contained in all these hyperplanes.

It remains to show that

$$\bigcap_{j \in J} \mathcal{H}^j \cap \bigcap_{k \in K} \mathcal{H}_k = \{V_J\}.$$

Let X be in the intersection. First, since $X \in \bigcap_{k \in K} \mathcal{H}_k$, we have

$$X = \sum_{j \in J} \mu_j C_j^o, \tag{4.8.25}$$

where (for non-redundancy) we may assume that $\sum_{j \in J} \mu_j \geq 1$ and $\mu_j \geq 0, j \in J$.

Now, fix $i \in J$, so that $X \in \mathcal{H}^i$. We thus have (4.8.22)–(4.8.24). Comparing these with (4.8.25), we see that $\alpha_k = 0$ for $k \in K$, so that in (4.8.22)–(4.8.24) I can be replaced by J. Moreover, comparing coefficients, we obtain

$$\mu_i = \alpha_i + \frac{\lambda_i}{1 + \lambda_i} \sum_{j \in J \setminus \{i\}} \frac{\alpha_j}{1 - \sigma_{\{i,j\}}}$$

and

$$\mu_j = \frac{\alpha_j}{1 - \sigma_{\{i,j\}}} \frac{\lambda_j}{1 + \lambda_j}, \quad j \in J \setminus \{i\}. \tag{4.8.26}$$

Solving for α_i, we also get

$$\alpha_i = \mu_i - \frac{\lambda_i}{1 + \lambda_i} \sum_{j \in J \setminus \{i\}} \frac{1 + \lambda_j}{\lambda_j} \mu_j. \tag{4.8.27}$$

In (4.8.23), expressing the α's in terms of the μ's in the use of (4.8.26) and (4.8.27), after a simple computation, we obtain

$$\mu_i - \sum_{j \in J \setminus \{i\}} \frac{\mu_j}{\lambda_j} = 1.$$

We now vary $i \in J$ and consider this as a system of equations for $\mu_j, j \in J$. We see that

$$\mu_i \left(1 + \frac{1}{\lambda_i}\right) = c,$$

where c is a constant, independent of i. The value of the constant can be determined by substitution:

$$c = \frac{1}{1 - \sigma_J}.$$

We finally arrive at

$$\mu_i = \frac{1}{1 - \sigma_J} \frac{\lambda_i}{1 + \lambda_i}.$$

Thus, we have $X = V_J$. Case (ii) follows.

The proof of the proposition is now complete.

Remark. Given $j \in I$, for $i \in I \setminus \{j\}$, we have $V_{I \setminus \{i\}} \in \mathcal{H}^j$. Since $V_I \in \mathcal{H}^j$, by Lemma 4.8.4, we also have $C_i \in \mathcal{H}^j$. We obtain that $\mathcal{H}^j = \langle V_I, \{C_i\}_{i \in I \setminus \{j\}} \rangle$. Thus, apart from the base $\langle \{C_i\}_{i \in I} \rangle$, the hyperplanes $\mathcal{H}^j$, $j \in I$, are bounding the simplex $[V_I, \{C_i\}_{i \in I}]$ in $\mathcal{X}_I$.

The following geometric picture emerges: $\mathcal{P}_I$ is the intersection of this simplex with the truncated cone $\mathcal{T}_I$.

For the next step, we define the "bulges" noted at the beginning of this section. For $I \subset \mathcal{I}$, we let $\mathcal{B}_I = \mathcal{T}_I \cap C$. For $I = \{i_1, \ldots, i_k\}$, we also write $\mathcal{B}_I = \mathcal{B}(C_{i_1}, \ldots, C_{i_k})$. $\mathcal{B}_I$ is called the "bulge" for the linear slice $C \cap \mathcal{X}_I$ over $[\{C_i^o\}_{i \in I}]$.

Theorem 4.8.10. $\mathcal{B}_I \subset \mathcal{P}_I$.

Proof. As usual we proceed by induction with respect to $|I|$. The theorem holds for $|I| = 2$ by inspection of Example 4.8.6.

We will show that $\mathcal{B}_I$ is on the same side of the hyperplane extension of each face of the covering polytope $\mathcal{P}_I$. Let $\mathcal{H}$ be a hyperplane extension of a face F of $\mathcal{P}_I$. By Proposition 4.8.9, $\mathcal{H} = \mathcal{H}_0$ or $\mathcal{H} = \mathcal{H}_i = \mathcal{X}_{I \setminus \{i\}}$, or $\mathcal{H} = \mathcal{H}^i$ for some $i \in I$. Since $\mathcal{B}_I \subset \mathcal{T}_I$, and the hyperplane extensions of the faces of $\mathcal{T}_I$ are $\mathcal{H}_0$ and $\mathcal{H}_i$, $i \in I$, we may assume that $\mathcal{H} = \mathcal{H}^j$ for some $j \in I$.

It is enough to show that the interior of $\mathcal{B}_I$ is on the *same side of $\mathcal{H}$ as the origin* 0.

Let $X \in \operatorname{int} \mathcal{B}_I$. Let $i \in I \setminus \{j\}$. Then $j \in I \setminus \{i\}$ so that $V_{I \setminus \{i\}} \in F \subset \mathcal{H}$. By Lemma 4.8.4, we also have $C_i \in \mathcal{H}$.

$C_i \notin \mathcal{T}_I$ since $C_i = -\lambda_i C_i^o$. Since $\mathcal{T}_I$ is convex, the line segment $[X, C_i]$ intersects the boundary of $\mathcal{T}_I$ at a unique point $X_i \in \partial \mathcal{T}_I$ (Corollary 1.1.9).

C_i and C_i^o are at opposite sides of the hyperplane $\mathcal{X}_{I \setminus \{i\}}$. Thus C_i and X are also on opposite sides of $\mathcal{X}_{I \setminus \{i\}}$. Hence the line segment $[X, C_i]$ intersects $\mathcal{X}_{I \setminus \{i\}}$ at a unique point $Y_i \in \mathcal{X}_{I \setminus \{i\}}$. Note that, by convexity, $X_i, Y_i \in C$.

Case (i): $X_i = Y_i$. Since $X_i \in \partial \mathcal{T}_I \cap \mathcal{X}_{I \setminus \{i\}}$, we also have $X_i \in \mathcal{B}_{I \setminus \{i\}}$. By the induction hypothesis, $\mathcal{B}_{I \setminus \{i\}} \subset \mathcal{P}_{I \setminus \{i\}}$, so that $X_i \in \mathcal{P}_{I \setminus \{i\}}$. Consider $\mathcal{H} \cap \mathcal{X}_{I \setminus \{i\}}$. This is a hyperplane extension of a face of $\mathcal{P}_{I \setminus \{i\}}$ in $\mathcal{X}_{I \setminus \{i\}}$ and it contains $V_{I \setminus \{i\}}$. We see that X_i and 0 are on the same side of $\mathcal{H} \cap \mathcal{X}_{I \setminus \{i\}}$ in $\mathcal{X}_{I \setminus \{i\}}$. Thus X_i and 0 are on the same side of $\mathcal{H}$. Since $X_i \in [C_i, X]$ and $C_i \in \mathcal{H}$, we obtain that X and 0 are on the same side of $\mathcal{H}$.

Case (ii): $X_i \neq Y_i$. We first claim that $X_i \in F_0$. Indeed, since $X_i \in \partial \mathcal{T}_I$, the only other possibility in this case would be $X_i \in \mathcal{X}_{I \setminus \{k\}}$ for some $k \in I \setminus \{i\}$. Write $X = \sum_{l \in I} \mu_l C_l^o$ with $\sum_{l \in I} \mu_l > 0$ and $\mu_l > 0$, $l \in I$. (Recall that X is in the interior of $\mathcal{B}_I$.) Then, by the definition of X_i, for some $0 < t < 1$, we have

$$X_i = (1 - t)X + tC_i = (1 - t) \sum_{l \in I} \mu_l C_l^o - t\lambda_i C_i^o.$$

If $X_i \in \mathcal{X}_{I \setminus \{k\}}$ then $(1 - t)\mu_k = 0$, a contradiction. The claim follows, and $X_i \in F_0$.

Since C_i^o and 0 are on the same side of $\mathcal{H}$ so is X_i. As before, X and 0 are on the same side of $\mathcal{H}$. The theorem follows.

We now return to the original setting of Theorem 4.8.3 with $\mathcal{C} \in \mathfrak{B}$ and $O \in$ int $\mathcal{C}$. As noted previously, we may assume that O is the origin 0 of $\mathcal{X}$. As usual, we assume (4.8.6) and (4.8.9) so that 0 is a regular point. We take a simplicial minimal configuration $\{C_0, \ldots, C_n\} \in \mathfrak{C}(0)$ with corresponding index set $\mathcal{I} = \{0, \ldots, n\}$. Letting $\lambda_i = \Lambda(C_i, 0)$, $i = 0, \ldots, n$, (4.8.9) rewrites as

$$1 < \sigma = \sum_{i=0}^{n} \frac{1}{\lambda_i + 1} < 1 + \frac{1}{n+1}. \qquad (4.8.28)$$

For $i = 0, \ldots, n$, we define

$$\sigma_i = \sum_{j=0; j \neq i}^{n} \frac{1}{\lambda_j + 1} = \sigma - \frac{1}{\lambda_i + 1}.$$

By (4.8.6) and (4.8.9) we have

$$\sigma_i + \frac{1}{n+1} \leq \sigma_i + \frac{1}{\lambda_i + 1} = \sigma < 1 + \frac{1}{n+1}$$

so that $\sigma_i < 1$. In addition, since $[C_0, \ldots, C_n]$ is an n-simplex with the origin 0 in its interior, $\{C_0, \ldots, \widehat{C}_i, \ldots, C_n\}$ is *linearly independent*. By (4.8.13) and the preceding discussion these were the very assumptions under which the construction of the polytope $\mathcal{P}_I$, $I = \mathcal{I} \setminus \{i\}$, was carried out. We obtain the polytope $\mathcal{P}_i = \mathcal{P}(C_0, \ldots, \widehat{C}_i, \ldots, C_n)$ *containing the bulge* $\mathcal{B}_i = \mathcal{B}(C_0, \ldots, \widehat{C}_i, \ldots, C_n)$ (Theorem 4.8.10), and the vertex $V_i = V(C_0, \ldots, \widehat{C}_i, \ldots, C_n)$. Thus, so far we have ·

$$\mathcal{C} = [C_0^o, \ldots, C_n^o] \cup \bigcup_{i=0}^{n} \mathcal{B}_i \subset [C_0^o, \ldots, C_n^o] \cup \bigcup_{i=0}^{n} \mathcal{P}_i \qquad (4.8.29)$$

Our next result is the following:

Theorem 4.8.11. *We have*

$$\mathcal{C} \subset [V_0, \ldots, V_n]. \qquad (4.8.30)$$

Before the proof we introduce a useful method that compares the geometry of $[V_0, \ldots, V_n]$ with the geometry of the inscribed simplex $\Delta = [C_0, \ldots, C_n]$.

Throughout, we let $0 \leq i \leq n$. Recall that $\lambda_i = \Lambda(C_i, 0)$ with our base point O set at the origin 0. For simplicity, we also set $\lambda_{\Delta,i} = \Lambda_\Delta(C_i, 0)$.

Since $\Delta \subset C$, we have

$$\lambda_i \leq \lambda_{\Delta,i}. \tag{4.8.31}$$

Since Δ is a simplex, by Lemma 4.1.2, we also have

$$\sum_{i=0}^{n} \frac{1}{\lambda_{\Delta,i} + 1} = 1, \quad \text{and} \quad \sum_{i=0}^{n} \frac{1}{\lambda_{\Delta,i} + 1} C_i = 0. \tag{4.8.32}$$

We substitute these into the defining formula (4.8.15) for V_i which, for convenience, we make explicit here:

$$V_i = -\frac{1}{1 - \sigma_i} \sum_{j=0; j \neq i}^{n} \frac{1}{\lambda_j + 1} C_j.$$

Using (4.8.31)–(4.8.32) and (4.8.9), a short computation gives

$$V_i = \frac{1}{\frac{1}{\lambda_i + 1} - \epsilon} \left(\frac{1}{\lambda_i + 1} C_i - \sum_{j=0}^{n} \epsilon_j C_j \right), \tag{4.8.33}$$

where

$$\epsilon_i = \frac{1}{1 + \lambda_i} - \frac{1}{1 + \lambda_{\Delta,i}} \geq 0,$$

and

$$\epsilon = \sum_{i=0}^{n} \epsilon_i = \sigma - 1 > 0.$$

(Note that $1/(\lambda_i + 1) - \epsilon = 1/(\lambda_i + 1) - \sigma + 1 = 1 - \sigma_i > 0$.)

As in (4.8.11) we finally define

$$\tilde{C} = \frac{1}{\epsilon} \sum_{i=0}^{n} \epsilon_i C_i \in \Delta.$$

With this (4.8.33) can be written as

$$V_i - \tilde{C} = \frac{1}{1 - \sigma_i} \frac{1}{\lambda_i + 1}(C_i - \tilde{C}). \tag{4.8.34}$$

From this it is immediately clear that $[V_0, \ldots, V_n]$ is an n-simplex.

Lemma 4.8.12. *We have* $C_i \in [V_0, \ldots, V_m]$.

Proof. Eliminating the denominators in (4.8.34), multiplying by ϵ_i and summing up with respect to $i = 0, \ldots, m$, the definition of $\tilde{C}$ gives

$$\sum_{i=0}^{n} \epsilon_i (1 - \sigma_i)(\lambda_i + 1)(V_i - \tilde{C}) = \sum_{i=0}^{n} \epsilon_i (C_i - \tilde{C}) = 0.$$

The coefficients are non-negative and their sum is positive. This means that $\tilde{C} \in [V_0, \ldots, V_n]$. Finally, (4.8.34) can be written as

$$\left(\frac{1}{\lambda_i + 1} - \epsilon\right) V_i + \epsilon \tilde{C} = \frac{1}{\lambda_i + 1} C_i.$$

Since the coefficients are positive, this means that $C_i \in [V_i, \tilde{C}] \subset [V_0, \ldots, V_n]$. The lemma follows.

Proof of Theorem 4.8.11. Recall that $V_i = V(C_0, \ldots, \widehat{C}_i, \ldots, C_n)$. Applying Lemma 4.8.4 inductively (first to $I = \{0, \ldots, \hat{i}, \ldots, n\}$) and using Lemma 4.8.12, we see that, for $J \subset I$, we have $V_J \in [V_0, \ldots, V_n]$. In the last step of the induction we obtain $C_j^o = V_{\{j\}} \in [V_0, \ldots, V_n]$, and so

$$[C_0^o, \ldots, C_n^o] \subset [V_0, \ldots, V_n].$$

Proposition 4.8.8 also gives

$$\mathcal{P}_i \subset [V_0, \ldots, V_n], \quad i = 0, \ldots, n.$$

Using (4.8.29), we obtain

$$\mathcal{C} \subset [C_0^o, \ldots, C_n^o] \cup \bigcup_{i=0}^{n} \mathcal{P}_i \subset [V_0, \ldots, V_n].$$

The theorem follows.

Proof of Theorem 4.8.3. The coefficient in (4.8.34) can be estimated as

$$\frac{1}{1 - \sigma_i \lambda_i + 1} = 1 + \frac{\sigma - 1}{\frac{1}{\lambda_i + 1} + 1 - \sigma} \leq 1 + \frac{\sigma - 1}{\frac{1}{n+1} + 1 - \sigma} = \frac{1}{1 - (n+1)(\sigma - 1)}.$$
(4.8.35)

By (4.8.10)–(4.8.12), we define

$$\tilde{C}_i = S_{\tilde{r}, \tilde{C}}(C_i) = \tilde{C} + \tilde{r}(C_i - \tilde{C}), \quad i = 0, \ldots, n,$$
(4.8.36)

and

$$\tilde{\Delta} = S_{\tilde{r}, \tilde{C}}(\Delta) = [\tilde{C}_0, \ldots, \tilde{C}_n].$$

By (4.8.34)–(4.8.36), we have

$$[V_0, \ldots, V_n] \subset \tilde{\Delta}.$$

This, combined with Theorem 4.8.11, gives Theorem 4.8.3.

Exercises and Further Problems

1.* Let $C \in \mathcal{B}$ and $O \in \text{int}\, C$. For $k \geq 1$, define

$$\Sigma_k(O) = \sup_{\{C_0, \ldots, C_k\} \in \mathcal{C}_k(O)} \sum_{i=0}^{k} \frac{1}{\Lambda(C_i, O) + 1}, \quad O \in \text{int}\, C.$$

(Compare this with (4.1.1).) Show that $\Sigma_k = (k+1)/2 - \sigma_k$.

2.* Show that, in a minimal n-configuration, there is *at least one* configuration point at which $\Lambda(., O)$ assumes its *global* maximum $\mathfrak{m}(O)$ over ∂C.

3.* Let $\bar{B} \subset \mathbb{R}^n$ be the closed unit ball. Use induction with respect to n to show that

$$\sigma(O) = 1 + (n-1)\frac{1 - |O|}{2}, \quad O \in \text{int}\, \bar{B}.$$

4. Derive the formula in (4.6.3) for the function σ on the half-disk C based on the discussion in Example 4.6.3.

5. Let C be a planar convex polygon and V_0, V_1, V_2, V_3 consecutive vertices of C. Let α_1 and α_2 be interior angles at V_1 and V_2. Assume that $\alpha_1 + \alpha_2 > \pi$. Using affine diameters, show that $\text{int}\, C \cap [V_1, V_2, W] \subset S$, where S is the singular

set and $W = [V_0, V_2] \cap [V_1, V_3]$. Use this to give a complete description of the regular and singular sets in convex quadrilaterals.

6.   Let $\mathcal{B}_n$, $n \geq 2$, be the Birkhoff polytope of doubly stochastic $n \times n$-matrices (Example 2.2.4). Show that

$$\sigma_{\mathcal{B}_n}(E_n) = \frac{(n-1)^2 + 1}{n},$$

where $E_n \in \mathcal{B}_n$ is the $n \times n$-matrix all of whose entries are equal to $1/n$.

7.*   Let $\mathcal{C} \in \mathfrak{B}$ with $\mathfrak{m}^* = n - 1$ and assume that the critical set $\mathcal{C}^*$ consists of a single point: $\mathcal{C}^* = \{O^*\}$. Show that O^* is singular if and only if $\mathcal{C}$ has a codimension 1 simplicial slice across O^*.

8.*   Let $\mathcal{C} \in \mathfrak{B}$ be symmetric with center of symmetry at O. Let $O' \in \operatorname{int}\mathcal{C}$, $O' \neq O$. Let $\langle O, O' \rangle \cap \mathcal{C} = [C, C^o]$ with $O \in [O', C]$. Show that $\Lambda(., O')$ attains its global maximum at C (and its global minimum at C^o). Moreover, for any affine plane $\mathcal{K}$ that contains $\langle O, O' \rangle$, the distortion is increasing along the boundary arcs of $\mathcal{K} \cup \partial \mathcal{C}$ from C^o to C.

9.*   Use the previous problem to give a simple proof of the second statement of Theorem 4.5.2: The interior of a symmetric convex body consists of singular points only.

10.   Let $\mathcal{C} = [\mathcal{C}_0, V] \subset \mathcal{X}$ be a cone with base $\mathcal{C}_0$ and vertex $V \in \mathcal{X} \setminus \mathcal{X}_0$, where $\mathcal{X}_0 = \langle \mathcal{C}_0 \rangle$ is a codimension 1 affine subspace in $\mathcal{X}$. Let $O_0 \in \operatorname{int}\mathcal{C}_0$ and $O_\lambda = (1 - \lambda)O_0 + \lambda V, 0 \leq \lambda < 1$. Show that $O_\lambda \in \mathcal{R}_\mathcal{C}$ implies $O_0 \in \mathcal{R}_{\mathcal{C}_0}$ and, in this case we have $\sigma_\mathcal{C}(O_\lambda) = \lambda + (1 - \lambda)\sigma_{\mathcal{C}_0}(O_0)$. (The converse is false; see Example 4.6.14.)

11.*   Theorem 4.8.10 can be interpreted in terms of the "bulging function" $\beta : F_0 \to \mathbb{R}$ (for a given $I \subset \mathcal{I}$ and applied to the slice $\mathcal{C} \cap \mathcal{X}_I$). It is defined, for $X \in F_0$, as the largest number $\beta(X)$ such that $\beta(X)X \in \partial\mathcal{C}$. Show that $1/\beta$ is a convex function. Define the bulging function $\beta_I : F_0 \to \mathbb{R}$ for the covering polytope $\mathcal{P}_I$ analogously for $X \in F_0$, as the largest number $\beta_I(X)$ with $\beta_I(X)X \in \partial\mathcal{P}_I$. By Theorem 4.8.10, we have $1 \leq \beta \leq \beta_I$. Show that $1/\beta_I$ is convex and *piecewise linear*. Derive the formula

$$\beta_I\left(\frac{1 - \sigma_J}{|J| - \sigma_J}V_J\right) = \frac{|J| - \sigma_J}{1 - \sigma_J}.$$

Conclude that the maximum bulging of the slice $\mathcal{C} \cap \mathcal{X}_I$ over F_0 is

$$\max_{F_0} \beta \leq \max_{F_0} \beta_I = \beta\left(\frac{1 - \sigma_I}{|I| - \sigma_I}V_I\right) = \frac{|I| - \sigma_I}{1 - \sigma_I}.$$

Apply this estimate to the "maximum bulging" of $\mathcal{C}$ over the inscribed *antipodal* simplex $[C_0^o, \ldots, C_n^o] \subset \mathcal{C}$ as follows. Denote by $\beta_i : [C_0^o, \ldots, \widehat{C_i^o}, \ldots, C_m^o] \to \mathbb{R}$ the bulging function of the ith face, and conclude

$$\max_{0\le i\le n} \max_{[C_0^o,\dots,\widehat{C_i^o},\dots,C_n^o]} \beta_i \le \max_{0\le i\le n} \frac{n-\sigma_i}{1-\sigma_i} = 1 + \frac{n-1}{1-\sigma+\min_{0\le i\le n}\frac{1}{1+\lambda_i}}$$

$$\le 1 + \frac{n^2-1}{1-(n+1)(\sigma-1)}.$$

12.* Let $C = \Delta \cup B \subset X$, where $\Delta = [C_0,\dots,C_n]$ is an n-simplex and the "bulging" B is a convex set with $\Delta \cap B = \partial\Delta \cap \partial B = [C_1,\dots,C_n]$ and the rest of the boundary of B is given by the graph of a smooth non-negative function $f : [C_1,\dots,C_n] \to \mathbb{R}$ measured as the distance along rays emanating from C_0 and passing through $[C_1,\dots,C_n]$. Assume that $B \setminus [C_1,\dots,C_n]$ is contained in the interior of the n-simplex $[C_1,\dots,C_n,V]$, where $V = (C_1+\dots+C_n-C_0)/(n-1)$. Show that $\mathcal{R} = \text{int }\Delta$. In addition, letting $O = (1-\lambda)C_0 + \lambda X \in \mathcal{R}, X \in \text{int}[C_1,\dots,C_n]$ and $0 < \lambda < 1$, derive the formula

$$\sigma(O) = 1 + \lambda \frac{f(X)}{|X| + f(X)}.$$

Note that, for non-zero f, this is an example of a non-simplicial convex body with n affinely independent flat cells and nearby regular points. (Compare with Theorem 4.6.1/(I).)

Appendix A
Moduli for Spherical $\mathcal{H}$-Maps

A.1 Spherical $\mathcal{H}$-Maps and Their Moduli

One of the original motivations to introduce the sequence of mean Minkowski measures $\{\sigma_k\}_{k\geq 1}$ to convex geometry comes from a specific classical problem in Riemannian geometry. For a given compact Riemannian manifold M and an eigenvalue λ of the Laplace–Beltrami operator Δ^M on M, one can consider maps with domain M into Euclidean spheres with components being eigenfunctions of Δ^M with eigenvalue λ. Such maps can naturally be parametrized by a compact convex body, called the moduli space. The fundamental problem is to measure how symmetric (or asymmetric) the moduli space is.

In this appendix we explore the connection between convex geometry and Riemannian geometry, construct the moduli space for these maps, and study how far they are from being symmetric. We assume basic knowledge of Riemannian geometry, and some elementary facts from the representation theory of certain compact Lie groups.

Let M be a compact *Riemannian (C^∞-)manifold* and $C^\infty(M)$ the space of smooth functions on M. We endow $C^\infty(M)$ with the L^2-scalar product (with respect to the Riemannian measure defined by the Riemannian metric g). Let V be a finite dimensional *Euclidean* vector space and $f : M \to V$ a smooth map. A *component* of f is given by composition with a linear functional $\alpha \in V^*$: $\alpha \circ f \in C^\infty(M)$. Precomposition by f defines a linear map $V^* \to C^\infty(M)$ onto the *space of components* $V_f = \{\alpha \circ f \,|\, \alpha \in V^*\}$.

We call $f : M \to V$ *full* if the linear span of the image of f is V. (Note that any map $f : M \to V$ can be made full by restricting the range of f to the linear span of its image.) Clearly, $f : M \to V$ is full if and only if the linear map $V^* \to V_f$ above is injective, that is, a linear isomorphism.

Two full smooth maps $f : M \to V$ and $f' : M \to V'$ are called *congruent* if there is a linear isometry $U : V \to V'$ such that $U \circ f = f'$. Congruence is an equivalence relation on the set of smooth full maps.

© Springer International Publishing Switzerland 2015
G. Toth, *Measures of Symmetry for Convex Sets and Stability*,
Universitext, DOI 10.1007/978-3-319-23733-6

Let $\mathcal{H} \subset C^\infty(M)$ be a fixed *finite dimensional* linear subspace. On $\mathcal{H} \subset C^\infty(M)$ we take the *scaled L^2-scalar product*

$$\langle \chi_1, \chi_2 \rangle = h \int_M \chi_1 \cdot \chi_2 \, v_M, \tag{A.1.1}$$

where $h = \dim \mathcal{H}$ and v_M is the Riemann measure associated to the Riemannian metric g on M scaled to $\int_M v_M = \text{vol}(M) = 1$. Whenever convenient, using this scalar product, we will identify $\mathcal{H}$ and its dual $\mathcal{H}^*$. In terms of an orthonormal basis $\{\chi_i\}_{i=1}^h \subset \mathcal{H}$, the identification $\mathcal{H}^* = \mathcal{H}$ is given by $\alpha \mapsto \sum_{i=1}^h \alpha(\chi_i)\chi_i, \alpha \in \mathcal{H}^*$.

A map $f : M \to V$ is called an *$\mathcal{H}$-map* if $V_f \subset \mathcal{H}$.

The archetype of an $\mathcal{H}$-map is the *Dirac delta map* $\delta_{\mathcal{H}} : M \to \mathcal{H}$. With respect to an orthonormal basis $\{\chi_i\}_{i=1}^h \subset \mathcal{H}$ as above, it is defined by $\delta_{\mathcal{H}}(x) = \sum_{i=1}^h \chi_i(x)\chi_i$. Clearly, the definition does not depend on the choice of the orthonormal basis. The Dirac delta map is *maximal* in the sense that $V_{\delta_{\mathcal{H}}} = \mathcal{H}$.

Remark. The Dirac delta map gets its name from the equivalent definition $\delta_{\mathcal{H}} : M \to \mathcal{H}^*$, $\delta_{\mathcal{H}}(x)(\chi) = \chi(x)$, $\chi \in \mathcal{H}$. Using the identification $\mathcal{H}^* = \mathcal{H}$ by the scalar product (A.1.1), this definition is equivalent to the one above since $\delta_{\mathcal{H}}(x) = \sum_{i=1}^h \delta_{\mathcal{H}}(x)(\chi_i)\chi_i = \sum_{i=1}^h \chi_i(x)\chi_i, x \in M$.

Let $f : M \to V$ be a full $\mathcal{H}$-map. We denote by $A : \mathcal{H} \to V$ the *adjoint* of the composition $V^* \to V_f \subset \mathcal{H} = \mathcal{H}^*$ (precomposition by f followed by inclusion). Clearly A is onto. In addition, we have $f = A \circ \delta_{\mathcal{H}}$, in particular, $\dim V \leq \dim \mathcal{H}$. Indeed, for $\alpha \in V^*$ and $x \in M$, we have

$$\alpha(A \circ \delta_{\mathcal{H}}(x)) = \langle \alpha \circ f, \delta_{\mathcal{H}}(x) \rangle = \sum_{i=1}^h \langle \alpha \circ f, \chi_i \rangle \chi_i(x) = (\alpha \circ f)(x).$$

We associate to f the symmetric linear endomorphism $\langle f \rangle = A^* \cdot A - I \in S^2(\mathcal{H})$. This association is well-defined on the *congruence classes* of full $\mathcal{H}$-maps. Indeed, with $f = A \circ \delta_{\mathcal{H}}$ as above, for any linear isometry $U : V \to V'$, we have

$$\langle U \circ f \rangle = (U \cdot A)^* \cdot (U \cdot A) - I = A^* \cdot (U^* \cdot U) \cdot A - I = A^* \cdot A - I = \langle f \rangle.$$

In addition, since $A^* \cdot A$ is automatically *positive semi-definite*, we see that, for any full $\mathcal{H}$-map $f : M \to V$, the associated symmetric endomorphism $\langle f \rangle$ of $\mathcal{H}$ belongs to the set

$$\mathcal{C}(\mathcal{H}) = \{C \in S^2(\mathcal{H}) \,|\, C + I \geq 0\}.$$

As noted in Example 4.1.6, positive semi-definiteness is a closed and convex condition, so that $\mathcal{C}(\mathcal{H})$ is a closed convex subset of $S^2(\mathcal{H})$. Note also that $\mathcal{C}(\mathcal{H})$ has non-empty interior consisting of those symmetric endomorphisms C of $\mathcal{H}$ for which $C + I > 0$. In particular, the Dirac delta map $\delta_{\mathcal{H}}$ corresponds to the origin: $\langle \delta_{\mathcal{H}} \rangle = 0$ as an interior point of $\mathcal{C}(\mathcal{H})$.

Finally, note that for a full $\mathcal{H}$-map $f : M \to V$, we have

$$\text{im}\,((\langle f \rangle + I) = \text{im}\,(A^* \cdot A) = \text{im}\,(A^*) = V_f. \tag{A.1.2}$$

Now let $f_i : M \to V_i$ be full $\mathcal{H}$-maps and $\lambda_i \in (0,1)$, $i = 0, \ldots, d$, with $\sum_{i=0}^{d} \lambda_i = 1$. Then for the $\mathcal{H}$-map $f : M \to V$, $V = \sum_{i=0}^{d} V_i$ given by $f = (\sqrt{\lambda_0} f_0, \ldots, \sqrt{\lambda_d} f_d)$, a simple computation using the definition of the parametrization gives

$$\langle f \rangle = \sum_{i=0}^{d} \lambda_i \langle f_i \rangle \in \mathcal{C}(\mathcal{H}). \tag{A.1.3}$$

(Note that f is not necessarily full.) Using (A.1.2)–(A.1.3), for the respective spaces of components, we obtain

$$V_f = \text{im}\,((\langle f \rangle + I) = \text{im}\left(\sum_{i=0}^{d} \lambda_i \langle f_i \rangle + I\right) = \sum_{i=0}^{d} \text{im}\,((\langle f_i \rangle + I) = \sum_{i=0}^{d} V_{f_i} \subset \mathcal{H}. \tag{A.1.4}$$

(The third equality is because, for any set of *symmetric positive semi-definite* endomorphisms $\{Q_i\}_{i=0}^{d} \subset S^2(\mathcal{H})$, we have $\text{im}(\sum_{i=0}^{d} Q_i) = \sum_{i=0}^{d} \text{im}\, Q_i$.)

We will show shortly that the association $f \mapsto \langle f \rangle$ gives rise to a *one-to-one* correspondence of the set of congruence classes of full $\mathcal{H}$-maps of M onto $\mathcal{C}(\mathcal{H})$, so that $\mathcal{C}(\mathcal{H})$ can be considered as a *parameter space* or *moduli space* for such maps.

This parameter space $\mathcal{C}(\mathcal{H})$ is non-compact, however. We therefore normalize our $\mathcal{H}$-maps, so that the corresponding reduced parameter space or *reduced moduli space* will be the compact slice

$$\mathcal{C}_0(\mathcal{H}) = \{C \in \mathcal{C}(\mathcal{H}) \,|\, \text{trace}\, C = 0\}.$$

Recall that $\mathcal{C}_0(\mathcal{H})$ is a convex body in $S_0^2(\mathcal{H}) = \{C \in S^2(\mathcal{H}) \,|\, \text{trace}\, C = 0\}$ analyzed in detail in Example 4.1.6.

We call an $\mathcal{H}$-map $f : M \to V$ *normalized* if

$$\int_M |f|_V^2 \, v_M = 1,$$

where the norm $|\cdot|_V$ is defined by the given scalar product on the Euclidean vector space V.

We first note that the Dirac delta map $\delta_{\mathcal{H}} : M \to \mathcal{H}$ is normalized. Indeed, in terms of an orthonormal basis $\{\chi_i\}_{i=1}^{h} \subset \mathcal{H}$, we have

$$\int_M |\delta_{\mathcal{H}}|^2 \, v_M = \int_M \left(\sum_{i=1}^{h} \chi_i^2\right) v_M = \sum_{i=1}^{h} \int_M \chi_i^2 \, v_M = \sum_{i=1}^{h} \frac{1}{h} = 1.$$

We now claim that $f : M \to V$ is normalized if and only if trace $\langle f \rangle = 0$. Indeed, letting $f = A \circ \delta_{\mathcal{H}}$ with $A : \mathcal{H} \to V$ as above, we have $\langle f \rangle = A^* \cdot A - I$. In terms of an orthonormal basis $\{\chi_i\}_{i=1}^h \subset \mathcal{H}$, we have

$$\int_M |f|_V^2 \, v_M = \int_M \langle f, f \rangle_V \, v_M = \int_M \langle (A^* \cdot A) \circ \delta_{\mathcal{H}}, \delta_{\mathcal{H}} \rangle \, v_M$$

$$= \sum_{i,j=1}^h \langle (A^* \cdot A) \chi_i, \chi_j \rangle \int_M \chi_i \cdot \chi_j \, v_M = \frac{1}{h} \sum_{i=1}^h \langle (A^* \cdot A) \chi_i, \chi_i \rangle$$

$$= \frac{1}{h} \operatorname{trace} (A^* \cdot A) = \frac{1}{h} \operatorname{trace} \langle f \rangle + 1.$$

The claim follows.

We obtain that, under $f \mapsto \langle f \rangle$, the congruence class of a full normalized $\mathcal{H}$-map corresponds to a traceless symmetric endomorphism of $\mathcal{H}$, an element in $\mathcal{C}_0(\mathcal{H})$.

Proposition A.1.1. *The correspondence $f \mapsto \langle f \rangle$ is one-to-one on the congruence classes of full $\mathcal{H}$-maps onto $\mathcal{C}(\mathcal{H})$. Under this correspondence, the congruence classes of full normalized $\mathcal{H}$-maps correspond to the convex body $\mathcal{C}_0(\mathcal{H}) \subset S_0^2(\mathcal{H})$.*

Proof. To show injectivity, let $f_1 : M \to V_1$ and $f_2 : M \to V_2$ be full $\mathcal{H}$-maps and assume that $\langle f_1 \rangle = \langle f_2 \rangle$. Setting $f_1 = A_1 \circ \delta_{\mathcal{H}}$ and $f_2 = A_2 \circ \delta_{\mathcal{H}}$ with $A_1 : \mathcal{H} \to V_1$ and $A_2 : \mathcal{H} \to V_2$ linear and surjective, we have

$$A_1^* \cdot A_1 = A_2^* \cdot A_2. \tag{A.1.5}$$

In particular, A_1 and A_2 have the same kernel. We may assume that this kernel is trivial since otherwise we would restrict both linear maps to its orthogonal complement. We now apply the polar decomposition to obtain $A_1 = U_1 \cdot Q_1$ and $A_2 = U_2 \cdot Q_2$, where $U_1 : \mathcal{H} \to V_1$ and $U_2 : \mathcal{H} \to V_2$ are linear isometries and Q_1 and Q_2 are symmetric positive definite endomorphisms of $\mathcal{H}$. Substituting these into (A.1.5) and taking the square root of both sides, we obtain $Q_1 = Q_2$. Hence $A_1 = U_1 \cdot U_2^{-1} \cdot A_2$, so that $f_1 = (U_1 \cdot U_2^{-1}) \circ f_2$. Congruence of f_1 and f_2 follows.

To show surjectivity of the parametrization, let $C \in \mathcal{C}(\mathcal{H})$. Since $C + I \geq 0$ we can define $A = (C + I)^{1/2} \in S^2(\mathcal{H})$. Restricting A to its image $V = \operatorname{im} A$, we obtain a surjective linear map $A : \mathcal{H} \to V$ (denoted by the same letter), and a full $\mathcal{H}$-map $f = A \circ \delta_{\mathcal{H}} : M \to V$. Working backwards, we have $\langle f \rangle = A^* \cdot A - I = C$. Surjectivity of the parametrization, and therefore the proposition follows.

From geometric point of view, general $\mathcal{H}$-maps $f : M \to V$ are uninteresting. On the other hand, $\mathcal{H}$-maps with specific geometric properties have long been of great interest in Riemannian geometry. A case of particular importance is when $\mathcal{H} = \mathcal{H}_\lambda$ is the eigenspace of the Laplace–Beltrami operator Δ^M of M (acting on the space of functions $C^\infty(M)$) corresponding to an eigenvalue $\lambda > 0$. In addition, we impose sphericality, that is, we assume that $f : M \to S_V(\subset V)$ maps into the unit sphere S_V of the Euclidean vector space V. With these, we arrive at the concept of λ-*eigenmap*, an $\mathcal{H}_\lambda$-map $f : M \to S_V$.

One of the principal reasons why such maps are important is due to the fact that a λ-eigenmap is a *harmonic map* (of constant energy-density) in the sense of [Eells–Sampson]. (See also [Eells–Lemaire].)

A great number of classical examples include the *Hopf map* Hopf : $S^3 \to S^2$, the (real) *Veronese maps* $\text{Ver}_k : S^2 \to S^{2k}$, $k \geq 2$, and their generalizations.

In addition, imposing conformality on λ-eigenmaps, one arrives at the concept of *spherical minimal immersion*, an isometric minimal immersion $f : M \to S_V$ (up to scaling of the Riemannian metric g of M).

A famous example is the *tetrahedral minimal immersion* Tet : $S^3 \to S^6$. Its name signifies that it defines an isometric minimal *embedding* of the *tetrahedral manifold* (the quotient of S^3 by the binary tetrahedral group) into S^6. Similarly, we have the *octahedral* and *icosahedral minimal immersions* Oct : $S^3 \to S^8$ and Ico : $S^3 \to S^{12}$. (See Problems 5–6.)

With this motivation in mind we now return to the general setting.

An $\mathcal{H}$-map $f : M \to V$ is called *spherical* if it maps M into the unit sphere S_V of V. In this case we write $f : M \to S_V$. To study spherical $\mathcal{H}$-maps, the Dirac delta map $\delta_{\mathcal{H}} : M \to \mathcal{H}$ needs to be spherical. To ensure this, we assume that M is *Riemannian homogeneous*, that is, M is endowed with a transitive action of a compact Lie group G of isometries. As usual, we write $M = G/K$, where $K \subset G$ is a closed subgroup.

The action of G induces a linear action on $C^\infty(M)$ by setting $g \cdot \chi = \chi \circ g^{-1}$, $\chi \in C^\infty(M)$, $g \in G$.

In addition, we also assume that the linear subspace $\mathcal{H} \subset C^\infty(M)$ is G-invariant. Since the scaled L^2-scalar product (A.1.1) is G-invariant, this means that $\mathcal{H}$ is an *orthogonal G-module*, that is, G acts on $\mathcal{H}$ via *orthogonal transformations*.

We claim that under these assumptions the Dirac delta map is spherical. We first show that $\delta_{\mathcal{H}} : M \to \mathcal{H}$ is equivariant with respect to the action of G on M and the orthogonal action on $\mathcal{H}$. Equivariance means that we have

$$g \cdot \delta_{\mathcal{H}}(x) = \delta_{\mathcal{H}}(g \cdot x), \quad g \in G, \ x \in M.$$

Indeed, with respect to an orthonormal basis $\{\chi_i\}_{i=1}^h \subset \mathcal{H}$, for $g \in G$, $x \in M$, we have

$$g \cdot \delta_{\mathcal{H}}(x) = g \cdot \left(\sum_{i=1}^h \chi_i(x) \, \chi_i \right) = \sum_{i=1}^h \chi_i(x) \, (g \cdot \chi_i)$$

$$= \sum_{i=1}^h \chi_i(x) \sum_{j=1}^h g_{ji} \chi_j = \sum_{j=1}^h \chi_j \sum_{i=1}^h (g^{-1})_{ij} \chi_i(x)$$

$$= \sum_{j=1}^h \chi_j \, (g^{-1} \cdot \chi_j)(x) = \sum_{j=1}^h \chi_j \chi_j(g \cdot x) = \delta_{\mathcal{H}}(g \cdot x),$$

where $(g_{ij})_{i,j=1}^h \in O(\mathcal{H})$ is the orthogonal matrix of $g \in G$ acting on $\mathcal{H}$ with respect to the fixed orthonormal basis. Equivariance of $\delta_{\mathcal{H}}$ follows.

Since G acts transitively on M, it follows that $\delta_{\mathcal{H}}$ maps M into a sphere in $\mathcal{H}$ with center at the origin and radius $r > 0$. Now, once again with respect to an orthonormal basis $\{\chi_i\}_{i=1}^h \subset \mathcal{H}$, for $x \in M$, we have

$$r^2 = |\delta_{\mathcal{H}}(x)|^2 = \sum_{i=1}^h \chi_i(x)^2$$

Integrating, and using orthonormality of the basis, we obtain

$$r^2 = \sum_{i=1}^h \int_M \chi_i^2 \, v_M = 1.$$

Sphericality of the Dirac delta map now follows: $\delta_{\mathcal{H}} : M \to S_{\mathcal{H}}$.

We now take an arbitrary full $\mathcal{H}$-map $f : M \to V$ with $f = A \circ \delta_{\mathcal{H}}$ and $A : \mathcal{H} \to V$ linear and onto. For $x \in M$, we calculate

$$
\begin{aligned}
|f(x)|^2 - 1 &= |f(x)|^2 - |\delta_{\mathcal{H}}(x)|^2 \\
&= \langle (A^* \cdot A - I) \circ \delta_{\mathcal{H}}(x), \delta_{\mathcal{H}}(x) \rangle \\
&= \langle \langle f \rangle, \delta_{\mathcal{H}}(x) \odot \delta_{\mathcal{H}}(x) \rangle = 0,
\end{aligned}
$$

where $\odot$ stands for the symmetric tensor product. As in Example 4.1.6, we use here the natural scalar product on $S^2(\mathcal{H})$ given by

$$\langle C_1, C_2 \rangle = \operatorname{trace}(C_1 \cdot C_2), \quad C_1, C_2 \in S^2(\mathcal{H}).$$

This scalar product satisfies

$$\langle C \cdot \chi_1, \chi_2 \rangle = \langle C, \chi_1 \odot \chi_2 \rangle, \quad C \in S^2(\mathcal{H}), \ \chi_1, \chi_2 \in \mathcal{H}.$$

We conclude that a full $\mathcal{H}$-map $f : M \to V$ is spherical if and only if the associated symmetric endomorphism $\langle f \rangle$ belongs to the *linear* subspace

$$\mathcal{E}(\mathcal{H}) = \{\delta_{\mathcal{H}}(x) \odot \delta_{\mathcal{H}}(x) \,|\, x \in M\}^{\perp} \subset S^2(\mathcal{H}),$$

where $\perp$ stands for the orthogonal complement.

Let $C \in \mathcal{E}(\mathcal{H})$. Using an orthonormal basis $\{\chi_i\}_{i=1}^h \subset \mathcal{H}$ and integrating, we have

$$
\begin{aligned}
\int_M \langle C, \delta_{\mathcal{H}}(x) \odot \delta_{\mathcal{H}}(x) \rangle v_M &= \int_M \langle C \circ \delta_{\mathcal{H}}(x), \delta_{\mathcal{H}}(x) \rangle v_M \\
&= \sum_{i,j=1}^h \langle C\chi_i, \chi_j \rangle \int_M \chi_i \cdot \chi_j \, v_M \\
&= \frac{1}{h} \sum_{i=1}^h \langle C\chi_i, \chi_i \rangle = \frac{1}{h} \operatorname{trace} C = 0.
\end{aligned}
$$

Thus, $\mathcal{E}(\mathcal{H}) \subset S_0^2(\mathcal{H})$. We now define

$$\mathcal{L}(\mathcal{H}) = \mathcal{C}(\mathcal{H}) \cap \mathcal{E}(\mathcal{H}) = \mathcal{C}_0(\mathcal{H}) \cap \mathcal{E}(\mathcal{H}) = \{C \in \mathcal{E}(\mathcal{H}) \mid C + I \geq 0\}.$$

In particular, since $\mathcal{C}_0(\mathcal{H})$ is compact, so is $\mathcal{L}(\mathcal{H})$. We obtain that $\mathcal{L}(\mathcal{H})$ is a convex body in $\mathcal{E}(\mathcal{H})$. Proposition A.1.1 gives the following;

Corollary A.1.2. *The correspondence $f \mapsto \langle f \rangle$ is one-to-one on the congruence classes of full spherical $\mathcal{H}$-maps onto $\mathcal{L}(\mathcal{H})$.*

The convex body $\mathcal{L}(\mathcal{H})$ is called the *moduli space* for spherical $\mathcal{H}$-maps. As before, the Dirac delta map $\delta_{\mathcal{H}}$ corresponds to the origin of $\mathcal{L}(\mathcal{H})$. The interior of $\mathcal{L}(\mathcal{H})$ consists of those symmetric endomorphisms $C \in \mathcal{E}(\mathcal{H})$ for which $C + I > 0$.

Let $f : M \to S_V$ be a full spherical $\mathcal{H}$-map. By definition, the corresponding parameter point is $\langle f \rangle = A^* \cdot A - I \in \mathcal{C}(\mathcal{H})$, where $f = A \circ \delta_{\mathcal{H}}$ and $A : \mathcal{H} \to V$ linear and onto. By (A.1.2), we have

$$\dim V = \dim V_f = \dim \operatorname{im}(\langle f \rangle + I) \leq \dim \mathcal{H} = h.$$

Equality holds if and only if $\langle f \rangle + I > 0$. By the above, this happens if and only if $\langle f \rangle$ is in the interior of $\mathcal{L}(\mathcal{H})$. We obtain that full $\mathcal{H}$-eigenmaps $f : M \to S_V$ that correspond to boundary points in $\partial \mathcal{L}(\mathcal{H})$ are exactly those for which $\dim V < \dim \mathcal{H}$. These $\mathcal{H}$-maps are said to be of *boundary type*.

Beyond the fact that it is convex, the geometry of the moduli $\mathcal{L}(\mathcal{H})$, even for the simplest Riemannian homogeneous spaces, is very complex. (For example, its dimension is known for compact rank one symmetric spaces $M = G/K$ only; see below.) As the simplest measure of symmetry, recall from Example 4.1.6 that the distortion $\Lambda(C, 0)$ at a boundary point $C \in \partial \mathcal{C}(\mathcal{H})$ is the *maximal eigenvalue* of C as a symmetric endomorphism of $\mathcal{H}$. Thus, $\mathfrak{m}(0) = \max_{\partial \mathcal{L}(\mathcal{H})} \Lambda(., 0)$ is the *maximal eigenvalue* of the symmetric endomorphisms corresponding to all spherical $\mathcal{H}$-maps of boundary type.

In this appendix we will study the sequence $\{\sigma_{\mathcal{L}(\mathcal{H}),k}(0)\}_{k \geq 1}$ (at the *origin* 0 which corresponds to the Dirac delta map $\delta_{\mathcal{H}}$).

We begin with the following elementary fact:

Proposition A.1.3. *Let $f : M \to S_V$ be a full spherical $\mathcal{H}$-map of boundary type. Then we have*

$$\frac{h}{\dim V} \leq \Lambda(\langle f \rangle, 0) + 1 \leq \frac{h}{\nu(\langle f \rangle)}, \tag{A.1.6}$$

where $\nu(\langle f \rangle)$ is the multiplicity of the maximal eigenvalue $\Lambda(\langle f \rangle, 0)$ of $\langle f \rangle$. Equality holds (in both places) if and only if $f : M \to S_V$ has L^2-orthogonal components of the same norm with respect to an orthonormal basis in V.

Proof. We let $C = \langle f \rangle = A^* \cdot A - I \in S^2(\mathcal{H})$, where $f = A \circ \delta_{\mathcal{H}}$ with $A : \mathcal{H} \to V$ linear and onto, and, for brevity, we set $\nu = \nu(C)$ and $\dim V = n$. As noted in

Example 4.1.6, the eigenvalues of C are contained in $[-1, h-1]$. Since $A : \mathcal{H} \to V$ is onto, we have $\text{rank}\,(C + I) = \text{rank}\,(A^* \cdot A) = \text{rank}\,A = \dim V = n$. Hence the multiplicity of the minimal eigenvalue -1 of C is equal to $h - n$. Thus, for the multiplicity ν of the maximal eigenvalue $\Lambda(C, 0)$, we must have $\nu \leq n$. Let $\lambda_1, \ldots, \lambda_{n-\nu}$ denote the non-minimal and non-maximal eigenvalues. With these, the condition that C is traceless can be written as

$$\nu\Lambda(C, 0) + \sum_{i=1}^{n-\nu} \lambda_i = h - n.$$

Now, using $-1 < \lambda_i < \Lambda(C, 0)$ in this equation, we obtain the two estimates in (A.1.6).

Finally, observe that $\nu(\langle f \rangle) = \dim V$ if and only if $f : M \to S_V$ has L^2-orthogonal components of the same norm with respect to an orthonormal basis in V.

A.2   The G-Structure of the Moduli

As in the previous section, we let $M = G/K$ be a compact Riemannian homogeneous space, with a compact Lie group G of isometries of M, $\mathcal{H} \subset C^\infty(M)$ a finite dimensional orthogonal G-module endowed with the scaled L^2-scalar product (A.1.1), and $\mathcal{L}(\mathcal{H}) \subset \mathcal{E}(\mathcal{H})(\subset S_0^2(\mathcal{H}))$ the moduli space parametrizing the full spherical $\mathcal{H}$-maps.

The action of G on $\mathcal{H}$ induces an action on $S^2(\mathcal{H})$ by conjugation $g \cdot C = g \circ C \circ g^{-1}$, $C \in S^2(\mathcal{H})$, $g \in G$. As easy computation shows, all the spaces $\mathcal{C}(\mathcal{H}) \subset S^2(\mathcal{H})$, $\mathcal{C}_0(\mathcal{H}) \subset S_0^2(\mathcal{H})$, and $\mathcal{L}(\mathcal{H}) \subset \mathcal{E}(\mathcal{H})$ are G-invariant, and the parametrization $f \to \langle f \rangle$ with $f : M \to V$ running on the respective sets of full $\mathcal{H}$-maps, is equivariant in the sense that $g \cdot \langle f \rangle = \langle f \circ g^{-1} \rangle$.

Proposition A.2.1. *Assume that $\mathcal{H}$ is irreducible as a G-module. Then G acts on $S_0^2(\mathcal{H})$ with no non-zero fixed points.*

PROOF. Let $C \in S_0^2(\mathcal{H})$. Since C is symmetric, it is diagonalizable, and $\mathcal{H}$ splits into an orthogonal sum of the eigenspaces of C.

Assume now that $C \in S_0^2(\mathcal{H})$ is G-fixed: $g \cdot C = C \cdot g$, for all $g \in G$. This implies that every eigenspace of C is G-invariant. By irreducibility, there can only be one eigenspace. Hence C is a constant multiple of the identity, thereby orthogonal to $S_0^2(\mathcal{H})$. We obtain $C = 0$.

The following proposition is the cornerstone of our calculations. We state this result in a more general setting as follows:

Proposition A.2.2. *Let $\mathcal{X}$ be an orthogonal G-module, and assume that G acts on $\mathcal{X}$ with no non-zero fixed points. If $C \subset \mathcal{X}$ is a G-invariant convex body with $0 \in \text{int}\,C$ then, for the mean Minkowski measure σ_C, we have*

$$\sigma_C(0) = \frac{\dim C + 1}{\mathfrak{m}_C(0) + 1}, \tag{A.2.1}$$

where $\mathfrak{m}_C$ is the maximal distortion.

Proof. Assume that the distortion function $\Lambda(.,0)$ attains its global maximum $\mathfrak{m}(0)$ at $C \in \partial \mathcal{C}$. Consider the convex hull $[G(C)] \subset \mathcal{C}$ of the orbit $G(C) \subset \partial \mathcal{C}$ passing through C. This convex hull is a G-invariant compact convex set. It contains its centroid which must be fixed by G. Since $\mathcal{X}$ has no non-zero G-fixed points, this centroid must be the origin 0. By Carathéodory's theorem (Section 1.3), there exists $\{C_0, \ldots, C_n\} \subset G(C)$, $n = \dim \mathcal{X}$, such that $\sum_{i=0}^{n} \lambda_i C_i = 0$, $\sum_{i=0}^{n} \lambda_i = 1$, $\{\lambda_0, \ldots, \lambda_n\} \subset [0,1]$. What we just concluded means that $\{C_0, \ldots, C_n\} \in \mathfrak{C}(0)$ is an n-configuration with respect to 0. Therefore, we have

$$\sigma(0) \leq \sum_{i=0}^{n} \frac{1}{\Lambda(C_i,0)+1} = \frac{n+1}{\mathfrak{m}(0)+1}.$$

On the other hand, by (4.1.8), the opposite inequality also holds. The proposition follows.

Remark. Note that connectedness of the Lie group G is not assumed. In particular, Proposition A.2.2 can be used to calculate the mean Minkowski measure $\sigma(O)$ for polytopes (with respect to the centroid O) possessing sufficiently large (finite) symmetry groups. Specific examples include the *regular* solids in any dimension, and the Birkhoff polytope $\mathcal{B}_n$ (Example 2.2.4).

Returning to our specific setting, we now apply Propositions A.2.1–A.2.2 to the moduli $\mathcal{L}(\mathcal{H}) \subset \mathcal{E}(\mathcal{H})$. Assuming *irreducibility* of $\mathcal{H}$, we obtain

$$\sigma_{\mathcal{L}(\mathcal{H})}(0) = \frac{\dim \mathcal{L}(\mathcal{H}) + 1}{\mathfrak{m}_{\mathcal{L}(\mathcal{H})}(0) + 1} \tag{A.2.2}$$

In addition, Proposition A.1.3 gives the following:

Proposition A.2.3. *Let $M = G/K$ be a compact Riemannian homogeneous space and $\mathcal{H} \subset C^\infty(M)$, $\dim \mathcal{H} = h$, a G-invariant irreducible linear subspace. Then we have*

$$\frac{\dim \mathcal{L}(\mathcal{H}) + 1}{h} \leq \sigma_{\mathcal{L}(\mathcal{H})}(0) \leq \frac{\dim \mathcal{L}(\mathcal{H}) + 1}{h} \dim V_{\min}, \tag{A.2.3}$$

where $f : M \to S_{V_{\min}}$ is a spherical $\mathcal{H}$-map with minimum range dimension. Equality holds in the upper estimate if and only if

$$\mathfrak{m}_{\mathcal{L}(\mathcal{H})}(0) + 1 = \frac{h}{\dim V_{\min}}.$$

In this case $f : M \to S_{V_{\min}}$ has L^2-orthogonal components with the same norm.

Proof. Proposition A.1.3 applied to a minimal range spherical $\mathcal{H}$-map $f : M \to S_{V_{\min}}$ gives

$$\frac{h}{\dim V_{\min}} \leq \Lambda(\langle f \rangle, 0) + 1 \leq \mathfrak{m}_{\mathcal{L}(\mathcal{H})}(0) + 1.$$

By (A.2.2), the upper estimate and the last statement of the proposition follow immediately.

The lower estimate holds because $\Lambda(C, 0)$ is the maximal eigenvalue (in our case) of $C \in \mathcal{L}(\mathcal{H})$ and, by Example 4.1.6, all eigenvalues are contained in the interval $[-1, h-1]$.

Remark. Although there are many known examples of spherical $\mathcal{H}$-maps which (up to scaling) have L^2-orthonormal components, the full classification of such maps, even in the simplest settings, is an unsolved problem. In addition, finding the lowest rangle dimension for certain classes of spherical (e.g., conformal) $\mathcal{H}$-maps is and old and difficult problem.

We now specialize $M = G/K$ to be a *compact rank one symmetric space.* (For a thorough account, see [Besse].) It is a classical fact that M is then the Euclidean sphere $S^m \subset \mathbb{R}^{m+1}$, or one of the real, complex, or quaternionic projective spaces $\mathbb{RP}^m$, $\mathbb{CP}^m$, $\mathbb{HP}^m$, or the (16-dimensional) Cayley projective plane $\mathbb{CAP}^2$.

We let $\mathcal{H} = \mathcal{H}_\lambda \subset C^\infty(M)$ be the eigenspace of the Laplace–Beltrami operator $\triangle^M$ corresponding to an eigenvalue $\lambda > 0$. (The trivial case $\lambda = 0$ is excluded as it corresponds to the one-dimensional eigenspace of constant functions.) It is also a classical fact that $\mathcal{H}_\lambda$ is an *irreducible* G-module [Helgason 1, Helgason 2, Helgason 3]. In fact, there are several explicit geometric representations of $\mathcal{H}_\lambda$; in particular, $h_\lambda = \dim \mathcal{H}_\lambda$ is known.

A spherical $\mathcal{H}_\lambda$-map $f : M \to S_V$ is called a λ-*eigenmap*. By Corollary A.1.2, the moduli space $\mathcal{L}_\lambda = \mathcal{L}(\mathcal{H}_\lambda)$ parametrizes the congruence classes of full λ-eigenmaps of M into various Euclidean spheres. For simplicity, we denote $\mathcal{E}_\lambda = \mathcal{E}(\mathcal{H}_\lambda)$, $\delta_\lambda = \delta_{\mathcal{H}_\lambda} : M \to S_{\mathcal{H}_\lambda}$, etc. Note that, in this case, the Dirac delta map is also known as the *standard eigenmap.*

For a compact rank one symmetric space $M = G/K$, the G-module structure of the *quotient* $S_0^2(\mathcal{H}_\lambda)/\mathcal{E}_\lambda$; in particular, $\dim \mathcal{E}_\lambda$ is known [Toth 3]. More precisely, the finite sums of products $\mathcal{H}_\lambda \cdot \mathcal{H}_\lambda$ of functions in $\mathcal{H}_\lambda$ is a G-submodule of $S^2(\mathcal{H}_\lambda)$, and

$$\mathcal{E}_\lambda = S^2(\mathcal{H}_\lambda)/\mathcal{H}_\lambda \cdot \mathcal{H}_\lambda. \tag{A.2.4}$$

In addition, if $\{\lambda_k\}_{k \geq 0}$ denotes the sequence of eigenvalues in increasing order, we have

$$\mathcal{H}_{\lambda_k} \cdot \mathcal{H}_{\lambda_k} = \begin{cases} \sum_{i=0}^k \mathcal{H}_{\lambda_{2i}} & \text{if } M = S^m \\ \sum_{i=0}^{2k} \mathcal{H}_{\lambda_i} & \text{if } M = \mathbb{RP}^m, \ \mathbb{CP}^m, \ \mathbb{HP}^m, \ \mathbb{CAP}^2 \end{cases} \tag{A.2.5}$$

As $h_\lambda = \dim \mathcal{H}_\lambda$ is known, combining (A.2.4)–(A.2.5), $\dim \mathcal{E}_\lambda = \dim \mathcal{L}_\lambda$ can be calculated. In summary, Propositions A.2.1–A.2.3 now give

$$\frac{\dim \mathcal{L}_\lambda + 1}{h_\lambda} \leq \sigma_{\mathcal{L}_\lambda}(0) = \frac{\dim \mathcal{L}_\lambda + 1}{\mathfrak{m}_{\mathcal{L}_\lambda}(0) + 1} \leq \frac{\dim \mathcal{L}_\lambda + 1}{h_\lambda} \dim V_{\min}, \tag{A.2.6}$$

where $f : M \to S_{V_{\min}}$ is a minimum range dimensional λ-eigenmap.

The most studied and explicit case is the Euclidean sphere $M = S^m \subset \mathbb{R}^{m+1}$ with $G = SO(m+1)$ and $K = SO(m)$. Calculating the kernel of the Euclidean Laplacian acting on k-homogeneous polynomials in $(m+1)$-variables, and comparing it with the spherical Laplacian on S^m, we obtain that the kth eigenvalue is $\lambda_k = k(k+m-1)$. In addition, the eigenfunctions corresponding to λ_k are the restrictions of *harmonic* k-homogeneous polynomials in $(m+1)$-variables to $S^m \subset \mathbb{R}^{m+1}$. (For details, see [Berger–Gauduchon–Mazet] or [Toth 4].) These are classically known as *spherical harmonics of order k*. This computation also shows that

$$h_{\lambda_k} = \dim \mathcal{H}_{\lambda_k} = \binom{k+m}{m} - \binom{k+m-2}{m}. \tag{A.2.7}$$

Now, (A.2.4), (A.2.5), and (A.2.7) give

$$\dim \mathcal{L}_{\lambda_k} = \dim \mathcal{E}_{\lambda_k} = \dim S^2(\mathcal{H}_{\lambda_k}) - \dim(\mathcal{H}_{\lambda_k} \cdot \mathcal{H}_{\lambda_k})$$

$$= \binom{h_{\lambda_k}+1}{2} - \sum_{i=0}^{2k} h_{\lambda_{2i}}$$

$$= \binom{\binom{k+m}{m} - \binom{k+m-2}{m} + 1}{2} - \binom{2k+m}{m}. \tag{A.2.8}$$

This shows, in particular, that the moduli space $\mathcal{L}_{\lambda_k}$ parametrizing λ_k-eigenmaps $f : S^m \to S_V$ is non-trivial if and only if $m \geq 3$ and $k \geq 2$. Note that triviality of the moduli for $m = 2$ is due to [Calabi], and it can be paraphrased by saying that a full λ_k-eigenmap $f : S^2 \to S_V$, $k \geq 1$, must be congruent to the Veronese map $\mathrm{Ver}_k : S^2 \to S^{2k}$ (with L^2-orthonormal components).

For $M = S^m$, in view of (A.2.6) and (A.2.8), to calculate $\sigma_{\mathcal{L}_{\lambda_k}}(0)$, $m \geq 3$ and $k \geq 2$, one needs to know the maximal distortion $\mathfrak{m}_{\mathcal{L}_{\lambda_k}}(0)$ for λ_k-eigenmaps. This is a difficult and largely unsolved problem [Escher–Weingart, Toth 4]. To obtain an upper bound for $\sigma_{\mathcal{L}_{\lambda_k}}(0)$, again by (A.2.6), one needs to know the minimal range dimension of such maps. This is the *DoCarmo Problem*. In general, to give bounds on the minimum range dimension is (once again) an old and difficult problem [DoCarmo–Wallach] (Remark 1.6) and [Moore, Toth 2, Toth–Ziller, Weingart, DeTurck–Ziller 1, DeTurck–Ziller 2, DeTurck–Ziller 3].

A.3 $SU(2)$-Equivariant Moduli

In the rest of this appendix we will treat the first non-trivial domain S^3.

Identifying $\mathbb{R}^4$ with $\mathbb{C}^2$ in the usual way makes the Lie group $SU(2)$ a (normal) subgroup of $SO(4)$. The (real) orthogonal transformation $(z, w) \mapsto (z, \bar{w})$, $z, w \in \mathbb{C}$, of $\mathbb{C}^2$ conjugates $SU(2)$ into another copy $SU(2)' \subset SO(4)$, and we have the splitting

$$SO(4) = SU(2) \cdot SU(2)' \quad \text{with} \quad SU(2) \cap SU(2)' = \{\pm I\}.$$

Since $\mathcal{L}_{\lambda_k} \subset \mathcal{E}_{\lambda_k}$ is a convex body, taking the fixed points sets by $SU(2)$, we see that $\mathcal{L}_{\lambda_k}^{SU(2)} = \mathcal{L}_{\lambda_k} \cap \mathcal{E}_{\lambda_k}^{SU(2)}$ is a convex body in the linear subspace $\mathcal{E}_{\lambda_k}^{SU(2)} \subset \mathcal{E}_{\lambda_k}$. Due to the splitting $SO(4) = SU(2) \cdot SU(2)'$, this linear subspace $\mathcal{E}_{\lambda_k}^{SU(2)}$ is actually an $SU(2)'$-module. Similarly, $\mathcal{L}_{\lambda_k}^{SU(2)'}$ is cut out from $\mathcal{L}_{\lambda_k}$ by the $SU(2)$-module $\mathcal{E}_{\lambda_k}^{SU(2)'}$. Moreover, by restriction, $\mathcal{E}_{\lambda_k}^{SU(2)}$ and $\mathcal{E}_{\lambda_k}^{SU(2)'}$ are orthogonal in $\mathcal{E}_{\lambda_k}$.

Finally, since the parametrization is equivariant, $\mathcal{L}_{\lambda_k}^{SU(2)}$ and $\mathcal{L}_{\lambda_k}^{SU(2)'}$ parametrize $SU(2)$- and $SU(2)'$-*equivariant* eigenmaps. Because of this they are called *equivariant moduli*. Since $SU(2)'$ is a conjugate of $SU(2)$, the module structures on the respective equivariant moduli are isomorphic via this conjugation.

The module structures on $\mathcal{E}_{\lambda_k}^{SU(2)}$ and $\mathcal{E}_{\lambda_k}^{SU(2)'}$ are known, in particular, we have [Toth–Ziller, Toth 4]

$$\dim \mathcal{L}_{\lambda_k}^{SU(2)} = \dim \mathcal{E}_{\lambda_k}^{SU(2)} = \dim \mathcal{L}_{\lambda_k}^{SU(2)'} = \dim \mathcal{E}_{\lambda_k}^{SU(2)'} = \left[\frac{k}{2}\right]\left(2\left[\frac{k}{2}\right] + 3\right).$$
$$(A.3.1)$$

Example A.3.1. For $k = 2$, a quick dimension computation in the use of (A.2.8) and (A.3.1) gives

$$\mathcal{E}_{\lambda_2} = \mathcal{E}_{\lambda_2}^{SU(2)} \oplus \mathcal{E}_{\lambda_2}^{SU(2)'}.$$

The corresponding first non-trivial (ten-dimensional) moduli $\mathcal{L}_{\lambda_2}$ is particularly simple as it is the convex hull of the five-dimensional slices $\mathcal{L}_{\lambda_2}^{SU(2)}$ and $\mathcal{L}_{\lambda_2}^{SU(2)'}$.

In fact, $\mathcal{L}_{\lambda_2}^{SU(2)}$ is the *convex hull* of the $SU(2)'$-orbit of the parameter point $\langle \text{Hopf} \rangle$ corresponding to the Hopf map Hopf $: S^3 \to S^2$. This orbit $SU(2)'(\langle \text{Hopf} \rangle)$, in turn, is the real projective plane $\mathbb{RP}^2$ embedded into a copy of the 4-sphere in $\mathcal{E}_{\lambda_2}^{SU(2)}$ as a Veronese surface, the image of the Veronese map $\text{Ver}_2 : S^2 \to S^4$. It now follows that the extremal set $\mathcal{L}_{\lambda_2}^{\wedge}$ consists of this orbit and its isomorphic copy $SU(2)(\langle \text{Hopf}' \rangle)$ in $\mathcal{L}_{\lambda_2}^{SU(2)'}$.

Since maximal distortion occurs at an extremal point (Corollary 2.2.3), we have $\Lambda_{\mathcal{L}_2}(\langle \text{Hopf} \rangle) = \mathfrak{m}_{\mathcal{L}_2}(0) = 2$, where the value 2 will be calculated in the theorem below. (Note that the Hopf map also has the lowest range dimension, even for topological reasons.)

By (A.2.6)–(A.2.7), we obtain

$$\sigma_{\mathcal{L}_{\lambda_2}}(0) = \frac{\dim \mathcal{L}_{\lambda_2} + 1}{\mathfrak{m}_{\lambda_2}(0) + 1} = \frac{11}{3}.$$

In addition, its explicit description shows that $\mathcal{L}_{\lambda_2}$ (actually, $\mathcal{L}_{\lambda_2}^{SU(2)}$) has a triangular slice across 0. ($\langle \text{Hopf} \rangle$ can be chosen as one of the vertices of the triangle. Its antipodal is the parameter point $\langle \text{Ver}^{\mathbb{C}} \rangle$, where $\text{Ver}^{\mathbb{C}} : S^3 \to S^5$, $\text{Ver}^{\mathbb{C}}(z, w) = (z^2, \sqrt{2}zw, w^2)$, $z, w \in S^3 \subset \mathbb{C}^2$, is the complex Veronese map. The latter is the center of a disk on the boundary of the moduli, and the boundary circle of the disk is on the orbit $SU(2)'(\langle\text{Hopf}\rangle)$.) Thus, equality holds in (4.1.11) of Theorem 4.1.5, and we obtain

$$\sigma_{\mathcal{L}_{\lambda_2,j}}(0) = \frac{j+1}{3}, \quad j \geq 2.$$

For the equivariant moduli, we have the following:

Theorem A.3.2. *For $k \geq 2$, we have*

$$\max_{\partial\mathcal{L}_{\lambda_k}^{SU(2)}} \Lambda(.,0) = \mathfrak{m}_{\mathcal{L}_{\lambda_k}^{SU(2)}}(0) = \begin{cases} k & \text{if } k \text{ is even} \\ \frac{k-1}{2} & \text{if } k \text{ is odd}. \end{cases} \tag{A.3.2}$$

The dimension d_{λ_k} of the largest simplicial slice of $\mathcal{L}_{\lambda_k}^{SU(2)}$ across 0 is equal to this maximal distortion, and we have

$$\sigma_{\mathcal{L}_{\lambda_k}^{SU(2)},j}(0) = \begin{cases} 1 & \text{if } j \leq d_{\lambda_k} \\ \frac{j+1}{d_{\lambda_k}+1} & \text{if } j > d_{\lambda_k}. \end{cases} \tag{A.3.3}$$

In particular, we have

$$\sigma_{\mathcal{L}_{\lambda_k}^{SU(2)}}(0) = \begin{cases} \frac{k+2}{2} & \text{if } k \text{ is even} \\ k & \text{if } k \text{ is odd}. \end{cases} \tag{A.3.4}$$

Remark. In the light of Theorem 4.1.1, it is instructive to compare (A.3.1) and (A.3.4). We see that $\sigma_{\mathcal{L}_{\lambda_k}^{SU(2)}}(0) = O(k)$ whereas $\dim \mathcal{L}_{\lambda_k}^{SU(2)} = O(k^2)$ as $k \to \infty$. This means that, for large k, the equivariant moduli $C_{\lambda_k}^{SU(2)}$ is far from being symmetric.

Before the proof, we need to recall a few facts from the representation theory of the group $SU(2)$. (See [Fulton–Harris, Börner, Knapp, Vilenkin, Vilenkin–Klimyk, Weingart].)

The irreducible *complex* $SU(2)$-modules are parametrized by their dimension, and they can be realized as submodules appearing in the (multiplicity one) decomposition of the $SU(2)$-module of complex homogeneous polynomials $\mathbb{C}[z, w]$ in two variables.

For $k \geq 0$, the kth $SU(2)$-submodule W_k, $\dim_{\mathbb{C}} W_k = k + 1$, comprises the homogeneous polynomials of degree k. With respect to the L^2-scalar product (suitably scaled), the standard orthonormal basis for W_k is

$$\left\{ \frac{z^{k-j}w^j}{\sqrt{(k-j)!j!}} \right\}_{j=0}^{k}.$$

For k odd, W_k is irreducible *as a real $SU(2)$-module.* For k even, the fixed point set R_k of the complex anti-linear self map

$$z^j w^{k-j} \mapsto (-1)^j \bar{z}^{k-j} \bar{w}^j, \quad j = 0, \ldots, k,$$

of W_k is an irreducible *real* submodule with $W_k = R_k \otimes_{\mathbb{R}} \mathbb{C}$.

Conforming with the splitting $SO(4) = SU(2) \cdot SU(2)'$, the $SO(4)$-module of real spherical harmonics can be written as

$$\mathcal{H}_{\lambda_k} = R_k \otimes R_k' \ (k \text{ even}) \text{ and } \mathcal{H}_{\lambda_k} = W_k \otimes W_k' \ (k \text{ odd}).$$

(Here if V is an $SU(2)$-module then V' denotes the $SU(2)'$-module obtained from V by conjugating $SU(2)'$ to $SU(2)$ as above.) Restriction to $SU(2)$ gives

$$\mathcal{H}_{\lambda_k}|_{SU(2)} = (k+1)R_k \ (k \text{ even}) \text{ and } \mathcal{H}_{\lambda_k}|_{SU(2)} = \frac{k+1}{2} W_k \ (k \text{ odd}).$$

Ignoring, for a moment, the second statement of Theorem A.3.2, and letting d_{λ_k} denote the right-hand side of (A.3.2), we see that (in both parities) $\mathcal{H}_{\lambda_k}|_{SU(2)}$ has $d_{\lambda_k} + 1$ irreducible components. Since, by (A.2.7), we have $\dim \mathcal{H}_{\lambda_k} = (k+1)^2$, each irreducible component is of dimension $(k+1)^2/(d_{\lambda_k} + 1)$.

Let $f : S^3 \to S_V$ be any full $SU(2)$-equivariant λ_k-eigenmap. Equivariance means that the parameter point $C = \langle f \rangle \in \mathcal{L}_{\lambda_k}$ is fixed by $SU(2)$, that is, C commutes with the action of $SU(2)$ on $\mathcal{H}_{\lambda_k}$. In particular, each eigenspace is $SU(2)$-invariant, therefore a multiple of R_k (k even) or a multiple of W_k (k odd).

Now, since trace $C = 0$ and the eigenvalues are ≥ -1 (as $C + I \geq 0$), the largest possible eigenvalue is d_k. We obtain $\mathrm{m}_{\mathcal{L}_{\lambda_k}^{SU(2)}}(0) \leq d_k$.

For the reverse inequality, let $V_0 \subset \mathcal{H}_{\lambda_k}|_{SU(2)}$ be an irreducible component. By the above, $V_0 = R_k$ (k even) or $V_0 = W_k$ (k odd). Mimicking the construction of the Dirac delta map, we let $f_0 : S^3 \to V_0$ be defined by $f_0(x) = \sum_{i=1}^{h} \chi_i(x)\chi_i, x \in S^3$, where $\{\chi_i\}_{i=1}^{h} \subset V_0$ is an orthonormal basis with respect to (A.1.1). (Note that $h = \dim R_k = k+1$ (k even) or $h = \dim W_k = 2(k+1)$ (k odd).) The argument used for the construction of the Dirac delta map goes through giving $SU(2)$-equivariance and sphericality of f_0. Thus, we obtain the full $SU(2)$-equivariant $\mathcal{H}_{\lambda_k}$-eigenmap $f_0 : S^3 \to S_{V_0}$. By construction, $\langle f_0 \rangle \in \mathcal{L}_{\lambda_k}^{SU(2)}$ has only two eigenvalues; the maximal eigenvalue with multiplicity $\dim V_0$ and the minimal eigenvalue -1. (See also (A.1.2).) Therefore, the maximal eigenvalue must be d_{λ_k}. Now (A.3.2) follows.

Next we apply Proposition A.2.2 to the $SU(2)'$-module $\mathcal{E}_{\lambda_k}^{SU(2)}$ and its convex body $\mathcal{L}_{\lambda_k}^{SU(2)}$ to obtain

$$\sigma_{\mathcal{L}_{\lambda_k}^{SU(2)}}(0) = \frac{\dim \mathcal{L}_{\lambda_k}^{SU(2)} + 1}{\mathrm{m}_{\mathcal{L}_{\lambda_k}^{SU(2)}}(0) + 1}.$$

Using the dimension formula (A.3.1) and (A.3.2) we arrive at (A.3.4).

It remains to prove (A.3.3). This will follow from the last statement of Theorem 4.1.5 provided that we show that equality holds in (4.1.11) there. For this, we will construct an explicit d_{λ_k}-simplicial slice of $\mathcal{L}_{\lambda_k}^{SU(2)}$.

As in the discussion above, we have $\mathcal{H}_{\lambda_k}|_{SU(2)} = \sum_{j=0}^{d_{\lambda_k}} V_j$ where all V_j, $j = 0, \ldots, d_{\lambda_k}$, are isomorphic as $SU(2)$-modules and $V_j = R_k$ (k even) or $V_j = W_k$ (k odd). Mimicking, once again, the construction of the Dirac delta map, we obtain a full $SU(2)$-equivariant $\mathcal{H}_{\lambda_k}$-eigenmap $f_j : S^3 \to S_{V_j}$ for each $j = 0, \ldots, d_{\lambda_k}$. We now let $f = ((d_{\lambda_k}+1)^{-1/2}f_0, \ldots, (d_{\lambda_k}+1)^{-1/2}f_{d_{\lambda_k}}) : S^3 \to S_{\mathcal{H}_{\lambda_k}}$. The orthonormal bases in V_j used to define f_j, $j = 0, \ldots, d_{\lambda_k}$, unite to an orthonormal basis in $\mathcal{H}_{\lambda_k}$ and we see that f is the Dirac delta map $\delta_{\mathcal{H}_{\lambda_k}}$. Since the latter corresponds to the origin $0 \in \mathcal{L}_{\lambda_k}$, by (A.1.3), we obtain

$$0 \in [\langle f_0 \rangle, \ldots, \langle f_{d_{\lambda_k}} \rangle].$$

The convex hull here is a d_{λ_k}-simplex whose faces are contained in the boundary of $\mathcal{L}_{\lambda_k}^{SU(2)}$. This follows from (A.1.4) since any interior point of the jth face (antipodal to $\langle f_j \rangle$) corresponds to a λ_k-eigenmap whose space of components does not contain the components of f_j. Therefore this λ_k-eigenmap must be of boundary type. We obtain that $[\langle f_0 \rangle, \ldots, \langle f_{d_{\lambda_k}} \rangle]$ is a d_{λ_k}-dimensional simplicial intersection of $\mathcal{L}_{\lambda_k}^{SU(2)}$. The formula in (A.3.3) and with this Theorem A.3.2 follows.

Remark. Using the terminology of Section 4.5, (A.3.3) can be paraphrased saying that the degree of singularity of the origin in $\mathcal{L}_{\lambda_k}^{SU(2)}$ is dim $\mathcal{L}_{\lambda_k}^{SU(2)} - d_{\lambda_k}$.

Exercises and Further Problems

1. Let $\xi = \sum_{j=0}^{k} c_j z^{k-j} w^j \in W_k$ be a non-zero polynomial. Define the *orbit map* $f_\xi : S^3 \to W_k$ through ξ by $f_\xi(g) = g \cdot \xi = \xi \circ g^{-1}$, $g \in SU(2)$. (a) Show that, for k odd, $f_\xi : S^3 \to W_k$ is full, and, for k even and $\xi \in R_k$, $f_\xi : S^3 \to R_k$ is full. (b) Use $SU(2)$-equivariance to verify that, up to scaling, f_ξ is a spherical λ_k-eigenmap, and that the scaling condition is $\sum_{j=0}^{k}(k-j)!j!|c_j|^2 = 1$. (c) Show that $\Lambda(\langle f_\xi \rangle, 0) = d_{\lambda_k}$. (d) Use (c) to conclude that f_ξ has orthogonal components with the same norm.
2. Calculate $f_\xi : S^3 \to S_{R_2}$ in Problem 1 explicitly, and verify that for $\xi(z, w) = izw$, up to congruence, we obtain the Hopf map Hopf : $S^3 \to S^2$:

$$\text{Hopf}(a, b) = (|a|^2 - |b|^2, 2a\bar{b}), \quad (a, b) \in S^3 \subset \mathbb{C}^2.$$

3. Generalize Problem 2 to calculate the orbit map $f_\xi : S^3 \to S_{R_{2k}}$ for $\xi(z, w) = (i^k/k!)z^k w^k$ to show that $f_\xi = \text{Ver}_k \circ \text{Hopf}$, where $\text{Ver}_k : S^2 \to S^{2k}$ is the real Veronese map.
4. Impose the condition of homothety on a λ-eigenmap $f : G/K \to S_V$ as

$$\langle f_*(X), f_*(Y) \rangle = \frac{\lambda}{m}\langle X, Y \rangle, \quad X, Y \in T(M),$$

where $T(M)$ is the tangent bundle of M and f_* is the differential of f. (A λ-eigenmap satisfying this homothety condition is called a spherical minimal immersion. For more information see [DoCarmo–Wallach, Wallach] and also [DeTurck–Ziller 1, DeTurck–Ziller 2, DeTurck–Ziller 3, Toth–Ziller, Toth 4, Weingart].)

Assume that $M = G/K$ is *isotropy irreducible* (that is, the isotropy group K acts on the tangent space $T_{\{K\}}(G/K)$ irreducibly). Show that the Dirac delta map $\delta_\lambda : M \to S_{\mathcal{H}_\lambda}$ is homothetic. Then, show that the moduli $\mathcal{M}_\lambda$ of the congruence classes of full homothetic λ-eigenmaps is the slice of $\mathcal{C}_\lambda$ by the linear subspace

$$\mathcal{F}_\lambda = \{f_*(X) \odot f_*(Y) \mid X, Y \in T(M)\}^{\perp} \subset S^2(\mathcal{H}_\lambda),$$

where tangent vectors in a vector space at any point are identified by ordinary vectors via parallel translation to the origin.

5.* Show that $f_\xi : S^3 \to S_{W_k}$ in Problem 1 is homothetic (as in Problem 4) if and only if the coefficients of ξ satisfy the following

$$\sum_{j=0}^{k}(2j-k)^2(k-j)!j!|c_j|^2 = \frac{k(k+2)}{3},$$

$$\sum_{j=0}^{k-2}(j+2)!(k-j)!c_j\bar{c}_{j+2} = 0,$$

$$\sum_{j=0}^{k-1}(k-2j-1)(j+1)!(k-j)!c_j\bar{c}_{j+1} = 0.$$

(See [Mashimo 1, Mashimo 2].)

6. Choose Klein's tetrahedral form $\Omega \in R_6$, $\Omega(z,w) = zw(z^4 - w^4)$ to verify that $\xi = \Omega/(4\sqrt{15})$ satisfies the conditions of conformality in Problem 5 and obtain a full conformal λ_6-eigenmap $f_{\Omega/(4\sqrt{15})} : S^3 \to S^6$. Show that this eigenmap factors through the binary tetrahedral group $\mathbf{T}^*$ (the lift of the tetrahedral group in $SO(3)$ to the 2-fold cover S^3) and gives an isometric minimal embedding of the tetrahedral manifold $S^3/\mathbf{T}^*$ to S^6. Do a similar analysis for the octahedral form $O \in R_8$, $O(z,w) = z^8 + 14zw + w^8$ to obtain the octahedral minimal immersion of $f_{O/(96\sqrt{21})} : S^3 \to S^8$ and for the icosahedral form $\mathcal{I} \in R_{12}$, $\mathcal{I}(z,w) = z^{11}w + 11z^6w^6 - zw^{11}$ to obtain the icosahedral minimal immersion $f_{\mathcal{I}/(3600\sqrt{11})} : S^3 \to S^{12}$. (For these and similar examples, see [DeTurck–Ziller 1, DeTurck–Ziller 2, DeTurck–Ziller 3], and for a study of the corresponding moduli, see [Toth–Ziller, Toth 4].)

Appendix B
Hints and Solutions for Selected Problems

Chapter 1.

1. See [Eggleston 1, p. 12].
5. See [Schneider 2, 1.8].
7. Let $C \in \mathcal{C}$ be one such point. Assuming $X \neq C$, consider the hyperplane $\mathcal{H} \subset \mathcal{X}$ that contains C and has normal vector $N = X - C$.
10. Let $\mathcal{C} \in \mathfrak{B}$. Define the *normal cone* $\mathcal{K}_{\mathcal{C}}(C)$ of $\mathcal{C}$ at $C \in \partial \mathcal{C}$ as the union of all the rays emanating from C and having direction vector as the outward normal vector of a hyperplane supporting $\mathcal{C}$ at C. (Here outward means that the normal vector points into the respective open half-space disjoint from $\mathcal{C}$.) Realize that C is a smooth point if and only if $\mathcal{K}_{\mathcal{C}}(C)$ is a single ray, and C is a vertex if and only if $\mathcal{K}_{\mathcal{C}}(C)$ has a non-empty interior in $\mathcal{X}$. Finally, show that for $C_1, C_2 \in \partial \mathcal{C}$, $C_1 \neq C_2$, the normal cones $\mathcal{K}_{\mathcal{C}}(C_1)$ and $\mathcal{K}_{\mathcal{C}}(C_2)$ are disjoint.
13. See [Danzer–Grünbaum–Klee, 2.3].
14. See [Danzer–Grünbaum–Klee, 2.4]. The key is to encode all the data in $\mathcal{X} \times \mathbb{R}$. Extend the scalar product of $\mathcal{X}$ to $\mathcal{X} \times \mathbb{R}$ in a natural way: $\langle (X, x), (Y, y) \rangle = \langle X, Y \rangle + xy$, $X, Y \in \mathcal{X}$, $x, y \in \mathbb{R}$. For $C \in \mathcal{X}$, define the two (complementary) open half-spaces $\mathcal{G}_C^{\pm} = \{(X, x) \in \mathcal{X} \times \mathbb{R} \mid \langle (X, x), (C, 1) \rangle \gtrless 0\}$. Assume that each $n + 2$ members of the family $\mathfrak{G} = \{\mathcal{G}_A^+ \mid A \in \mathcal{A}\} \bigcup \{\mathcal{G}_B^- \mid B \in \mathcal{B}\}$ have a non-empty intersection. Apply Helly's theorem (Section 1.4) (dim $\mathcal{X} \times \mathbb{R} = n + 1$) to get a point $(X_0, x_0) \in \bigcap \mathfrak{G}$ and show that the hyperplane $\{X \in \mathcal{X} \mid \langle (X_0, x_0), (X, 1) \rangle = 0\}$ separates $\mathcal{A}$ and $\mathcal{B}$.
15. The following proof of the theorem of Steinitz is due to Valentine and Grünbaum. We may assume that $\mathcal{A}$ is finite. Setting O at the origin, let $\mathcal{D} \subset \mathcal{X}$ be the union of the linear spans of all the subsets of $\mathcal{A}$ with $\leq n - 1$ elements. Since $\mathcal{A}$ is finite, $\mathcal{D}$ is a finite union of proper linear subspaces of $\mathcal{X}$. Since the origin is an interior point of the convex polytope $[\mathcal{A}]$, there exists a chord $[C, C'] \subset [\mathcal{A}]$, $C, C' \in \partial[\mathcal{A}]$, with $0 \in (C, C')$ such that the line extension $\langle C, C' \rangle$ meets $\mathcal{D}$ only at the origin. Let $\mathcal{H}$ and $\mathcal{H}'$ be hyperplanes supporting $[\mathcal{A}]$ at C and C'. By Carathéodory's theorem (Section 1.3) and since $\langle C, C' \rangle$ avoids

© Springer International Publishing Switzerland 2015

G. Toth, *Measures of Symmetry for Convex Sets and Stability*,
Universitext, DOI 10.1007/978-3-319-23733-6

$D \setminus \{0\}$, the point $C \in [\mathcal{A}] \cap \mathcal{H}$ can be expressed as a linear combination with *exactly n* elements $\{C_1, \ldots, C_n\} \subset \mathcal{A} \cap \mathcal{H}$. In particular, $[C_1, \ldots, C_n] \subset \mathcal{H}$ is a convex body with C in its interior relative to $\mathcal{H}$. In a similar vein, $[C'_1, \ldots, C'_n] \subset \mathcal{H}'$ is a convex body with C' in its interior relative to $\mathcal{H}'$. We obtain $0 \in \text{int}[C_1, \ldots, C_n, C'_1, \ldots, C'_n]$.

17. See [Berger, 9.11.5].

19. See [Rockafellar–Wets, IV].

20. See [Roberts–Varberg, pp. 3–7]. (a) For a closed subinterval $[u, v] \subset \text{int}\, I$, $u < v$, $M = \max(f(u), f(v))$ is obviously an upper bound of f on $[u, v]$. For a lower bound, apply the defining inequality for convexity to the interval $[(u + v)/2 - w, (u + v)/2 + w]$, $|w| \leq (u - v)/2$, and its midpoint $(u + v)/2$, and estimate $f((u + v)/2 + w)$ from below. For Lipschitz continuity, let $[u, v] \subset \text{int}\, I$, $u < v$, as above, and choose $\epsilon > 0$ such that $[u - \epsilon, v + \epsilon] \subset I$. Given $a, b \in [u, v]$, $a < b$, apply the defining inequality for convexity to the interval $[a, b + \epsilon]$ and its interior point b to obtain $f(b) - f(a) \leq \lambda(f(b + \epsilon) - f(a))$, where $\lambda = (b - a)/(b - a + \epsilon) < (b - a)/\epsilon$. Estimate and obtain $f(b) - f(a) \leq \frac{M-m}{\epsilon}(b - a)$, where M and m are upper and lower bounds of f on $[u - \epsilon, v + \epsilon]$. For estimating $f(a) - f(b)$ use the interval $[a - \epsilon, b]$ and its interior point a to obtain $f(a) - f(b) \leq \frac{M-m}{\epsilon}(b - a)$ (b) The chain of inequalities follows from various rearrangements of the defining inequality for convexity. Then, setting $a = x_3$, we have

$$\frac{f(x_2) - f(a)}{x_2 - a} \leq \frac{f(x_4) - f(a)}{x_4 - a}.$$

Now the chain of inequalities shows that the left-hand side here increases as $x_2 \to a^-$ and the right-hand side decreases as $x_4 \to a^+$. Thus, both one-sided derivatives exist and $f'_-(a) \leq f'_+(a)$, $a \in \text{int}\, I$. Moreover, again by the above, for $a = x_1$ and $b = x_2$, we have

$$f'_+(a) \leq \frac{f(x_2) - f(a)}{x_2 - a} \leq \frac{f(x_3) - f(b)}{x_3 - b} \leq f'_-(b).$$

This, combined with the previous inequality gives monotonicity of $f'_\pm$. (c) By (b) and continuity of f, we have

$$\lim_{x_2 \to x_1^+} f'_+(x_2) \leq \lim_{x_2 \to x_1^+} \frac{f(x_3) - f(x_2)}{x_3 - x_2} = \frac{f(x_3) - f(x_1)}{x_3 - x_1}.$$

Letting $x_1 = a$ and $x_2 = x, x_3 \to a^+$, we obtain $\lim_{x \to a^+} f'_+(x) \leq f'_+(a)$. The opposite inequality is because of monotonicity of f'_+ established in (b). The first limit relation in (c) follows. The proofs of the remaining three relations are similar.

Chapter 2.

1. See [Soltan, Hammer 2, Busemann 2]. First, reduce the statement to the planar case (by intersecting C with a plane through O). Second, set O at the origin, and let $\rho : S \to \mathbb{R}$ denote the (radial) distance of the boundary ∂C from 0 as the function of the polar angle on the unit circle S. Similarly, let $\sigma : S \to \mathbb{R}$ be the radial distance for the boundary of the convex body reflected in the origin. Clearly, $\rho(\theta + \pi) = \sigma(\theta)$, $\theta \in S$. Now, the crux is that the condition that every chord through the origin is an affine diameter gives $\rho'/\rho = \sigma'/\sigma$ wherever the derivatives exist. As noted in the text, ρ and σ are Lipschitz continuous, therefore absolutely continuous so that the derivatives exist almost everywhere (except a countable subset [Berger]). Since $\ln \rho$ and $\ln \sigma$ are also Lipschitz continuous, integrating $(\ln \rho)' = (\ln \sigma)'$, we obtain $\rho(\theta)/\rho(\theta_0) = \sigma(\theta)/\sigma(\theta_0)$, $\theta, \theta_0 \in S$. Since the boundary of C and that of the reflected convex body intersect (for θ_0, say) we obtain $\rho = \sigma$ identically on S. Thus, any plane intersection of C through the origin is symmetric, so that C itself is symmetric.

2. $m(O) > 1$ can be assumed. (Otherwise $m(O) = 1$ and C is symmetric with respect to O with any chord through O being an affine diameter.) By Corollary 2.2.2, there is an affine diameter through O with ratio $m(O)$ which must then be different from the given one.

3. (2) Assume that the cycle $a_{i_1 j_1}, a_{i_1 j_2}, a_{i_2 j_2}, \ldots, a_{i_m j_m}, a_{i_m j_{m+1}}$ is minimal. Then we have $i_1 = i_m$ and $j_1 = j_{m+1}$, and the cycle $a_{i_2 j_2}, a_{i_2 j_3}, a_{i_3 j_3}, \ldots, a_{i_m j_m}, a_{i_1 j_2}$ contains less elements; a contradiction.

6. See [Klee 2]. For $O_0 \in \operatorname{int} C_0$ and $0 < \lambda < 1$, set $O_\lambda = (1 - \lambda)O_0 + \lambda V$. First, use Lemma 2.1.5 to show that $1/(\Lambda_C(C, O_\lambda) + 1) = (1 - \lambda)/(\Lambda_{C_0}(C, O_0) + 1)$, $C \in \partial C_0$. Observe that the extremal points of C are those of C_0 and the vertex V. Second, use Corollary 2.2.3 to show that $1/(m_C(O_\lambda) + 1) = \min((1 - \lambda)(m_{C_0}(O_0) + 1), \lambda)$. Then maximize both sides for $0 < \lambda < 1$ and $O_0 \in \operatorname{int} C_0$. Realize that, for fixed $O_0 \in \operatorname{int} C_0$, maximum occurs on the right-hand side in $0 < \lambda < 1$ if the two entries in the minimum are equal to the common value $1/(m_{C_0}(O_0) + 2)$. Conclude that $1/(m_C^* + 1) = 1/(m_{C_0}^* + 2)$.

8. Fix a vertex V_0 and the opposite edge E_0. Deleting V_0, the rest of the vertices split into two subsets according to which side of E_0 they are located. Show that these two sets have the same number of vertices.

12. Clearly $C \subset \bigcap_{X \in C} \bar{B}_d(X)$, $d = D_C$ holds. Assume that $C \in \bar{B}_d(X)$, for all $X \in C$. If $C \notin C$ the use Problem 8 at the end of Chapter 1 to get a contradiction.

13. Let $\mathcal{H}_i$, $i = 0, \ldots, n$, be the affine span of the ith face $[V_0, \ldots, \hat{V}_i, \ldots, V_n]$ of Δ opposite to V_i. For $i = 0, \ldots, n$, let $\mathcal{G}_i$ be the half-space with boundary $\mathcal{H}_i$ which does not contain O. Finally, let $\mathcal{D}_i = \partial(\bar{B}_d(V_i)) \cap \mathcal{G}_i$. By construction, for $C \in \partial C$, we have $C \in \mathcal{D}_i$ for some $i = 0, \ldots, n$. By symmetry, we may assume that $i = 0$. Let $X \in \mathcal{D}_i$, $\alpha = \angle OV_0 X$ and $\beta = \angle OV_0 V_1$. Clearly, we have $0 \leq \alpha \leq \beta \leq \pi/2$. Use the law of cosines for the triangles $[O, V_0, X]$ and $[O, V_0, V_1]$ and the equality $d(X, V_0) = d(V_1, V_0) = d$ to conclude that $d(X, O) \leq d(V_1, O) = R$.

14. Argue by contradiction, and assume that $C^\sharp$ is *not complete*. Let $\tilde{C}$ be a completion of $C^\sharp$, so that $C \subset C^\sharp \subset \tilde{C}$, and $\tilde{C}$ is complete with $D_{\tilde{C}} = D_{C^\sharp} = d$, in particular, $\tilde{C}$ is of constant width d. Since $C^\sharp$ is maximal in $\mathfrak{F}$, we have $\tilde{C} \notin \mathfrak{F}$; in particular, $\tilde{C} \not\subset \bar{B}_R(O)$. Let $Z \in \partial\tilde{C}$ be a point with maximal distance ($> R$) from O, and let $Y \in \mathcal{X}$ be the unique point such that $O \in [Y,Z]$ and $d(Y,O) = d - R (> 0)$. Notice that $Y \notin C^\sharp$, even $Y \notin \tilde{C}$, since $d(Y,Z) = d(Y,O)+d(Z,O) = d-R+d(Z,O) > d$, and $\tilde{C}$ is of constant width d. On the other hand, $Y \in \bar{B}_R(O)$ since $d(Y,O) = d-R = D_C - R_C \le R_C = R$. Conclude that the convex hull $[C^\sharp, Y] \subset \bar{B}_R(O)$ properly contains $C^\sharp$. Finally, for $X \in C^\sharp$, $d(X,Y) \le d(X,O)+d(Y,O) \le R+d-R = d$ so that the diameter of $[C^\sharp, Y]$ is still equal to d. Thus, $[C^\sharp, Y] \in \mathfrak{F}$, a contradiction to the maximality of $C^\sharp$ in $\mathfrak{F}$.

Chapter 3.

3. We claim $C_0 \subset C \subset m_C^* \cdot C$. Since the first inclusion is obvious, we need to show that $\partial C \subset m_C^* \cdot C_0$. Set $O^* = 0$, the origin. For $C \in \partial C \setminus C'$, there exists $0 < \mu < 1$ such that $\mu C \in \partial C'(\cap C \subset C_0)$. This means that $-\mu C \in \partial C$ so that $1/\mu = \Lambda(C,0) \le m_C^*$.

5. (3) Let $A_0 = (A+A')/2$ and $Z_0 = (Z+Z')/2$, so that $\mathcal{E}_0 = A_0\bar{B}+Z_0$ is also an ellipsoid of maximal volume in C. Observe that $\det(A) = \det(A') = \det(A_0)$. Then calculate as

$$\det(A_0)^{1/n} = \frac{1}{2}\det(A + A')^{1/n} \ge \frac{1}{2}\det(A)^{1/n} + \frac{1}{2}\det(A')^{1/n} = \det(A_0)^{1/n}.$$

(The middle inequality is a variant of the Brunn–Minkowski inequality.) Noting that equality holds if and only if $A' = cA$ for some $c \in \mathbb{R}$, conclude that $c = 1$. (4) If $Z \ne Z'$ then $\mathcal{E}$ and $\mathcal{E}'$ are translates, and their convex hull contains an ellipsoid having larger volume than $\mathrm{vol}(\mathcal{E})$.

6. (1) Δ can be assumed to be regular. Use the symmetries of Δ to show that the John's ellipsoid is the inball of Δ.

8. See [Grünbaum 2, 6.1]. Let $O' \in \mathrm{int}\,C'$, $O'' \in \mathrm{int}\,C''$ so that $O = O' + O'' \in \mathrm{int}\,C$, where $C = C' + C''$. Let $\mathcal{H}' \ni O'$, $\mathcal{H}'' \ni O''$ and $\mathcal{H} \ni O$ be parallel hyperplanes. First claim

$$\mathcal{R}_C(\mathcal{H},O) \le \max(\mathcal{R}_{C'}(\mathcal{H}',O'), \mathcal{R}_{C''}(\mathcal{H}'',O'')).$$

All the supporting hyperplanes defining the ingredients in this inequality are parallel. Taking the respective ratios, the claim follows from the elementary inequality

$$\frac{a' + a''}{b' + b''} \le \max\left(\frac{a'}{b'}, \frac{a''}{b''}\right), \quad a',a'',b',b'' > 0.$$

Now choose $O' \in C'$ and $O'' \in C''$ with $\mathrm{m}_{C'}(O') = \mathrm{m}_{C'}^*$ and $\mathrm{m}_{C''}(O'') = \mathrm{m}_{C''}^*$. Take the suprema with respect to $\mathcal{H}' \ni O'$ and $\mathcal{H}'' \ni O''$ separately, and obtain

$$\mathcal{R}_C(\mathcal{H}, O) \leq \max(\mathrm{m}_{C'}^*, \mathrm{m}_{C''}^*).$$

Since this holds for all hyperplanes $\mathcal{H} \ni O$, finally arrive at

$$\mathrm{m}_C^* \leq \mathrm{m}_C(O) = \sup_{\mathcal{H}(\ni O)} \mathcal{R}_C(\mathcal{H}, O) \leq \max(\mathrm{m}_{C'}^*, \mathrm{m}_{C''}^*).$$

9. By convexity, $g(C) \in C$. Use Corollary 1.2.3 to rule out the boundary.

12. (1) For $\mathcal{A}_1$ and $\mathcal{A}_2$ open rectangular parallelepipeds, concavity is a simple application of the comparison of geometric and arithmetic means. (2) If $\mathcal{A}_1 = \bigcup_{i=1}^k \mathcal{A}_1^i$ and $\mathcal{A}_2 = \bigcup_{j=1}^l \mathcal{A}_2^j$ are disjoint unions of open rectangular parallelepipeds then use induction with respect to $k + l$ as follows: Choose a hyperplane $\mathcal{H}$ which separates *at least two members in the decomposition of $\mathcal{A}_1$*. Let $\mathcal{G}'$ and $\mathcal{G}''$ be the two closed half-planes with common boundary $\mathcal{H}$. Taking intersections with $\operatorname{int} \mathcal{G}'$ and $\operatorname{int} \mathcal{G}''$, obtain $\mathcal{A}_1 = \mathcal{A}_1' \cup \mathcal{A}_1''$ and $\mathcal{A}_2 = \mathcal{A}_2' \cup \mathcal{A}_2''$. The number of participants in the respective decomposition of $\mathcal{A}_1'$ and $\mathcal{A}_1''$ are both $< k$, while the corresponding number for $\mathcal{A}_2'$ and $\mathcal{A}_2''$ are $\leq l$. Now apply the induction hypothesis. (3) Finally use an approximation argument.

13. This is an elementary but surprisingly technical integration.

15. $1 = \rho(\mathcal{B}, \mathcal{B}) \leq \rho(\mathcal{B}, C)\rho(C, C')\rho(C', \mathcal{B})$ and $\rho(C, C') \leq \rho(C, \mathcal{B})\rho(\mathcal{B}, C')$.

16. (1) See [Eggleston 1, 5.4]. (a) Elementary geometry. (b) Use Fubini's theorem. (c) We indicate a geometric and an analytic proof. The analytic proof also covers the equality case.

   Geometric Proof: A one-parameter family $\{C_\mu\}_{\mu \in [0,1]} \subset \mathfrak{B}$ is a *concave array* if, for any $\mu_0, \mu_1 \in [0, 1]$, we have $(1 - \lambda)C_{\mu_0} + \lambda C_{\mu_1} \subset C_{(1-\lambda)\mu_0 + \lambda\mu_1}$, $\lambda \in [0, 1]$. Given a hyperplane $\mathcal{H} \subset X$ and a concave array $\{C_\mu\}_{\mu \in [0,1]} \subset \mathfrak{B}$, show that the Steiner symmetrized array $\{C_\mu[\mathcal{H}]\}_{\mu \in [0,1]} \subset \mathfrak{B}$ is also concave. (Construct the convex body $C_X \subset X \times \mathbb{R}$ by $C_X \cap (X \times \{\lambda\}) = C_\lambda$ if $\lambda \in [0, 1]$, and $C_X \cap (X \times \{\lambda\}) = \emptyset$ if $\lambda \notin [0, 1]$. Then consider the Steiner symmetrization of C_X with respect to the hyperplane $\mathcal{H} \times \mathbb{R}$ in $X \times \mathbb{R}$.) Now, given $C \in \mathfrak{B}$ and $B \subset X$ a closed unit ball *with center on* $\mathcal{H}$, show that the array $\{(1 - \mu)C + \mu B\}_{\mu \in [0,1]}$ is concave. After Steiner symmetrization use the definition of concavity of the array $\{((1 - \mu)C + \mu B)[\mathcal{H}]\}_{\mu \in [0,1]}$ (for $\mu_0 = 0$ and $\mu_1 = 1$) and $B[\mathcal{H}] = B$ to obtain $(1-\lambda)C[\mathcal{H}] + \lambda B \subset ((1-\lambda)C + \lambda B)[\mathcal{H}]$, $\lambda \in [0, 1]$. Letting $r = \lambda/(1 - \lambda)$, $\lambda \in [0, 1)$, conclude that $\overline{C[\mathcal{H}]}_r \subset \bar{C}_r[\mathcal{H}]$, $r \geq 0$ where the subscript means closed r-neighborhood (Section 1.1/A). Take volumes and use (b) to arrive at $\operatorname{vol}(\overline{C[\mathcal{H}]}_r) - \operatorname{vol}(C[\mathcal{H}]) \leq \operatorname{vol}(\bar{C}_r) - \operatorname{vol}(C)$, $r \geq 0$. Dividing by r and letting $r \to \infty$ conclude that $\operatorname{vol}_{n-1}\partial(C[\mathcal{H}]) \leq \operatorname{vol}_{n-1}\partial C$. (For Minkowski's definition of the surface area, see the end of Section 1.2.)

Analytic Proof: Introduce a coordinate system $x = (x_1, \ldots, x_n)$ in $\mathcal{X}$ such that $\mathcal{H} = \{x_n = 0\}$. Represent $\mathcal{C}$ as the convex body between the graphs of a concave–convex pair of functions $f, g : \mathcal{C}_0 \to \mathbb{R}$, where $\mathcal{C}_0$ is the (orthogonal) projection of $\mathcal{C}$ to $\mathcal{H}$; that is, $\mathcal{C} = \{x \mid g(x_1, \ldots, x_{n-1}) \leq x_n \leq f(x_1, \ldots, x_{n-1}), (x_1, \ldots, x_{n-1}) \in \mathcal{C}_0\}$. Assume that f and g are *piecewise* C^1-*functions on* $\mathcal{C}_0$. Split $\partial \mathcal{C}$ into three parts: The graph of f, the graph of g, and the "side" of $\mathcal{C}$ consisting of vertical line segments that project to $\partial \mathcal{C}_0$. Realize that the area of the side does not change under Steiner symmetrization (Fubini's theorem). Use the calculus formula for the surface area of a graph to obtain

$$\mathrm{vol}_{n-1}\partial(\mathcal{C}[\mathcal{H}]) - \mathrm{vol}_{n-1}\partial\mathcal{C} = 2\int_{\mathcal{C}_0}\left(1 + \left|\nabla\left(\frac{f-g}{2}\right)\right|^2\right)^{1/2} dx_1 \ldots dx_{n-1}$$
$$- \int_{\mathcal{C}_0}\left(\left(1+|\nabla f|^2\right)^{1/2} + \left(1+|\nabla g|^2\right)^{1/2}\right) dx_1 \ldots dx_{n-1}.$$

(Note that the gradients exist on $\mathcal{C}_0$ almost everywhere.) Apply the triangle inequality (to the vectors $(1, \nabla f)$ and $(1, -\nabla g)$ almost everywhere) to conclude that the right-hand side is ≤ 0, and equality holds if and only if $\nabla f + \nabla g = 0$ on $\mathcal{C}_0$ almost everywhere. Observe that the last equality means that $f + g$ is constant on $\mathcal{C}_0$, or equivalently, $\mathcal{C}$ is symmetric with respect to a hyperplane parallel to $\mathcal{H}$. Finally, use an approximation argument to remove the smoothness condition on f and g. (Note that the surface area and the Steiner symmetrization are continuous with respect to d_H on $\mathfrak{B}$; see again the discussion at the end of Section 1.2.) (d) Elementary geometry. (2) Let $\{\mathcal{C}_{i_k}\}_{k \geq 0}$ be a subsequence which converges to $\mathcal{C}_\infty$ (in the Hausdorff metric). Since there are finitely many remainders $\mathrm{mod}(n-1)$, observe that there exists $0 \leq r \leq n-2$ such that the congruence $x \equiv r(\mathrm{mod}(n-1))$ has infinitely many solutions in the sequence $\{i_k\}_{k \geq 1}$. Selecting this subsequence of solutions and changing the notation, assume that $i_k \equiv r(\mathrm{mod}(n-1))$, $k \geq 1$. Conclude that $\mathcal{C}_\infty$ is symmetric with respect to $\mathcal{H}_{r-1(\mathrm{mod}(n-1))}$. Assume that $\mathcal{C}_\infty$ is *not symmetric* with respect to $\mathcal{H}_r$ so that $\mathrm{vol}_{n-1}(\partial\mathcal{C}_\infty[\mathcal{H}_r]) < \mathrm{vol}_{n-1}(\partial\mathcal{C}_\infty)$. Observe that the subsequence $\{\mathcal{C}_{i_k+1}\}_{k \geq 1}$ converges to $\mathcal{C}_\infty[\mathcal{H}_r]$. Use (1/c) to obtain a contradiction with $\mathrm{vol}_{n-1}(\partial\mathcal{C}_{i_k+1}) \leq \mathrm{vol}_{n-1}(\partial\mathcal{C}_{i_k+1})$ as $k \to \infty$. Continue to augment r and repeat this process.

Chapter 4.

1. For any k-configuration $\{C_0, \ldots, C_k\} \in \mathfrak{C}_k(O)$, use the antipodal configuration $\{C_0^o, \ldots, C_k^o\} \in \mathfrak{C}_k(O)$ as in the remark before Corollary 4.1.3.
2. This is obvious if O is a singular point. If O is regular and $\{C_0, \ldots, C_n\} \in \mathfrak{C}(O)$ is a minimal simplicial configuration then, as in the previous exercise, take the antipodal simplicial configuration $\{C_0^o, \ldots, C_n^o\} \in \mathfrak{C}(O)$. The ith face $[C_0^o, \ldots, \widehat{C_i^o}, \ldots, C_n^o]$ of the corresponding n-simplex $[C_0, \ldots, C_n]$, projected to the boundary $\partial\mathcal{C}$ from O is a closed domain $\mathcal{D}_i \subset \partial\mathcal{C}$ which clearly contains C_i in its relative interior. Restricted to $\mathcal{D}_i$, the function $\Lambda(., O)$ assumes its

global maximum at C_i. (Otherwise C_i in the minimal configuration could be replaced by a point in $\mathcal{D}_i$ at which $\Lambda(.,O)$ assumes its global maximum over $\mathcal{D}_i$, a contradiction.) Now use $\bigcup_{i=0}^{n} \mathcal{D}_i = \partial \mathcal{C}$.

3. By (the last statement of) Theorem 4.5.2, we have $\mathcal{S} = \mathrm{int}\,\mathcal{C}$ and by Example 2.1.4, $1/(\mathrm{m}(O) + 1) = (1 - |O|)/2$, $O \in \mathrm{int}\,\mathcal{C}$. The case $n = 2$ is clear. For the general induction step use (4.1.4) along with the geometry of the codimension 1 slices of $\mathcal{C}$ across O.

7. First use (2.4.12) in Corollary 2.4.11 to prove that $O^* \in [\mathcal{M}^*]$, and so $\sigma(O^*) = 1 + 1/n$. Then use sub-arithmeticity.

8. Apply Lemma 4.6.6.

9. Let $O, O' \in \mathrm{int}\,\mathcal{C}$ and $C, C^o \in \partial \mathcal{C}$ be as in the previous exercise. Assume that O' is a regular point. Let $\{C_0, \dots, C_n\} \in \mathfrak{C}(O')$ be a simplicial minimal configuration. Let C_i, $i = 0, \dots, n$ be any of the configuration points, different from C and C^o. Let $\mathcal{K} = \langle O, O', C_i \rangle$. Then C_i can be moved along one of the boundary arcs of $\mathcal{K} \cap \partial \mathcal{C}$ from C^o to C with increasing $\Lambda(.,O')$. By minimality, $\Lambda(.,O')$ must stay constant. Since O' is regular, the condition $O' \in [C_0, \dots, C_n]$ stays intact when replacing C_i with a moved point. Hence C_i can be moved to C. This is a contradiction.

11. See [Toth 8, Section 3].

12. The assumption on the bulging $\mathcal{B}$ implies that no tangent plane of the (relative interior of the) graph of f is parallel to any of the sides of $[C_0, \dots, C_n]$ other than $[C_1, \dots, C_n]$. In particular, no affine diameter emanates from the (relative interior of the) graph of f except those that end in C_0. Thus, for $O \in \mathcal{B}$, the distortion $\Lambda(.,O)$ cannot have local maxima on the graph of f. Hence the interior of $\mathcal{B}$ consists of singular points. Since the singular set is closed, $\mathcal{R} \subset \mathrm{int}\,\Delta$ follows. To show that equality holds, given $O \in \mathrm{int}\,\Delta$, write $O = \lambda X + (1 - \lambda)C_0$ with $0 < \lambda < 1$ and $X \in \mathrm{int}[C_1, \dots, C_n]$. Setting $C_0 = 0$, write $X = \sum_{i=1}^{n} \lambda_i C_i$, where $\sum_{i=1}^{n} \lambda_i = 1, 0 < \lambda_i < 1, i = 1, \dots, n$. Since $1/(\Lambda(C_i, O) + 1) = \lambda \lambda_i$, $i = 1, \dots, n$, (Lemma 2.1.5) conclude that $\Lambda_\Delta(C_0, O) = \lambda/(1 - \lambda)$.

Assume that the extension of the line segment $[C_0, X]$ beyond X intersects the simplex $[C_1, \dots, C_n, V]$ at the side opposite to C_i. This intersection point is $X/(1 - \lambda_i)$. Rewrite the assumption on $\mathcal{B}$ as $f(X)/|X| + 1 < 1/(1 - \lambda_i)$. Eliminate λ_i to obtain $\Lambda(C_i, O) < (1/\lambda)|X|/f(X) - 1$, $i = 1, \dots, n$. Notice that since $\mathcal{C}$ has a simplicial intersection across O, to prove that O is a regular point, it is enough to show that $\sum_{i=0}^{n} 1/(\Lambda(C_i, O) + 1) < 1 + 1/(\mathrm{m}(O) + 1)$. Calculating the sum on the left-hand side (via comparison with $\sigma_\Delta(O) = 1$), reduce this to $\lambda f(X)/(|X| + f(X)) < 1/(\mathrm{m}(O) + 1)$.

The maximum of the distortion $\Lambda(.,O)$ occurs either at one of the vertices C_i, $i = 0, \dots, n$, or at a point on the graph of the function f. To show that the latter cannot happen replace the graph with the boundary of the covering simplex $[C_1, \dots, C_n, V]$ (except the base $[C_1, \dots, C_n]$). Since maximum distortion occurs at a vertex, it is enough to show that the distortion at V cannot be greater than the maximum. Calculate: $1/(\Lambda(V, O) + 1) = (n - 1)\lambda \min_{1 \le i \le n} \lambda_i \ge \lambda \lambda_j = 1/(\Lambda(C_j, O) + 1)$, where $\max_{1 \le i \le n} \lambda_i = \lambda_j$. Conclude that O is regular.

To calculate $\sigma(O)$, show that the only minimizing configuration with extremal configuration points is $\{C_0, \ldots, C_n\}$ (Proposition 4.5.4) and evaluate.

Appendix A.

1. (b) For harmonicity of the components (showing that f_ξ is a λ_k-eigenmap), first write f_ξ in coordinates:

$$f_\xi(a,b)(z,w) = \xi(\bar{a}z + \bar{b}w, -bz + aw), \quad a, b \in \mathbb{C}, \ |a|^2 + |b|^2 = 1, \ z, w \in \mathbb{C}.$$

Here $g = (a,b) \in S^3 \subset \mathbb{C}^2$ is identified with $\begin{bmatrix} a & -\bar{b} \\ b & \bar{a} \end{bmatrix} \in SU(2)$, and the inverse $g^{-1} = (\bar{a}, -b)$ acts on (z,w) by multiplication. Then apply the Laplace operator $\Delta = 4(\partial^2/\partial a \partial \bar{a} + \partial^2/\partial b \partial \bar{b})$ to verify that the components are harmonic polynomials. (c) Use $SU(2)$-equivariance of f_ξ and argue as in the proof of Theorem A.3.2 to conclude that $\langle f_\xi \rangle \in \mathcal{L}_{\lambda_k}$ commutes with the action of $SU(2) \subset SO(4)$ by restriction on $\mathcal{H}_{\lambda_k}$.

5. By $SU(2)$-equivariance, it is enough to check conformality at the tangent space $T_1(S^3)$. Now calculate the differential $(f_\xi)_*$ on a specific orthonormal basis.

Bibliography

Aubin–Frankowska. J.-P. Aubin, H. Frankowska, *Set-Valued Analysis* (Birkhäuser, Boston, 1990)

Bajmóczy–Bárány. E.G. Bajmóczy, I. Bárány, A common generalization of Borsuk's and Radon's theorem. Acta Math. Hungar. **34**, 347–350 (1979)

Ball 1. K. Ball, Ellipsoids of maximal volume in convex bodies. Geom. Dedicata **41**(2), 241–250 (1992)

Ball 2. K. Ball, Volume ratios and the reverse isoperimetric inequality. J. Lond. Math. Soc. **44**, 351–359 (1991)

Barthe 1. F. Barthe, Inégalités de Brascamp–Lieb et convexité. C. R. Acad. Sci. Paris **324**, 885–888 (1997)

Barthe 2. F. Barthe, Inégalités fonctionelles et géométriques obtenues par transport de mesures. Thèse de Doctorat, Université Marne-la-Vallee, 1997

Belloni–Freund. A. Belloni, R.M. Freund, On the symmetry function of a convex set. Math. Program. Ser. B **111**, 57–93 (2008)

Berger. M. Berger, *Geometry I-II* (Springer, New York, 1987)

Berger–Gauduchon–Mazet. M. Berger, P. Gauduchon, E. Mazet, *Le Spectre d'une variété Riemannienne*. Springer Lecture Notes, vol. 194 (Springer, Berlin/New York, 1971)

Besse. A. Besse, *Manifolds All of Whose Geodesics are Closed* (Springer, New York, 1978)

Birch. B.J. Birch, On $3N$ points in a plane. Proc. Camb. Philos. Soc. **55**, 289–293 (1959)

Birkhoff. G. Birkhoff, Three observations on linear algebra. Univ. Nac. Tacumán Rev. Ser. A **5**, 147–151 (1946)

Blaschke 1. W. Blaschke, *Kreis und Kugel*, 2nd edn. (W. de Gruyter, Berlin, 1956)

Blaschke 2. W. Blaschke, *Vorlesungen über Differentialgeometrie, II, Affine Differentialgeometrie* (Springer, Berlin, 1923)

Blaschke 3. W. Blaschke, Über affine Geometrie IX: Verschiedene Bemerkungen und Aufgaben. Ber. Verh. Sächs. Akad. Wiss. Leipzig. Math.-Nat. Kl. **69**, 412–420 (1917)

Bohnenblust. F. Bohnenblust, Convex regions and projections in Minkowski spaces. Ann. Math. **39**, 301–308 (1938)

Boltyanskiĭ–Martini. V.G. Boltyanskiĭ, H. Martini, Carathéodory's theorem and H-convexity. J. Combin. Theory Ser. A **93**(2), 292–309 (2001)

Bonnesen-Fenchel. T. Bonnesen, W. Fenchel, *Theorie der konvexen Körper*. Ergebn. Math., Bd. 3 (Springer, Berlin, 1934) [English translation: Theory of Convex Bodies (BCS, Moscow, 1987)]

Börner. H. Börner, *Representations of Groups* (North-Holland, Amsterdam, 1963)

Böröczky 1. K. Böröczky Jr., Around the Rogers–Shephard inequality. Mathematica Pannonica **7**, 113–130 (1996)

© Springer International Publishing Switzerland 2015   267
G. Toth, *Measures of Symmetry for Convex Sets and Stability*,
Universitext, DOI 10.1007/978-3-319-23733-6

Böröczky 2. K. Böröczky Jr., The stability of the Rogers–Shephard inequality and of some related inequalities. Adv. Math. **190**, 47–76 (2005)

Busemann 1. H. Busemann, The foundations of Minkowskian geometry. Commentarii Mathematici Helvetici **24**, 156–187 (1950)

Busemann 2. H. Busemann, *The Geometry of Geodesics* (Academic, New York, 1955)

Calabi. E. Calabi, Minimal immersions of surfaces in Euclidean spheres. J. Differ. Geom. **1**, 111–125 (1967)

Carathéodory. C. Carathéodory, Über den Variabilitätsbereich der Koeffizienten von Potenzreihen, die gegebene Werte nicht annehmen. Math. Ann. **64**, 95–115 (1907)

Chakerian–Groemer. G.D. Chakerian, H. Groemer, Convex bodies of constant width, in *Convexity and Its Applications*, ed. by P.M. Gruber, J.M. Willis, (Birkhäuser, Basel, 1983), pp. 49–96

Chakerian–Stein. G.D. Chakerian, S.K. Stein, On the measures of symmetry of convex bodies. Can. J. Math. **17**, 497–504 (1965)

Danzer–Grünbaum–Klee. L. Danzer, B. Grünbaum, V. Klee, Helly's theorem and its relatives, in *Convexity, Proceedings of Symposium in Pure Mathematics*, vol. 7 (American Mathematical Society, Providence, 1963), pp. 101–179

DeTurck–Ziller 1. D. DeTurck, W. Ziller, Minimal isometric immersions of spherical space forms in spheres. J. Differ. Geom. **1**, 111–125 (1967)

DeTurck–Ziller 2. D. DeTurck, W. Ziller, Minimal isometric immersions of spherical space forms into spheres. Commentarii Mathematici Helvetici **67**, 428–458 (1992)

DeTurck–Ziller 3. D. DeTurck, W. Ziller, Spherical minimal immersions of spherical space forms. Proc. Symp. Pure Math. Part I **54**, 111–120 (1993)

Dinghas. A. Dinghas, Über das verhalten der Enfernung zweier Punktmengen bei gleichzeitiger Symmetrisierung derselben. Arch. Math. **8**, 46–51 (1957)

Diskant. V.I. Diskant, Stability of the solution of a Minkowski equation. Sibirsk Math. Z. **14**, 669–673 (1973) [English translation: Siberian Math. J. **14**, 466–473 (1974)]

DoCarmo–Wallach. M. DoCarmo, N. Wallach, Minimal immersions of spheres into spheres. Ann. Math. **93**, 43–62 (1971)

Dol'nikov. V.L. Dol'nikov, On a question of Grünbaum, Issled. po teorii funktsii mnogih veschestv. Peremennih, Jaroslav. Gos. University, Yarolslavl' (1976), pp. 34–35 [Russian]

Dvoretzky 1. A. Dvoretzky, A theorem on convex bodies and applications to Banach spaces. Proc. Natl. Acad. Sci. U.S.A. **45**, 223–226 (1959)

Dvoretzky 2. A. Dvoretzky, Some results on convex bodies and Banach spaces, in *Proceedings of the International Symposium on Linear Spaces* (Jerusalem Academic, Jerusalem, 1960/Pergamon, Oxford, 1961), pp. 123–160

Eckhoff. J. Eckhoff, Helly, Radon, and Carathéodory type theorems, in *Handbook of Convex Geometry* (North-Holland, Amsterdam, 1993), pp. 389–448

Eggleston 1. H.G. Eggleston, *Convexity* (Cambridge University Press, Cambridge, 1958)

Eggleston 2. H.G. Eggleston, Sets of constant width in finite dimensional Banach spaces. Israel J. Math. **3**, 163–172 (1965)

Eells–Lemaire. J. Eells, L. Lemaire, A report on harmonic maps. Bull. Lond. Math. Soc. **10**, 1–68 (1978)

Eells–Sampson. J. Eells, J.H. Sampson, Harmonic mappings of Riemannian manifolds. Am. J. Math. **86**(1), 109–160 (1964)

Ehrhart. E. Ehrhart, Sur les ovales et les ovoides. C. R. Acad. Sci. Paris **240**, 583–585 (1955)

Escher–Weingart. Ch. Escher, G. Weingart, Orbits of $SU(2)$-representations and minimal isometric immersions. Math. Ann. **316**, 743–769 (2000)

Estermann. T. Estermann, Über den Vektorenbereich eines konvexen Körpers. Math. Z. **28**, 471–475 (1928)

Falconer. K.J. Falconer, Applications of a result on spherical integration to the theory of convex sets. Am. Math. Mon. **90**, 690–693 (1983)

Fulton–Harris. W. Fulton, J. Harris, *Representation Theory, a First Course* (Springer, New York, 1991)

Funk 1. P. Funk, Über Flächen mit lauter geschlossenen geodätischen Linien. Math. Ann. **74**, 278–300 (1913)

Funk 2. P. Funk, Über eine geometrische Anwendung der Abelschen Intergralgleichung. Math. Ann. **77**, 129–135 (1915)

Gardner. R.J. Gardner, The Brunn–Minkowski inequality. Bull. Am. Math. Soc. **39**, 355–405 (2002)

Giannopoulos–Perissinaki–Tsolomitis. A. Giannopoulos, I. Perissinaki, A. Tsolomitis, John's theorem for an arbitrary pair of convex bodies. Geom. Dedicata. **84**, 63–79 (2001)

Gluskin. E. Gluskin, Norms of random matrices and diameters of finite dimensional sets. Matematicheskii Sbornik **120**, 180–189 (1983)

Groemer 1. H. Groemer, Stability theorems for two measures of symmetry. Discrete Comput. Geom. **24**, 301–311 (2000)

Groemer 2. H. Groemer, Stability of geometric inequalities, in *Handbook of Convex Geometry* (North-Holland, Amsterdam, 1993), pp. 125–150

Groemer 3. H. Groemer, *Geometric Applications of Fourier Series and Spherical Harmonics* (Cambridge University Press, Cambridge, 1996)

Groemer 4. H. Groemer, On complete convex bodies. Geom. Dedicata **20**, 319–334 (1986)

Groemer 5. H. Groemer, On the Brunn–Minkowski theorem. Geom. Dedicata **27**, 357–371 (1988)

Groemer–Wallen. H. Groemer, L.J. Wallen, A measure of asymmetry for domains of constant width. Beiträge zur Algebra und Geometrie **42**(2), 517–521 (2001)

Gruber. P.M. Gruber, *Convex and Discrete Geometry* (Springer, New York, 2007)

Gruber–Wills. P.M. Gruber, J.M. Wills (eds.), *Handbook of Convex Geometry* (North-Holland, Amsterdam, 1993)

Grünbaum 1. B. Grünbaum, *Convex Polytopes* (Springer, New York, 2003)

Grünbaum 2. B. Grünbaum, Measures of symmetry for convex sets. Proc. Symp. Pure Math. **VII**, 233–270 (1963)

Grünbaum 3. B. Grünbaum, Partitions of mass-distributions and of convex bodies by hyperplanes. Pac. J. Math. **10**, 1257–1261 (1960)

Grünbaum 4. B. Grünbaum, The dimension of intersections of convex sets. Pac. J. Math. **12**(1), 197–202 (1962)

Guo 1. Q. Guo, Stability of the Minkowski measure of asymmetry for convex bodies. Discrete Comput. Geom. **34**, 351–362 (2005)

Guo 2. Q. Guo, On p-measures of asymmetry for convex bodies. Adv. Geom. **12**, 287–301 (2012)

Guo–Jin. Q. Guo, H. Jin, On a measure of asymmetry for Reuleaux polygons. J. Geom. **102**, 73–79 (2011)

Guo–Kaijser 1. Q. Guo, S. Kaijser, On asymmetry of some convex bodies. Discrete Comput. Geom. **27**, 239–247 (2002)

Guo–Kaijser 2. Q. Guo, S. Kaijser, Approximation of convex bodies by convex bodies. Northeast Math. J. **19**(4), 323–332 (2003)

Guo–Kaijser 3. Q. Guo, S. Kaijser, On the distance between convex bodies. Northeast Math. J. **15**(3), 323–331 (1999)

Guo–Toth. Q. Guo, G. Toth, Dual mean Minkowski measures and the Grünbaum conjecture for affine diameters. Preprint, at http://math.camden.rutgers.edu/files/Dual-Mean.pdf

Hadwiger. L. Hadwiger, *Vorlesungen über Inhalt, Oberfläche und Isoperimetrie*, Grundlehren der Mathematischen Wissenschaften, vol. 93 (Springer, New York, 1957)

Hammer 1. P.C. Hammer, The centroid of a convex body. Proc. Am. Math. Soc. **2**, 522–525 (1951)

Hammer 2. P.C. Hammer, Convex bodies associated with a convex body. Proc. Am. Math. Soc. **2**, 781–793 (1951)

Hammer 3. P.C. Hammer, Diameters of convex bodies. Proc. Am. Math. Soc. **5**, 304–306 (1954)

Hammer–Sobczyk. P.C. Hammer, A. Sobczyk, Critical points of a convex body, Abstract 112. Bull. Am. Math. Soc. **57**, 127 (1951)

Harazišvili. A.B. Harazišvili, Affine diameters of convex bodies. Soobsch. Akad. Nauk. Gruzin SSR **90**, 541–544 (1978)

Hausdorff. F. Hausdorff, *Grundzüge der Mengenlehre* (Veit & Comp., Leipzig 1914); [English translation: Chelsea, New York (1962)]

Helgason 1. S. Helgason, The Radon transform on Euclidean spaces, compact two-point homogeneous spaces and Grassmann manifolds. Acta Math. **113**, 153–180 (1965)

Helgason 2. S. Helgason, *Geometric Analysis on Symmetric Spaces*, Mathematical Surveys and Monographs, vol. 39 (American Mathematical Society, Providence, 1994)

Helgason 3. S. Helgason, Groups and geometric analysis. *Integral geometry, invariant differential operators, and spherical functions*, Mathematical Surveys and Monographs, vol. 83 (American Mathematical Society, Providence, 2000)

Helly. E. Helly, Über Mengen konvexer Körper mit gemeinschaftlichen Punkten. Jahrb. Deut. Math. Verein. Vol. 32 175–176 (1923)

Hug–Schneider. D. Hug, R. Schneider, A stability result for a volume ratio. Israel J. Math. **161**, 209–219 (2007)

Hurlbert. G. Hurlbert, *Linear Optimization: The Simplex Workbook* (Springer, New York, 2009)

Jin–Guo 1. H. Jin, Q. Guo, Asymmetry of convex bodies of constant width. Discrete Comput. Geom. **47**, 415–423 (2012)

Jin–Guo 2. H. Jin, Q. Guo, On the asymmetry for convex domains of constant width. Commun. Math. Res. **26**(2), 176–182 (2010)

Jin–Guo 3. H. Jin, Q. Guo, A note on the extremal bodies of constant width for the Minkowski measure. Geom. Dedicata **164**, 227–229 (2013)

Jin–Guo 4. H. Jin, Q. Guo, The mean Minkowski measures for convex bodies of constant width. Taiwan J. Math. **18**(4), 1283–1291 (2014)

John. F. John, Extremum problems with inequalities as subsidiary conditions, in *Courant Anniversary Volume* (Interscience, New York, 1948), pp. 187–204

Jung. H.W.E. Jung, Über die kleinste Kugel, die eine räumliche Figur einschliesst. J. Reine Angew. Math. **123**, 241–257 (1901) [ebenda **137**, 310–313 (1910)]

Klee 1. V.L. Klee, On certain intersection properties of convex sets. Can. J. Math. **3**, 272–275 (1951)

Klee 2. V.L. Klee, The critical set of a convex body. Am. J. Math. **75**(1), 178–188 (1953)

Klee 3. V.L. Klee, *Unsolved Problems in Intuitive Geometry, Mimeographed Notes* (University of Washington, Seattle 1960)

Knapp. A.W. Knapp, *Representation Theory of Semisimple Groups* (Princeton University Press, Princeton, 1986)

Kneser–Süss. H. Kneser, W. Süss, Die Volumina in linearen Scharen konvexer Körper. Mat. Tidsskr. B **1**, 19–25 (1932)

Knothe. H. Knothe, Contributions to the theory of convex bodies. Michigan Math. J. **4**, 39–52 (1957)

Koziński 1. A. Koziński, On involution and families of compacta. Bull. Acad. Polon. Sci. Cl. III **5**, 1055–1059 (1954)

Koziński 2. A. Koziński, On a problem of Steinhaus. Fund. Math. **46**, 47–59 (1958)

König 1. D. König, Über konvexe Körper. Math. Z. **14**, 208–210 (1922)

König 2. D. König, *Theorie der Endlichen und Unendlichen Graphen* (Akademische Verlags Gesellschaft, Leipzig, 1936)

Krasnosel'skiĭ. M.A. Krasnosel'skiĭ, On a proof of Helly's theorem on sets of convex bodies with common points. Tr. Voronezhsk. Gos. Univ. **33**, 19–20 (1954)

Krein-Milman. M. Krein-Milman, D. Milman, On extreme points of regularly convex sets. Studia Math. **9**, 133–138 (1940)

Kubota. T. Kubota, Einige Probleme über konvex-geschlossene Kurven und Flächen. Tôhoku Math. J. **17**, 351–362 (1920)

Kuiper. N.H. Kuiper, Double normals of convex bodies. Israel J. Math. **2**, 71–80 (1920)

Kuratowski. C. Kuratowski, *Topologie I. Monografje Matematyczne* (Warszawa-Lwów, 1933) (in French) [English edition: Academic, New York, 1966]

Lanchand-Robert–Oudet. T. Lanchand-Robert, É. Oudet, Bodies of constant width in arbitrary dimension. Math. Nachr. **280**(7), 740–750 (2007)

Leichtweiss 1. K. Leichtweiss, Zwei Extremalprobleme der Minkowski-Geometrie. Math. Z. **62**, 37–49 (1955)

Leichtweiss 2. K. Leichtweiss, Über die affine Exzentrizität konvexer Körper. Arch. Math. **10**, 187–199 (1959)

Lassak. M. Lassak, Approximation of convex bodies by centrally symmetric convex bodies. Geom. Dedicata **72**, 63–68 (1998)

Levi. F.W. Levi, On Helly's theorem and the axioms of convexity. J. Indian Math. Soc. N.S. Part A **15**, 65–76 (1951)

Macbeath. A.M. Macbeath, A compactness theorem for the affine equivalence-classes of convex regions. Can. J. Math. **3**, 54–61 (1951)

Maehara. H. Maehara, Convex bodies forming pairs of constant width. J. Geom. **22**, 101–107 (1984)

Martini–Swanepoel–Weiss. H. Martini, K. Swanepoel, G. Weiss, The geometry of Minkowski spaces - a survey. I. Expo. Math. **19**, 97–142 (2001)

Mashimo 1. K. Mashimo, Minimal immersions of 3-dimensional spheres into spheres. Osaka J. Math. **2**, 721–732 (1984)

Mashimo 2. K. Mashimo, Homogeneous totally real submanifolds in S^6. Tsukuba J. Math. **9**, 185–202 (1985)

Matoušek 1. J. Matoušek, *Radon's Lemma and Helly's Theorem. Lectures on Discrete Geometry.* Graduate Texts in Mathematics, vol. 212 (Springer, New York, 2002), pp. 9–12

Matoušek 2. J. Matoušek, *Nonembeddability Theorems: An Introduction Using the Borsuk–Ulam Theorem.* Lectures on Topological Methods in Combinatorics and Geometry (Springer, New York, 2003), pp. 88–92

Meissner. F. Meissner, Über Punktmengen konstanter Breite, Vierteljahresschr. Naturforsch. Ges. Zürich **56**, 42–50 (1911)

Melzak. Z.A. Melzak, A note on sets of constant width. Proc. Am. Math. Soc. **11**, 493–497 (1960)

Meyer–Schütt–Werner 1. M. Meyer, C. Schütt, E. Werner, Dual affine invariant points. Indiana J. Math. (to appear). arXiv:1310.0128v1

Meyer–Schütt–Werner 2. M. Meyer, C. Schütt, E. Werner, Affine invariant points. Israel J. Math. (to appear)

Meyer–Schütt–Werner 3. M. Meyer, C. Schütt, E. Werner, New affine measures of symmetry for convex bodies. Adv. Math. **228**, 2920–2942 (2011)

Milman. V.D. Milman, A new proof of A. Dvoretzky's theorem on cross-sections of convex bodies. Funkcional. Anal. i Prilozhen. **5**(4), 28–37 (1971) [Russian]

Minkowski 1. H. Minkowski, Allgemeine Lehrsätze über konvexe Polyeder. Nachr. Ges. Wiss. Göttingen, 198–219, 1897 [also in Gesamelte Abhandlungen, vol. 2 (Teubner, Leipzig-Berlin, 1911), pp. 103–121]

Minkowski 2. H. Minkowski, *Gesammelte Abhandlungen* (Teubner, Leipzig-Berlin, 1911)

Minkowski 3. H. Minkowski, Über eine Erweiterung des Begriffs der konvexen Funktionen, mit einer Anwendung auf die Theorie der konvexen Körper. S.-B. Akad. Wiss. Wien. **125**, 241–258 (1916)

Moore. J.D. Moore, Isometric immersions of space forms into space forms. Pac. J. Math. **40**, 157–166 (1972)

Moreno–Schneider 1. J.P. Moreno, R. Schneider, Local Lipschitz continuity of the diametric completion mapping. Houston J. Math. **38**, 1207–1223 (2012)

Moreno–Schneider 2. J.P. Moreno, R. Schneider, Lipschitz selections of the diametric completion mapping in Minkowski spaces. Adv. Math. **233**, 248–267 (2013)

Muto. Y. Muto, The space W_2 of isometric minimal immersions of the three-dimensional sphere into spheres. Tokyo J. Math. **7**, 337–358 (1984)

Neumann. B.H. Neumann, On some affine invariants of closed convex regions. J. Lond. Math. Soc. **14**, 262–272 (1984)

Nguyên. Nguyên. M.H. Nguyên, Affine diameters of convex bodies, Ph.D. Thesis, Moldova State University (1990)

Painlevé. M. Painlevé, C. R. Acad. Sci. Paris **148**, 1341–1344 (1909)

Palmon. O. Palmon, The only convex body with extremal distance from the ball is the simplex. Israel J. Math. **80**, 337–349 (1992)

Petty. C.M. Petty, Centroid surfaces. Pac. J. Math. **11**, 1535–1547 (1961)

Pompéiu. D. Pompéiu, Sur la continuité des foncitions de variables complexes. Ann. Toul. **2**(7), 264–315 (1905)

Price. G.B. Price, On the extreme points of convex sets. Duke Math. J. **3**, 56–67 (1937)

Rademacher–Schoenberg. H. Rademacher, I.J. Schoenberg, Helly's theorem on convex domains and Tchebycheff's approximation problem. Can. J. Math. **2**, 245–256 (1950)

Rado. R. Rado, Theorems on the intersection of convex sets of points. J. Lond. Math. Soc. **27**, 320–328 (1952)

Radon. J. Radon, Mengen konvexer Körper, die einen gemeinsamen Punkt enthalten. Math. Ann. **83**, 113–115 (1921)

Roberts–Varberg. A. Roberts, D. Varberg, *Convex Functions* (Academic, New York, 1973)

Rockafellar–Wets. R.T. Rockafellar, R.J.-B. Wets, *Variational Analysis* (third corrected printing edn.). Grundlehren der mathematischen Wissenschaften, vol. 317 (Springer, Berlin, 2005)

Rogers–Shephard. C.A. Rogers, G.C. Shephard, The difference body of a convex body. Arch. Math. **8**, 220–233 (1957)

Rudelson. M. Rudelson, Distance between non-symmetric convex bodies and the MM^*-estimate. Positive **4**(2), 161–178 (2000)

Sallee. G.T. Sallee, Sets of constant width, the spherical intersection property and circumscribed balls. Bull. Aust. Math. Soc. **33**, 369–371 (1986)

Sandgren. L. Sandgren, On convex cones. Math. Scand. **2**, 19–28 (1954) [corr. **3**, 170 (1955)]

Schneider 1. R. Schneider, Stability for some extremal properties of the simplex. J. Geom. **96**, 135–148 (2009)

Schneider 2. R. Schneider, *Convex Bodies, the Brunn–Minkowski Theory*, 2nd edn. (Cambridge University Press, Cambridge, 2014)

Schneider 3. R. Schneider, Functional equations connected with rotations and their geometric applications. Enseign. Math. **16**, 297–305 (1970)

Schneider 4. R. Schneider, *Simplices*. At http://home.mathematik.uni-freiburg.de/rschnei/ Simplices.pdf

Scott. P.R. Scott, Sets of constant width and inequalities. Q. J. Math. Oxford **2**(32), 345–348 (1981)

Shephard–Webster. G.C. Shephard, R.J. Webster, Metrics for sets of convex bodies. Mathematica **12**, 73–88 (1965)

Sierpiński. W. Sierpiński, *General Topology* (University Toronto, Toronto, 1934)

Schmuckenschlänger. M. Schmuckenschlänger, An extremal property of the regular simplex, in *Convex Geometric Analysis*, vol. 34, ed. by K.M. Ball, V. Milman. MSRI Publications (Cambridge University Press, Cambridge, 1998), pp. 199–202

Soltan. V. Soltan, Affine diameters of convex bodies - a survey. Expo. Math. **23**, 47–63 (2005)

Soltan–Nguyên. V. Soltan, M.H. Nguyên, On the Grünbaum problem on affine diameters. Soobsch. Akad. Nauk. Gruzin SSR **132**, 33–35 (1988) [Russian]

Steinhagen. P. Steinhagen, Über die gröste Kugel in einer konvexen Punktmenge. Abh. aus dem Mathematischen Seminar de Hamburg Univ. **1**, 15–26 (1922)

Steinitz. E. Steinitz, Bedingt konvergente Reihen und konvexe Systeme, I-II-III. J. Reine Angew. Math. **143**, 128–175 (1913); **144**, 1–40 (1914); **146**, 1–52 (1916)

Süss 1. W. Süss, Über eine Affininvariant von Eibereichnen. Arch. Math. **1**, 127–128 (1948–1949)

Süss 2. W. Süss, Über Eibereiche mit Mittelpunkt. Math-Phys. Semesterber. **1**, 273–287 (1950)

Takahashi. T. Takahashi, Minimal immersions of Riemannian immersions. J. Math. Soc. Jpn. **18**, 380–385 (1966)

Tomczak-Jaegerman. N. Tomczak-Jaegerman, *Banach–Mazur Distances and Finite-Dimensional Operator Ideals*. Encyclopedia of Mathematics and its Applications, vol. 44 (Cambridge University Press, Cambridge, 1993)

Toth 1. G. Toth, Eigenmaps and the space of minimal immersions between spheres. Indiana Univ. Math. J. **46**(2), 637–658 (1997)

Toth 2. G. Toth, Infinitesimal rotations of isometric minimal immersions between spheres. Am. J. Math. **122**, 117–152 (2000)

Toth 3. G. Toth, Moduli for spherical maps and spherical minimal immersions of homogeneous spaces. J. Lie Theory **12**, 551–570 (2000)

Toth 4. G. Toth, *Finite Möbius Groups, Minimal Immersions of Spheres, and Moduli* (Springer, New York, 2002)

Toth 5. G. Toth, Simplicial intersections of a convex set and moduli for spherical minimal immersions. Michigan Math. J. **52**, 341–359 (2004)

Toth 6. G. Toth, On the shape of the moduli of spherical minimal immersions. Trans. Am. Math. Soc. **358**(6), 2425–2446 (2006)

Toth 7. G. Toth, On the structure of convex sets with applications to the moduli of spherical minimal immersions. Beitr. Algebra Geom. **49**(2), 491–515 (2008)

Toth 8. G. Toth, Convex sets with large distortion. J. Geom. **92**, 174–192 (2009)

Toth 9. G. Toth, On the structure of convex sets with symmetries. Geom. Dedicata **143**, 69–80 (2009)

Toth 10. G. Toth, A measure of symmetry for the moduli of spherical minimal immersions. Geom. Dedicata **160** 1–14 (2012)

Toth 11. G. Toth, Notes on Schneider's stability estimates for convex sets. J. Geom. **104**, 585–598 (2013)

Toth–Ziller. G. Toth, W. Ziller, Spherical minimal immersions of the 3-sphere. Comment. Math. Helv. **74**, 84–117 (1999)

Tverberg. H. Tverberg, A generalization of Radon's theorem. J. Lond. Math. Soc. **41**, 123–128 (1966)

Valentine. F.A. Valentine, *Convex Sets* (McGraw-Hill, New York, 1964)

Vilenkin. N.I. Vilenkin, *Special Functions and the Theory of Group Representations*. AMS Translations of Mathematical Monographs, vol. 22 (American Mathematical Society, Providence, 1968)

Vilenkin–Klimyk. N.I. Vilenkin, A.U. Klimyk, *Representations of Lie Groups and Special Functions*, vols. 1–3 (Kluwer, Dordrecht, 1991)

Vincensini. P. Vincensini, Sur une extension d'un théorème de M. J. Radon sur les ensembles de corps convexes. Bulletin de la Société Mathématique de Frances **67**, 115–119 (1939)

von Neumann. J. von Neumann, A certain zero-sum two-person game equivalent to an optimal assignment problem. Ann. Math. Stud. **28**, 5–12 (1953)

Vrećica. S. Vrećica, A note on the sets of constant width. Publ. Inst. Math. (Beograd) (Nouvelle Série) **43**(29), 289–291 (1981)

Wallach. N. Wallach, Minimal immersions of symmetric spaces into spheres, in *Symmetric Spaces* (Dekker, New York, 1972), pp.1–40

Webster. R.J. Webster, The space of affine-equivalence classes of compact subsets of Euclidean space. J. Lond. Math. Soc. **40**, 425–432 (1965)

Weingart. G. Weingart, Geometrie der Modulräume minimaler isometrischer Immersionen der Sphären in Sphären. Bonner Mathematische Schriften, Nr. 314, Bonn (1999)

Whyburn. G.T. Whyburn, Analytic topology. Am. Math. Soc. Colloq. Publ. **28**, 300–320 (1942)

Yaglom–Boltyanskiĭ. M. Yaglom, V.G. Boltyanskiĭ, *Convex Figures* (GITTL, Moscow, 1951) [Russian; English translation: Holt, Rinehart and Winston, New York, 1961; German translation: VEB Deutscher Verlag der Wissenschaften, Berlin, 1956]

Index

© Springer International Publishing Switzerland 2015
G. Toth, *Measures of Symmetry for Convex Sets and Stability*,
Universitext, DOI 10.1007/978-3-319-23733-6

Printed in the United States
By Bookmasters